Biophysics for the Life Sciences

Series Editor
Norma Allewell

For further volumes:
http://www.springer.com/series/10230

Jonathan D. Dinman

Editor

Biophysical Approaches to Translational Control of Gene Expression

Springer

Editor
Jonathan D. Dinman
Cell Biology and Molecular Genetics
University of Maryland
College Park, MD, USA

ISBN 978-1-4614-3990-5   ISBN 978-1-4614-3991-2 (eBook)
DOI 10.1007/978-1-4614-3991-2
Springer New York Heidelberg Dordrecht London

Library of Congress Control Number: 2012946734

Printed on acid-free paper

Springer is part of Springer Science+Business Media (www.springer.com)

Foreword

Biophysics of Protein Synthesis:
An Historical Perspective

Abstract The first evidence of the existence of the particles we now know as ribosomes appeared in the literature about 75 years ago, and ribosomes have been vigorously pursued by molecular biologists, biochemists, and biophysicists for the last 60 of those 75 years. This essay provides a brief history of the field that begins with a description of how the ribosome was discovered in the first place and ends with the announcement of atomic resolution crystal structures of the ribosome in 2000.

Introduction

The goal of the biophysical chemist, like that of every other kind of biologist, is to change the way people think about the living world for the better. Consequently, students must find it hard to understand why in the 1950s so many biophysical chemists concentrated on molecules like bovine serum albumin, myoglobin, hemoglobin, and tobacco mosaic virus. Why work on myoglobin when you could be working on, say, RNA polymerase II? Every biological scientist over the age of 50 knows the answer. Many biophysical techniques consume prodigious amounts of material, especially when they are first being developed, and hence, prior to the invention of methods for over-producing macromolecules, the choice confronted by biophysical chemists was often myoglobin, or nothing.

Biologically, in the 1950s, the ribosome was by far the most important macromolecule that you could prepare in gram quantities. Not surprisingly, therefore, virtually every physical technique that could be used to study ribosomes was used to study them as soon as it was invented, and, in fact, physical observations played an important role in the discovery of the ribosome.

The first observations relevant to the discovery of the ribosome were made in the 1930s by cytologists interested in characterizing biochemically the structures that can be visualized in eukaryotic cells using the light microscope, the most ancient of physical instruments used by biologists. By the early 1940s, it was clear that RNA is found primarily in the cytoplasms of eukaryotic cells, and that the more RNA there is in a eukaryotic cell, the more active it is likely to be in protein synthesis (Brachet 1941; Caspersson 1941). In that same era, using another physical instrument, the centrifuge, Claude identified a cellular fraction he dubbed "microsomes" (Claude 1941), and a few years later Brachet presciently suggested that the nucleoprotein particles abundant in microsomes might be involved in protein synthesis (Claude 1943; Brachet 1952).

The experiments that confirmed Brachet's hypothesis were carried out in the 1950s. Using two physical instruments that had just become commercialized, the electron microscope and the preparative ultracentrifuge, Porter, Palade, and Siekevitz discovered the endoplasmic reticulum (23–25), and proved that Claude's microsomes are fragments of endoplasmic reticulum and demonstrated that Claude's nucleoprotein particles are found free in the cytoplasms as well as associated with the membranous component microsomes. In 1954, Zamecnik and coworkers published the results of the elegant series of biochemical experiments done with radiolabeled amino acids (another post-war technology) proving that Claude's particles are indeed the sites where proteins are made in the cell (Keller et al. 1954). Experiments done with the analytical ultracentrifuge led to molecular weight estimates for these particles, confirmed that they are found in all cells, and showed that they are all 1:1 complexes of two nonequivalent ribonucleoprotein particles, the larger being about twice the molecular weight of the smaller (Chao and Schachman 1956; Chao 1957; Tissieres and Watson 1958). The name "ribosome" was bestowed on these particles in 1958 (Roberts 1958).

From the mid-1950s until 2000, the contributions made by biophysical chemists to our understanding of protein synthesis and the ribosome were overshadowed by those provided by biochemists and molecular biologists. Prominent among the many highlights of that four decade period were: (a) the discovery of tRNA (Hoagland et al. 1958; Crick 1958), (b) the discovery of messenger RNA (Brenner et al. 1961; Gros et al. 1961), (c) the elucidation of the code (Crick 1966), (d) the discovery that bacterial ribosomes can be reconstituted in vitro (Traub and Nomura 1968), (e) the elucidation of the protein composition of the ribosome (Waller 1964; Traut et al. 1967; Hardy 1975; Wittmann-Liebold 1986), and (f) the sequencing of the large ribosomal RNAs (Brosius et al. 1978; Noller and Woese 1981). [Many interesting accounts of this history have been written: e.g., (Zamecnik 1969; Tissieres 1974; Nomura 1990; Rheinberger 2004). Rheinberger's article is particularly useful because it was written by a professional historian of science, and the accounts it gives of events I know about are both fair and accurate.]

It would be incorrect to conclude from what has just been said that no biophysical research of consequence was done on the ribosome between ~1955 and 2000. (For a critical evaluation of much of that work see Moore 2011). For example, starting in the late 1960s, a lot of experiments were done on the ribosome using small angle X-ray (SAXS) and neutron scattering (SANS) techniques that provided a more refined sense of the overall shapes of the ribosome and its two subunits. The most complicated of these experiments were the SANS experiments that ultimately led to the so-called neutron map of the positions of proteins in the small ribosomal subunit from *Escherichia coli* (Capel et al. 1987). In those same years, an attempt was made to determine the locations of proteins in the ribosome by fluorescence energy transfer (FRET) (Huang et al. 1975), an approach to the study of protein synthesis that has come of age in the last decade or so.

Electron microscopy continued to contribute to our understanding of the ribosome, and of protein synthesis more generally. The first images of negatively stained ribosomes appeared in 1960 (Huxley and Zubay 1960), and by the mid-1970s, an accurate understanding of the shapes of the two ribosomal subunits, and their relationship in the intact ribosome had emerged (Lake 1976), as had a lot of information about the locations of specific proteins (Oakes et al. 1990). A highlight of this work was the discovery of the exit tunnel in 1982 (Bernabeu and Lake 1982; Milligan and Unwin 1986). EM studies of the ribosome done in that era also contributed significantly to the development of single particle reconstruction techniques, which are having a growing impact in all areas of structural biology today. Over the last 15 years, single particle reconstructions done with ribosome specimens maintained at liquid nitrogen temperatures (or below), i.e., cryo-EM, have become an increasingly important source of information about the conformational changes that occur as ribosomes function in protein synthesis (Frank et al. 1995).

Crystallographers too were hard at work. The first ribosome-related crystal structures solved were those of isolated ribosomal proteins (Leijonmarck et al. 1980). Over the years, roughly a dozen crystal structures and NMR structures were obtained for proteins from the bacterial ribosome. In addition, the structures of a number of ribosomal RNA fragments and ribosomal protein/RNA fragments were also determined. It was in this same era that crystallographic investigations of intact ribosomes and ribosomal subunits got underway. The first crystals of ribosomes large enough to work with were prepared in the late 1970s in Wittmann's laboratory in Berlin (Yonath et al. 1980). They diffracted poorly, as did all the other crystals of ribosomes and ribosomal subunits produced for many years thereafter. However, as experience deepened, resolution slowly improved (von Bohlen et al. 1991; Hope et al. 1989). By the early 1990s, the resolution of the diffraction patterns being obtained from ribosome crystals was so high that it was clear that atomic resolution structures of ribosomes would emerge if reliable strategies could be found for phasing them. That last hurdle was overcome in 1998 (Ban et al. 1998), and in the summer of 2000, as everyone knows, the landscape in the protein synthesis field was transformed by the publication of the first, atomic resolution structures of ribosomal subunits (Ban et al. 2000; Wimberly et al. 2000; Schluenzen et al. 2000).

The crystal structures of ribosomes that appeared in 2000 not only transformed the protein synthesis field, they re-energized it. Since 2000, a field that had for decades suffered from a dearth of atomic resolution information has been reveling in an abundance of riches. Crystal structures of 70S ribosomes trapped in different conformational states appear regularly. Cryo-electron microscopy continues to deliver an ever more detailed account of the structural dynamics of the ribosome. This structural information has enabled biochemists and molecular biologists to design and execute powerful experiments that are delivering important mechanistic insights, as well as stimulating renewed activity in the area of kinetics. Finally, single molecule studies are now underway that are likely to further transform the way we think about protein synthesis. The field is now in a golden age, and the chapters in this book, which of course will fill in much of the background missing from this sketch, will provide its readers with a sense of what the future is likely to bring.

New Haven, CT, USA   Peter B. Moore

References

Ban N, Freeborn B, Nissen P, Penczek P, Grassucci RA, Sweet R, Frank J, Moore PB, Steitz TA (1998) A 9 A resolution X-ray crystallographic map of the large ribosomal subunit. Cell 93:1105–1115

Ban N, Nissen P, Hansen J, Moore PB, Steitz TA (2000) The complete atomic structure of the large ribosomal subunit at 2.4 A resolution. Science 289:905–920

Bernabeu C, Lake JA (1982) Nascent polypeptide chains emerge from the exit domain of the large ribosomal subunit: immune mapping of the nascent chain. Proc Natl Acad Sci USA 79:3111–3115

Brachet J (1941) La detection histochimique et le microdosage des acides pentosenucleique. Enzymologia 10:87–96

Brachet J (1952) The role of the nucleus and cytoplasm in synthesis and morphogenesis. Symp Soc Exp Biol 6:173–200

Brenner S, Jacob F, Meselson M (1961) An unstable intermediate carrying information from genes to ribosomes for protein synthesis. Nature 190:567–581

Brosius J, Palmer ML, Kennedy PJ, Noller HF (1978) Complete nucleotide sequence of a 16S ribosomal RNA gene from *Escherichia coli*. Proc Natl Acad Sci USA 75:4801–4805

Capel MS, Engelman DM, Freeborn BR, Kjeldgaard M, Langer JA, Ramakrishnan V, Schindler DG, Schneider DK, Schoenborn BP, Sillers I-Y, Yabuki S, Moore PB (1987) A complete mapping of the proteins in the small ribosomal subunit of *E. coli*. Science 238:1403–1406

Caspersson T (1941) The protein metabolism of the cell. Naturwissenschaften 29:33–43

Chao FC (1957) Dissociation of macromolecular ribonucleoprotein of yeast. Arch Biochem Biophys 70:426–431

Chao FC, Schachman HK (1956) The isolation and characterization of a macromolecular ribonucleoprotein from yeast. Arch Biochem Biophys 61:220–230

Claude A (1941) Particulate components of cytoplasm. Cold Spring Harb Symp Quant Biol 9:263–271

Claude A (1943) The constitution of protoplasm. Science 97:451–455

Crick FHC (1958) On protein synthesis. Soc Exp Biol Symp London 12:138–163

Crick FHC (1966) The genetic code—yesterday, today and tomorrow. Cold Spring Harb Symp Quant Biol 31:3–9

Frank J, Zhu J, Penczek P, Li Y, Srivastava S, Verschoor A, Rademacher M, Grassucci R, Lata RK, Agrawal RK (1995) A model for protein synthesis based on cryo-electron microscopy of the *E. coli* ribosome. Nature 376:441–444

Gros F, Hiatt H, Gilbert G, Kurland CG, Risebrough RW, Watson JD (1961) Unstable ribonucleic acd revealed by pulse labelling of *E. coli*. Nature 190:581–585

Hardy SJS (1975) The stoichiometry of the ribosomal proteins of *Escherichia coli*. Mol Gen Genet 140:253–274

Hoagland MB, Stephenson ML, Scott JF, Hecht LI, Zamecnik PC (1958) A soluble ribonucleic acid intermediate in protein synthesis. J Biol Chem 231:241–257

Hope H, Frolow F, von Bohlen K, Makowski I, Kratky C, Halfori Y, Danz H, Webster P, Bartles KS, Wittmann HG, Yonath A (1989) Cryocrystallography of ribosomal particles. Acta Cryst B45:190–199

Huang KH, Fairclough RH, Cantor CR (1975) Singlet energy transfer studies of the arragnement of proteins in the 30S *Escherichia coli* ribosome. J Mol Biol 97:443–470

Huxley HE, Zubay G (1960) Electron microscopic observations on the structure of microsomal particles from *Escherichia coli*. J Mol Biol 2:10–18

Keller EB, Zamecnik PC, Loftfield RB (1954) The role of microsomes in the incorporation of amino acids into proteins. J Histochem Cytochem 2:378–386

Lake JA (1976) Ribosome structure determined by electron microscopy of *Escherichia coli* small subunits, large subunits and monomeric ribosomes. J Mol Biol 105:131–159

Leijonmarck M, Eriksson S, Liljas A (1980) Crystal structure of a ribosomal component at 2.6A resolution. Nature 286:824–826

Milligan RA, Unwin PNT (1986) Location of the exit channel for nascent protein in 80S ribosome. Nature 319:693–695

Moore PB (2011) Structure determination without crystals: the ribosome, 1970–2000. Cryst Growth Des 11:627–631

Noller HF, Woese CR (1981) Secondary structure of 16S ribosomal RNA. Science 212:403–411

Nomura M (1990) History of ribosome research: a personal account. In: Hill WE, Dahlberg A, Garrett RA, Moore PB, Schlessinger D, Warner JD (eds) The Ribosome: structure, function and evolution. Americal Society for Microbiology, Washington, DC, pp 3–55

Oakes MI, Scheinman A, Atha T, Shankweiler G, Lake JA (1990) Ribosome structure: three-dimensional locations of rRNA and proteins. In: Hill WE, Dahlberg AE, Garrett RA, Moore PB, Schlessinger D, Warner JR (eds) The Ribosome: structure, function and evolution. American Society for Microbiology, Washington, DC, pp 180–193

Rheinberger H-J (2004) A history of protein biosynthesis and ribosome research. In: Nierhaus KH, Wilson DN (eds) Protein Synthesis and Ribosome Structure. Wiley-VCH Verlag, Weinheim, pp 1–51

Roberts RB (1958) Microsomal particles and protein synthesis. Pergamon Press, New York

Schluenzen F, Tocilj A, Zarivach R, Harms J, Gluehmann M, Janell D, Bashan A, Bartles H, Agmon I, Franceschi F, Yonath A (2000) Structure of functionally activated small ribosomal subunit at 3.3 A resolution. Cell 102:615–623

Tissieres A (1974) Ribosome research: historical background. In: Nomura M, Tissieres A, Lengyel P (eds) Ribosomes. Cold Spring Harbor Laboratory, Cold Spring Harbor, NY, pp 3–12

Tissieres A, Watson JD (1958) Ribonulceoprotein particles from *Escherichia coli*. Nature 182:778–780

Traub P, Nomura M (1968) Structure and function of *E. coli* ribosomes V. Reconstitution of functionally active 30S ribosomal particles from RNA and proteins. Proc Natl Acad Sci USA 59:777–784

Traut RR, Moore PB, Delius H, Noller HF, Tissieres A (1967) Ribosomal proteins of *E. coli* I. Demonstration of primary structure differences. Proc Natl Acad Sci USA 57:1294–1301

von Bohlen K, Makowski I, Hansen HAS, Bartels H, Berkovitch-Yellin Z, Zaytzev-Bushan A, Meyer S, Paulke C, Franscheschi F, Yonath A (1991) Characterization and preliminary attempts for derivitization of crystals of large ribosomal subunits from *Haloarcula marismortui* diffracting to 3 A resolution. J Mol Biol 222:11–15

Waller JP (1964) Fractionation of the ribosomal protein from *Escherichia coli*. J Mol Biol 10:319–336

Wimberly BT, Brodersen DE, Clemons WM, Morgan-Warren RJ, Carter AP, Vonrhein C, Hartsch T, Ramakrishnan V (2000) Structure of the 30S ribosomal subunit. Nature 407:327–339

Wittmann-Liebold B (1986) Ribosomal proteins: their structure and evolution. In: Hardesty B, Kramer G (eds) Structure, function, and genetics of ribosomes. Springer-Verlag, New York, pp 326–361

Yonath A, Mussig J, Tesche B, Lorenz S, Erdmann VA, Wittmann HG (1980) Crystallization of the large ribosomal subunits from *Bacillus stearothermophilus*. Biochem Internat 1:428–435

Zamecnik PC (1969) An historical account of protein synthesis, with current overtones—a personalized view. Cold Spring Harb Symp Quant Biol 34:1–16

Contents

Chapter 1
X-Ray Analysis of Prokaryotic and Eukaryotic Ribosomes

**Lasse B. Jenner, Adam Ben-Shem, Natalia Demeshkina,
Marat Yusupov, and Gulnara Yusupova**

1.1 X-Ray Crystallography and the Ribosome

1.1.1 X-Ray Methodology

X-ray crystallography is an experimental technique that takes advantage of the fact that X-rays are scattered by electrons. Using electromagnetic radiation to visualize objects by scattering requires the wavelength of the radiation to be comparable to the smallest features to be resolved. Since the atomic bond lengths most commonly found in biological materials are in the 1–2 Angstrom (Å) range, the X-rays produced by in-house rotating anodes and large-scale facilities such as synchrotrons are well suited for this purpose. While scattering from one molecule is too weak to be measured, diffraction from a crystal containing millions of molecules all oriented in a regularly repeated manner is detectable. The diffraction data acquired by X-ray scattering off the periodic assembly of molecules in the crystal can be used to reconstruct the electron density. Electron distributions observed this way provides the locations of the atomic nuclei. An atomic model is iteratively constructed and refined into the observed electron density leading to a rather accurate molecular structure.

L.B. Jenner • A. Ben-Shem • N. Demeshkina • M. Yusupov (✉) • G. Yusupova (✉)
Département de Biologie et de Génomique Structurales, Institut de Génétique et de
Biologie Moléculaire et Cellulaire, CNRS, UMR7104, University of Strasbourg,
Strasbourg, Illkirch 67400, France
e-mail: Marat.Yusupov@igbmc.fr; gula@igbmc.u-strasbg.fr

J.D. Dinman (ed.), *Biophysical Approaches to Translational Control of Gene Expression*,
Biophysics for the Life Sciences 1, DOI 10.1007/978-1-4614-3991-2_1,
© Springer Science+Business Media New York 2013

1.1.2   Challenges for X-Ray Studies of Ribosomal Complexes

Crystallographic methods can shed light on many structure-related issues, from overall molecular conformations and ternary and quaternary interactions to secondary structure information and details about atomic bonds. In contrast to NMR and Cryo-EM approaches, there is no limitation to the size of molecule or assembly to be studied.

The main bottleneck in crystallographic studies is that a well-diffracting crystal must be found, and thus the information gleaned about the dynamic nature of the molecules to be studied will be very limited from only a single diffraction experiment. In other words, the price to pay for the high accuracy of X-ray crystallographic structures is that the method is very time-consuming.

For the ribosome as a huge complex consisting more than 50 components, it is very important to ensure that the samples are homogenous for crystallization to succeed. In our studies we have achieved this by two different strategies: For the prokaryotic studies we have chosen to work with a thermophilic bacteria because the ribosomes isolated from this organism are more robust and resistant to degradation. For the eukaryotic ribosome very gentle isolation protocols were developed to ensure that all the ribosomal components are intact and present. We exploited the observations that glucose starvation of the growing yeast cells leads to inhibition of initiation and accumulation of very homogenous ribosomes without any ligands (Ashe et al. 2000).

A further complication arises since ribosome crystals, as typically seen in RNA crystallography, diffract only poorly which results in electron density maps that are imprecise and difficult to interpret. Therefore special care has to be taken during post-crystallization treatment to avoid damaging the crystals (i.e., when transferring cryo-protection) and even for the freezing process itself we only use the most robust methods of freezing directly in the gaseous N_2 stream at 100 K rather than plunging into liquid N_2, ethane, or propane as is common practice in X-ray structural projects. A combination of severe radiation decay and generally weak diffracting power limits the amount of data that can be collected from each crystal making it necessary to merge data collected on different crystals to obtain complete datasets which invariantly degrades the data quality.

1.2   Crystal Structures of Prokaryotic Ribosome Complexes

1.2.1   Introduction

Translation of nucleotide sequence information in the form of mRNA codons into amino acids lies at the heart of protein biosynthesis. This process is accomplished by tRNA molecules that act as adaptors between the mRNA codon and the amino acids they code for. For accurate protein synthesis, the ribosome is required to position the

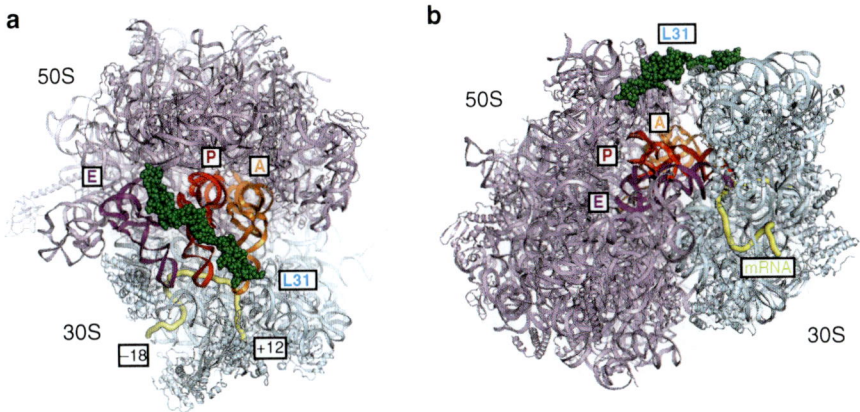

Fig. 1.1 Overall view of the 70S ribosome in elongation state. A, P, and E tRNA are shown in *orange*, *red*, and *magenta*, respectively, and 60-mer mRNA (position −18 to +12 visible) is shown in *gold*. Ribosomal proteins and RNA of the small and large subunits are shown in *light blue* and *violet*, respectively. The new intersubunit bridge formed by the protein L31 is shown in *green*. (**a**) *Top* view; (**b**) view from the E site

tRNAs in such a way that the reading frame of the mRNA (each codon consists of three consecutive nucleotides) is maintained throughout the translation process. The elongating ribosome contains three binding sites for tRNAs: The aminoacyl (A) site to which a cognate aminoacyl-tRNA is delivered such that it base pairs with the appropriate mRNA codon; The peptidyl (P) site where the tRNA carrying the nascent peptide chain is located. When a cognate aminoacyl-tRNA enters the A site the peptidyl transferase reaction takes place and the peptidyl chain carried by the P tRNA is added to the aminoacyl tRNA essentially adding one amino acid to the growing peptide chain; Last is the exit (E) site from which deacylated tRNA that has completed its role in translation is released.

After each peptide bond reaction, the ribosome must rearrange its contacts with mRNA and tRNA to allow translocation along the mRNA by a single three nucleotides codon. The ribosome controls the positioning of mRNA and tRNAs during the translation process through a number of direct intermolecular contacts. These interactions not only help to stabilize the binding of tRNA to the ribosome but are involved directly in functional processes such as mechanisms for discrimination of aminoacyl-tRNAs to increase the accuracy of tRNA selection; maintenance of the correct reading frame to avoid frame-shifting errors; and translocational movement of the tRNAs and mRNA within the ribosome. We are striving to understand these mechanisms by studying how the main substrates of protein synthesis such as mRNA and tRNA interact with the ribosome.

In order to shed light on these issues we recently determined high-resolution crystal structures of *Thermus thermophilus* 70S ribosomal complexes with different mRNA constructs and naturally modified tRNAs (Fig. 1.1) (Jenner et al. 2010a, b).

Crystals of the ribosome modeling the elongation state were obtained from ribosome complexes prepared with a 60 nucleotides long poly (U) mRNA containing a

Shine–Dalgarno (SD) sequence, UUU (Phe) codons in the A and P sites, and tRNA[Phe]. The structure of this complex was determined at 3.1 Å resolution. Crystals of the ribosome complex modeling the initiation state were prepared with a 27 nucleotide long mRNA comprising the SD sequence with AUG (Met) codon and initiator tRNA[fMet] in the P site. The structure of this complex was determined at 3.5 Å resolution.

1.2.2 A Novel Intersubunit Bridge Formed by Protein L31 May Regulate Swiveling of the 30S Head

Inspection of the electron density map corresponding to our elongation state revealed well-defined density for a novel element not fully seen in previous high-resolution structures. This element was the ribosomal protein L31 (Figs. 1.1 and 1.2) (Jenner et al. 2010a, b).

Protein L31 displays a considerable degree of similarity among bacteria (Fig. 1.2a), and is comprised of a three β-sheets Zn-binding domain followed by a loop area and an α-helix at its C-terminal. Interestingly, L31 crosses the intersubunit space yoking together the central protuberance of the 50S subunit and the head domain of the 30S subunit. At the 30S subunit head, L31 interacts with the two highly conserved proteins S13 and S19 that are known to form a loose hetero-dimer (Brodersen et al. 2002) (Fig. 1.2b).

Protein S13 is part of the B1a and B1b intersubunit bridges and has a C-terminal that approaches the P site (Yusupov et al. 2001). The central part of L31 (amino acids 32–52) interacts directly with S13 mostly through electrostatic interactions (Fig. 1.2c) while the interaction surface between S19 and L31 is not only of polar but also hydrophobic nature (Fig. 1.2d). The majority of the interacting residues of proteins S13, S19, and L31 are conserved. Overall, protein L31 clips together the globular N-domains of S13 and S19, presumably tightening their association.

The biological relevance of this intersubunit bridge composed of protein L31 may lie in regulating and safeguarding the swiveling of the 30S subunit head domain. It might function as a safety belt, delimiting the extent of 30S head rotation, in the ratchet-like motion supposed to happen during translocation (Spahn et al. 2004a; Frank and Agrawal 2000; Gao et al. 2003).

1.2.3 The Path of Messenger RNA Through the Ribosome

In the initial high-resolution (2.8–3.6 Å) structures of 70S ribosomal complexes, the mRNA was visualized only from positions −4 to +7 (Selmer et al. 2006; Weixlbaumer et al. 2007a, b; Korostelev et al. 2008; Laurberg et al. 2008) although earlier medium resolution structures (4.5–5.5 Å) had previously indicated the entire mRNA path (Jenner et al. 2007, 2005; Yusupova et al. 2006, 2001). In our recent high-resolution

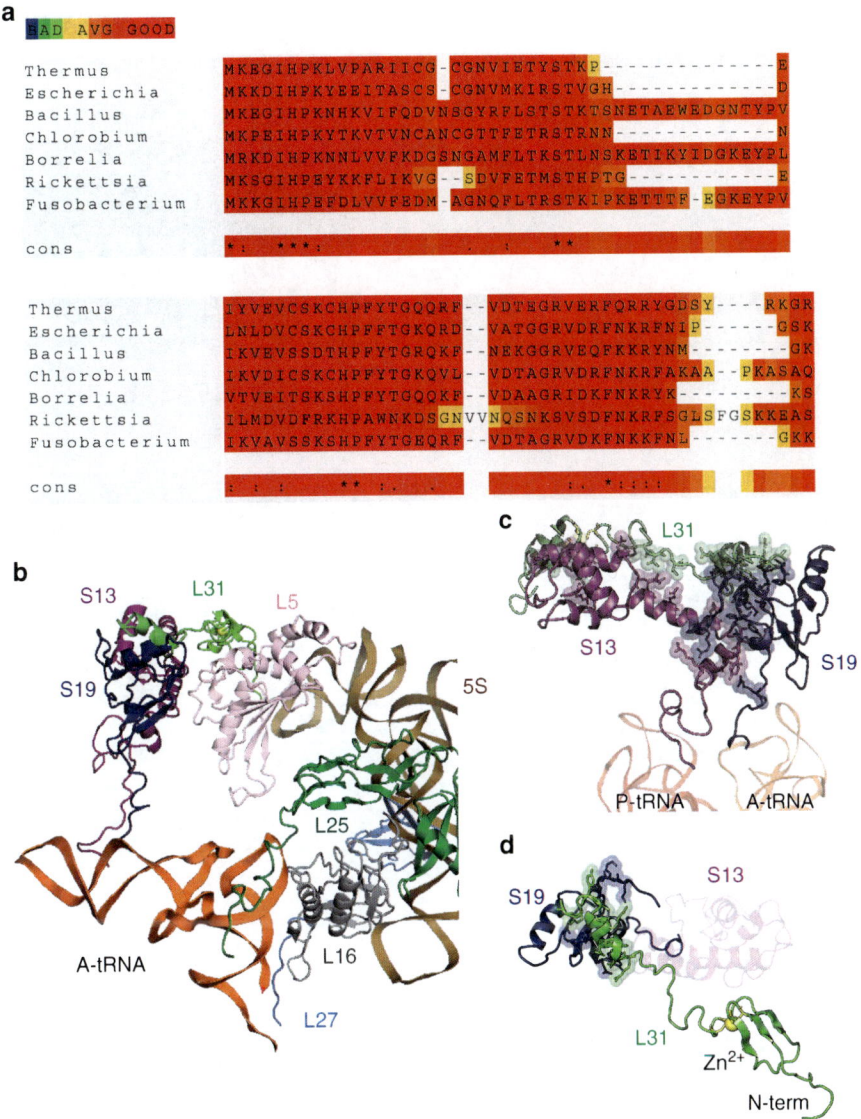

Fig. 1.2 Novel intersubunit bridge formed by protein L31. (**a**) Sequence alignment for L31. The degree of similarity among the major bacterial classes is color coded and specified in the *top* bar. The *bottom* color line indicates the conserved residues. (**b**) Protein L31 closes the circle of interactions around A-site tRNA. (**c**) Interactions between L31, S13, and S19. (**d**) Close-up view of the interactions between L31 and S19. The Zn²⁺ atom is shown as a *yellow sphere*

structure (3.1 Å) of a ribosomal complex in the elongation state we were again able to confidently model the mRNA from positions −18 to +12, which delimit the ribosome boundaries (Fig. 1.3a).

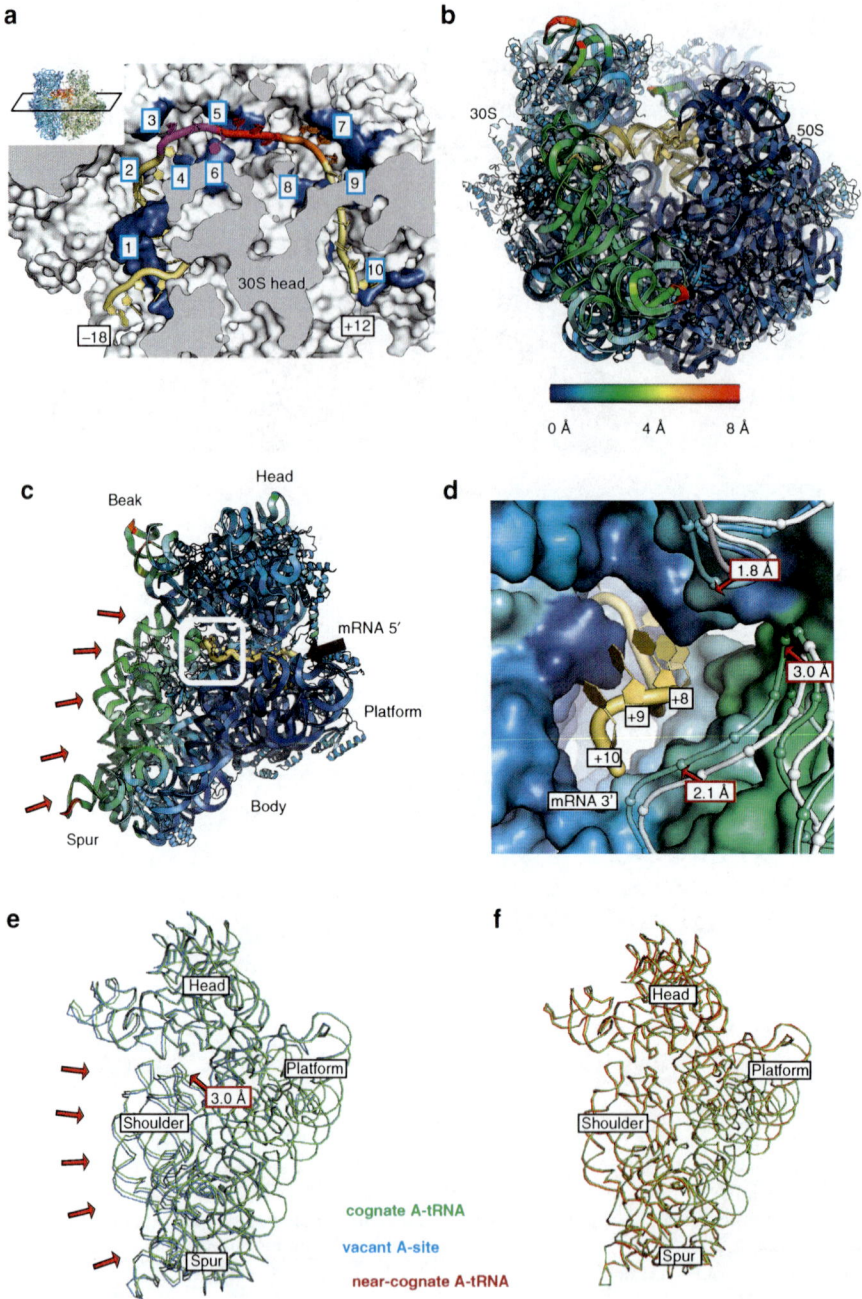

Fig. 1.3 View of mRNA and tRNA interactions with the ribosome. (a) Cross-section of the ribosome at the level of mRNA showing interactions between mRNA and the following ribosomal elements: (1) Shine–Dalgarno sequence of the 3′ of 16S rRNA; (2) ribosomal proteins S11 and S18; (3) loop of helix 23b (16S rRNA); (4) A1507 of 16S rRNA; (5) interaction with modified

The 3′ end of the mRNA enters the ribosome through a tunnel formed by ribosomal proteins S3, S4, and S5 (mRNA nucleotides +10 to +12) after which the mRNA passes a layer formed by 16S rRNA elements that are capable of contracting around the mRNA (mRNA nucleotides +7 to +9). The A- (+4 to +6), P- (+1 to +3), and E-site (−3 to −1) codons interact with the respective tRNAs on the interface between the ribosomal subunits. Finally the mRNA emerges on the platform of the 16S subunit where the 5′ end of the mRNA upstream of the E-site codon along with 3′-terminal tail of 16S rRNA forms the SD duplex (Shine and Dalgarno 1974).

1.2.4   Domain Closure

Comparison of two crystal structures of the ribosome modeling the initiation and elongation states reveals that, upon transition from initiation to elongation, the 30S subunit undergoes a conformational change whereupon helices 15–18 from the body of the 30S subunit contract towards the 30S neck (Fig. 1.3b, c) (Jenner et al. 2010b).

The remaining part of the body and most of the head of the 30S subunit remain immobile. This domain closure results in a contraction by 1–2 Å of the mRNA tunnel immediately downstream of the A-site codon causing it to grip the template more tightly in the elongation state than in the initiation state (Fig. 1.3d). A similar conformational change was seen in studies of the isolated 30S subunit with an anti-codon stem-loop bound in the A site (Ogle et al. 2002). From those studies it was hypothesized that domain closure occurs only when a cognate tRNA is bound in order to signal correct decoding. Unexpectedly, our results with the full functional

◄ ───

Fig. 1.3 (continued) nucleotide 37 of the P-site tRNA through $Mg(H_2O)_6^{2+}$ and stabilization of the mRNA kink between the P- and A-site codons via interactions with nucleotides from h44 (16S rRNA); (6) stacking of the base of mRNA position −1 with G926 from h28 (16S rRNA); (7) the mRNA A codon interactions with nucleotides G530, A1492, and A1493 (16S rRNA); (8) C1397 from 16S RNA; (9) aromatic stacking network between mRNA and U1196 and C1054 (16S rRNA); (10) ribosomal proteins S3, S4, and S5. (**b, c**) Conformational changes of the 70S ribosome. The 30S structure is colored according to the difference between phosphate and C_α positions in the initiation and elongation complexes, ranging from *blue* (0 Å difference) to *red* (8 Å difference). *Arrows* indicate the direction of movement of the domain closure during transition from the initiation to elongation state. The downstream mRNA tunnel has been marked with a white outline. From the superposition it is clear that only the shoulder of the 30S subunit moves, whereas the other parts of the 30S subunit remain immobile, and that the resulting movement leads to a contraction of the mRNA tunnel downstream of the A-site codon. (**d**) Detailed view of the RNA-layer part of the downstream mRNA tunnel seen from the solvent side of the 30S subunit. The RNA chains with the largest movements are shown in *white* (initiation) and color (elongation) with difference vectors marking the changes in position. The contraction of the downstream mRNA tunnel leads to a narrowing of the tunnel diameter by 1–3 Å, tightening the ribosome grip on the mRNA. (**e**) Superpostion of cognate A-tRNA (*green*) and vacant A-site (*blue*) states. The *black arrows* indicate the general movement of the 30S subunit domain towards the neck region of the subunit. (**f**) Superposition of cognate A-tRNA (*green*) and near-cognate A-tRNA (*red*) states

70S ribosome demonstrated the exact same domain closure happening upon binding of near-cognate tRNAs to the A site (Fig. 1.3e, f) [(Jenner et al. 2010a) and unpublished data]. The dissimilarities with the previous studies have led us to conclude that domain closure happens upon binding of any substrate in the A site and does not play an active role in decoding.

Examination of mRNA shows that the third nucleotide of the A-site codon interacts with the nucleotide base of C1397 from the neck region of 16S rRNA (h28). C1397 which nearly intercalates between the mRNA bases at positions +6 and +7 in the initiation complex (Fig. 1.4a) (Jenner et al. 2010b). Nucleotide C1397 protrudes from the side of the mRNA tunnel and seems to be able to adopt different conformations, depending on the state of the ribosome and the presence or absence of tRNA in the A site (Fig. 1.4a, b). Furthermore, in the initiation complex, the sugar moiety of mRNA nucleotide +9 forms a hydrogen bond with Gln162 of protein S3. However, in the elongation complex contraction of the downstream mRNA tunnel triggers the formation of an intricate network of interactions between 16S RNA, protein S3, and the mRNA adjacent to the A-site codon (Fig. 1.4c, d). Nucleotides +8 and +9 of the mRNA are held in place by a combination of hydrogen bonding and continuous aromatic base stacking with Gln162 of protein S3 and nucleotides U1196 and C1054 from helix 34 of 16S rRNA. Finally, C1054 interacts with G34 of A-tRNA as seen previously (Ogle et al. 2002).

This network of interactions between mRNA and the head of the 30S may align the mRNA immediately downstream of the A-site codon before its movement into the A site, such that the codon approaching the decoding center is pre-oriented for the interaction with the tRNA. We suggest that the ribosome preserves this network of interactions during translocation in order to strongly and accurately safeguard the mRNA. Thus, the mRNA reading frame is maintained not only by codon–anticodon interactions but also by this network in the downstream tunnel during swiveling of the 30S head in the course of the ratchet-like movement of the small ribosomal subunit relative to the large ribosomal subunit that accompanies translocation (Fig. 1.4e). After translocation, the mRNA interactions with h34 of 16S rRNA and protein S3 must be disrupted, and the 30S subunit head returns to its initial position.

$\longrightarrow$

Fig. 1.4 (continued) position +8 and U1196 of 16S rRNA further stabilizing the network. Finally nucleotide C1054 interacts with the "wobble" nucleotide G34 of the A-site tRNA. (**e**) Interface view of part of the 30S subunit, with mRNA colored according to codon: *magenta* (E), *red* (P), *orange* (A), and *yellow* (downstream of A codon). tRNAs and 50S subunit have been removed for clarity. Ribosomal elements of the 30S subunit head interacting (hydrogen bond or hydrophobic or electrostatic interactions) with either mRNA or tRNA have been colored correspondingly. We propose that when the head of the 30S subunit swivels (indicated by the *arrow*) in the course of the ratchet-like movement of the small ribosomal subunit relative to the large ribosomal subunit, the ribosome translocates the mRNA by maintaining the shown interactions

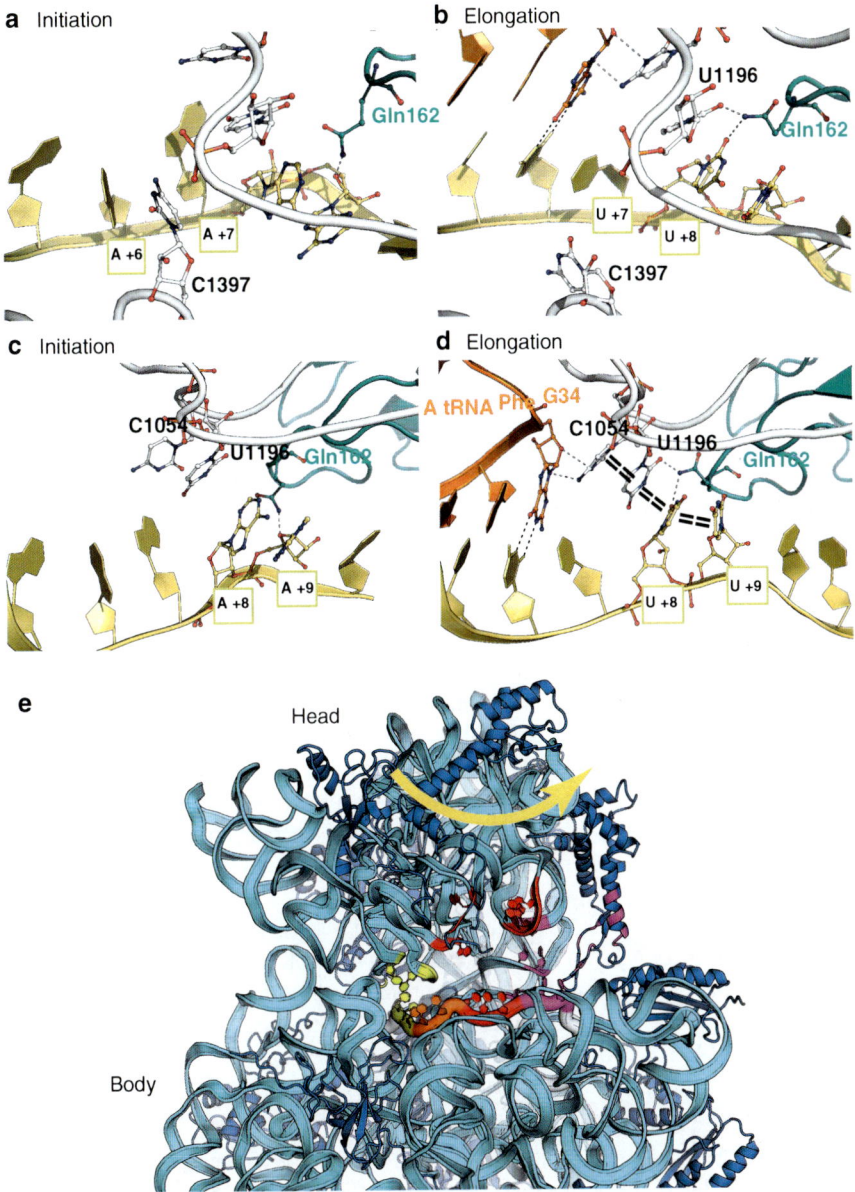

Fig. 1.4 mRNA–ribosome interactions downstream of the A codon. (**a**) In the initiation complex nucleotide C1397 (h28) from the neck of 16S rRNA interacts with the third nucleotide (+6) of the A codon. (**b**) View of the interactions downstream of the A codon in the elongation complex: nucleotide C1397 (h28) from the neck of 16S rRNA is tucked away from mRNA. (**c**) The sugar backbone of nucleotide +9 is held in place by hydrogen bonding with Gln162 of ribosome protein S3 in the initiation complex. (**d**) In the elongating ribosome a continuous base-stacking network is formed between nucleotides at positions +9 and +8 which in turn stack on 16S rRNA nucleotides U1196 and C1054. Gln162 of protein S3 forms hydrogen bonds with mRNA nucleotide at

1.2.5   tRNA Modifications Stabilize mRNA–tRNA Interactions

Most frame-shift errors introduce entirely incorrect amino acid sequences or premature stop codons and are as such destructive to the production of a functional protein. In contrast, missense errors are not necessarily fatal to protein synthesis. The error frequency associated with frame-shift events has been estimated to be not higher than one event per 30,000 amino acids incorporated (Jorgensen and Kurland 1990). Biochemical and genetic studies have shown that natural posttranscriptional modifications of tRNAs play a large role in maintaining the correct reading frame and decreasing the frequency of frame shifting (Bouadloun et al. 1986; Urbonavicius et al. 2001; Konevega et al. 2006; Gustilo et al. 2008). At present more than 80 different modified nucleosides have been characterized (Rozenski et al. 1999). Although they are present in tRNAs from all organisms and located at different positions within the tRNA, the majority are located in the anticodon region, especially at positions 34 (the wobble position) and 37 (3′ adjacent to the anticodon) with the type of modification varying with codon specificity. Crystals of isolated 30S subunits carrying an anticodon stem-loop in the A site indicated that a modification at position 34 (uridine 5-oxyacetic acid) not only increases the stability of codon–anticodon interaction but also expands the decoding capacity of the tRNA (Weixlbaumer et al. 2007a).

In our study (Jenner et al. 2010b) the elongation complex was formed with *E. coli* tRNA^{Phe}, which contains the hypermodified nucleotide 2-methylthio-N6-isopentenyl adenosine (ms^{2}i^{6}A37) in the anticodon loop. We described how the sulfur atom of ms^{2}i^{6}A37 in the A, P, and E sites stabilizes the codon–anticodon duplex by forming cross-strand stacking interactions with the bases of the first nucleotide in the respective mRNA codon (Fig. 1.5a–d). This enhanced interaction greatly increases the normal triplet interaction strength in the A and P sites.

Formation of a network of interactions around the first nucleotide of the P-site codon and the third nucleotide of the E-site codon is triggered by the presence of the methylthio-group of ms2i6A37 in P-site tRNA^{Phe} (Fig. 1.5e). This network involves a hydrated magnesium ion not found in the initiation complex where this modification is not present. The magnesium ion is coordinated by the sulfur atom of ms^{2}i^{6}A37 of the P-site tRNA, the phosphates of the mRNA nucleotides U+1 and U−1, and the bases of nucleotides A790 and U789 of 16S RNA. Nucleotide U−1 flips out of the A-helical conformation of the E-site codon and stacks directly on top of nucleotide G926 of 16S rRNA in the neck-helix 28. Such anchoring of the mRNA, which is not sequence dependent, probably increases the resistance against peptidyl-tRNA-slippage and mRNA frame shift. Translational errors due to modification deficiencies of ms^{2}i^{6}A37 in tRNAs have been studied extensively in vivo and in vitro, but the molecular mechanism of the ms^{2}i^{6}A37 contribution to reading frame maintenance was until now unclear (Urbonavicius et al. 2001, 2003; Wilson and Roe 1989).

Now we can rationalize previous biochemical and genetic data indicating that the absence of the methyl-thio group of ms^{2}i^{6}A37 significantly decreased the efficiency of tRNA^{Phe} binding to the ribosome (Menichi and Heyman 1976; Hoburg et al. 1979) and increased the frame-shift frequency. Seen in the light of our results, slippage of

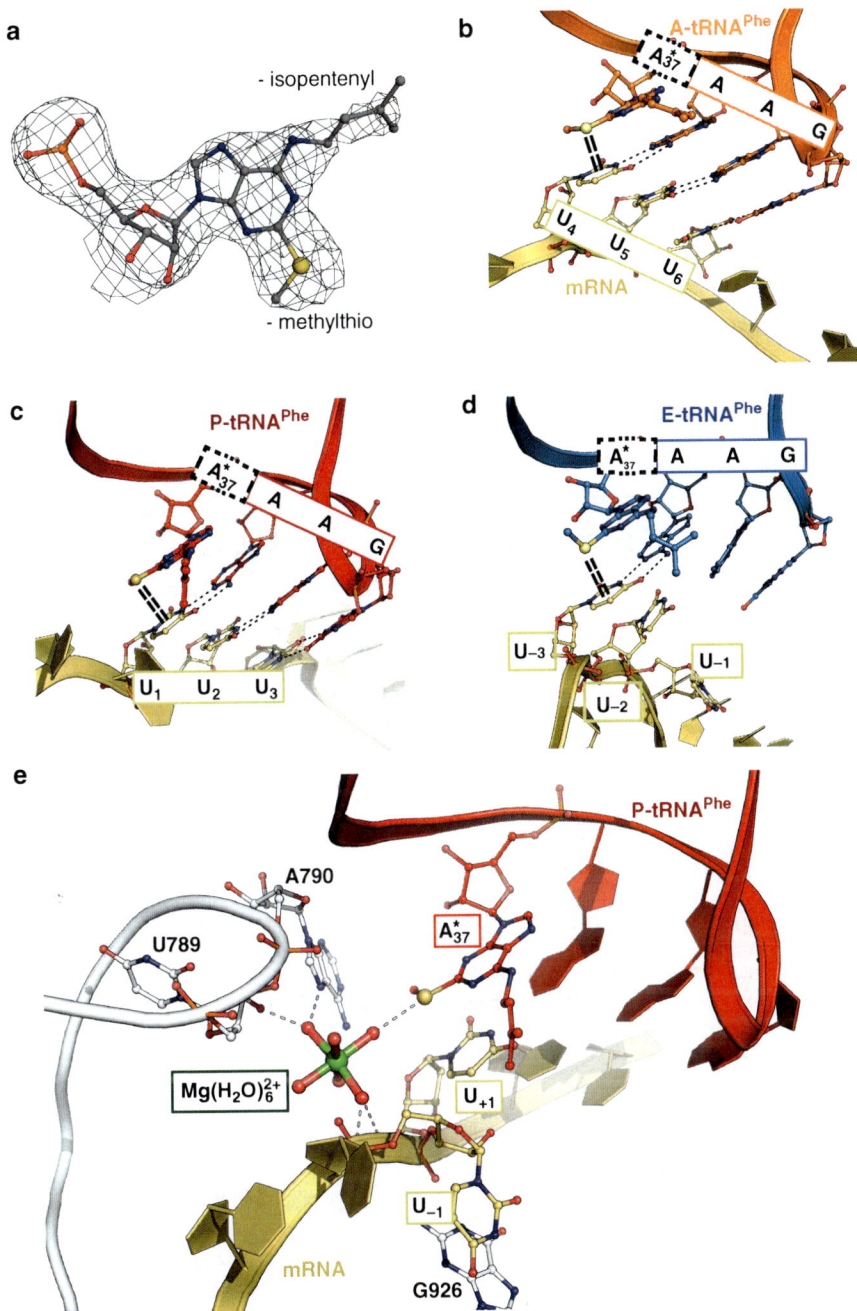

Fig. 1.5 Stabilization of codon–anticodon interactions by cross-strand stacking of hypermodified nucleotide 37 (ms^{2}i^{6}A37) in the A, P, and E sites. (**a**) 2F$_o$–F$_c$ density contoured at 1.5σ for ms^{2}i^{6}A37 of P-tRNA$_{GAA}$^{Phe}. (**b, c**) Cross-strand stacking and tRNA–codon interactions in the A (**b**) and P (**c**) sites in the elongation complex containing tRNA$_{GAA}$^{Phe}. (**d**) Base pairing at the first position of the E codon and stabilization by cross-strand stacking of the methylthio group of hypermodified nucleoside ms^{2}i^{6}A37 of tRNA$_{GAA}$^{Phe} in the elongation complex. (**e**) Anchoring of tRNA in the P site. A hydrated magnesium ion is coordinated by the sulfur atom of ms^{2}i^{6}A37 of tRNA$_{GAA}$^{Phe} in the P site, the mRNA backbone phosphates of nucleotides at positions +1 and −1 and bases A790 and U789 of 16S rRNA

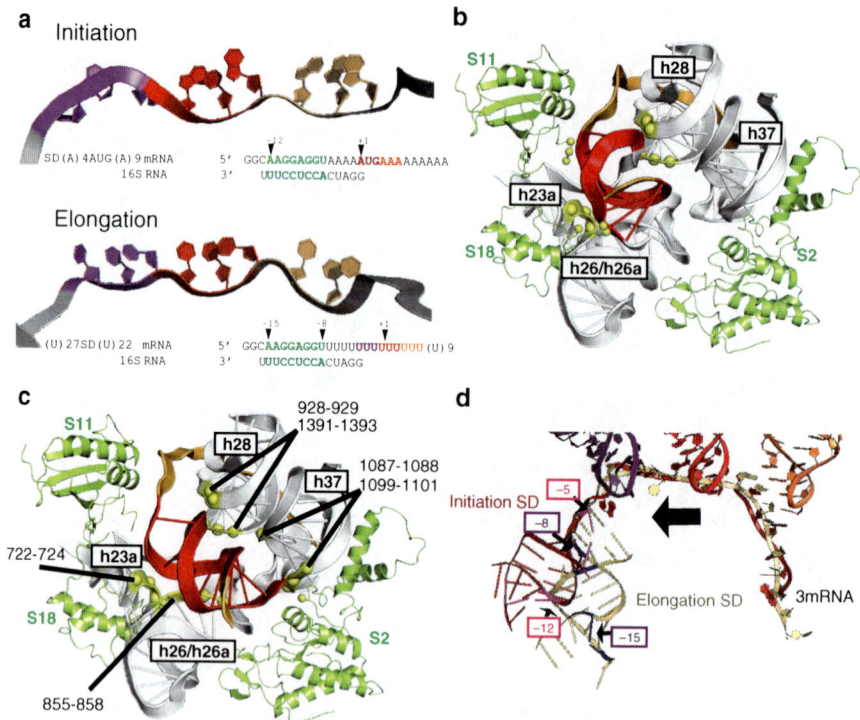

Fig. 1.6 The upstream mRNA is anchored on the platform. (**a**) Structure of E, P, and A codons in initiation (*top*) and elongation (*bottom*) complexes, respectively. The mRNA sequences and Shine–Dalgarno duplex (in *red*) are shown below. (**b, c**) Detailed view of the ribosomal environment of the SD duplex in the initiation (**c**) and elongation (**d**) complexes. (**d**) Positional shift of the Shine–Dalgarno duplex. Movement of the mRNA upon transition from initiation (*red*) to the elongation state (*yellow*) gives rise to a positional shift of the SD duplex. 5′ and 3′-ends of the SD sequence are marked in *pink* (initiation complex) and *magenta* (elongation complex)

tRNA in the P site in the direction of the E site (−1) or the A site (+1) is reduced when the additional stabilization of the P- and E-site codon region by ms^2i^6A37 is present.

1.2.6 The Upstream mRNA is Anchored on the Platform

In the elongation state, there is also an increased interaction between the mRNA and ribosomal elements upstream of the P codon (Fig. 1.3a). Previous ribosome studies (Selmer et al. 2006; Jenner et al. 2007; Yusupova et al. 2006) that investigated the nature of codon–anticodon pairing in the E site used mRNAs in a "tense" conformation with a minimal distance (seven nucleotides) between the core adenosine (−8) of the SD sequence and the first nucleotide of the P-site codon (+1) (Fig. 1.6a)

(Schurr et al. 1993; Ma et al. 2002). The E-site codons in those complexes and in the present initiation complex were found in a conformation not favorable for codon–anticodon interaction. These results agree fully with the accepted concept of translation initiation where deacylated tRNA should not appear in the E site until after the first translocation event. In the initiation complex, we observed that the SD duplex anchors the 5′ end of the mRNA onto the platform of the 30S subunit, where the SD is positioned close to proteins S11 and S18 on the platform (Fig. 1.6d).

Contrary to the initiation state, the SD duplex assumes a different position on the platform in the elongation state, where it contacts ribosomal protein S2 (Fig. 1.6c). The mRNA is in a "relaxed" conformation where the distance between the SD core adenosine (-11) and the P-site codon ($+1$) is increased from seven to ten nucleotides, resulting in an enlargement of the SD duplex to 12 nucleotides, although classical Watson–Crick base-pairing in the double helix occurs in only nine pairs (Fig. 1.6a, d). Thus, almost all nucleotides (from 1,531 to 1,542) of the single-stranded 3′ end of 16S rRNA are involved in the formation of the SD duplex. As a consequence of the mRNA relaxation, the E-site codon adopts a conformation that is closer to the classic A-helical conformation (Fig. 1.6a), which leads to formation of a base pair between the first position of the E-site codon and position 36 of the cognate E-site tRNA^{Phe} (Fig. 1.5d). This base pair is stabilized by G693 of 16S rRNA and the 2-methylthio group of the ms^{2}i^{6}A37 tRNA modification. Whether the observed codon–anticodon interaction in the E site would exist without stabilization by the modification remains uncertain. No base pairs are observed for the second and third positions of the E-site codon which are too distant from the E-site tRNA to form standard Watson–Crick pairs. Additionally, the base of mRNA nucleotide -4 interacts with A1507 of 16S rRNA in the elongation state immediately upstream of the E-site codon.

The "tense" and "relaxed" mRNA conformations may delimit the extent of the movement that the SD duplexes can undergo on the platform of the ribosome, so that the SD helix can take intermediate positions between these two extremes. In support of this notion, crystal structures of the 30S subunit (Kaminishi et al. 2007) and the 70S ribosome (Korostelev et al. 2006) report positions of the SD duplex that are between or near the ones observed here for the initiation and elongation states.

1.2.7   mRNA Movement on the Ribosome

During the past years we have employed structural methods to study how the mRNA interacts with the ribosome at different translational states. These various structures represent snapshots of the working ribosome and as such can be used to propose a trajectory for the process. We suggest the following simplified scheme for how the mRNA moves on the ribosome at the different states of translation (Fig. 1.7) (Jenner et al. 2010b, 2007; Yusupova et al. 2006).

In the first step of initiation complex formation, the mRNA binds to the ribosome and establishes the SD duplex between the 5′-untranslated region and the

a

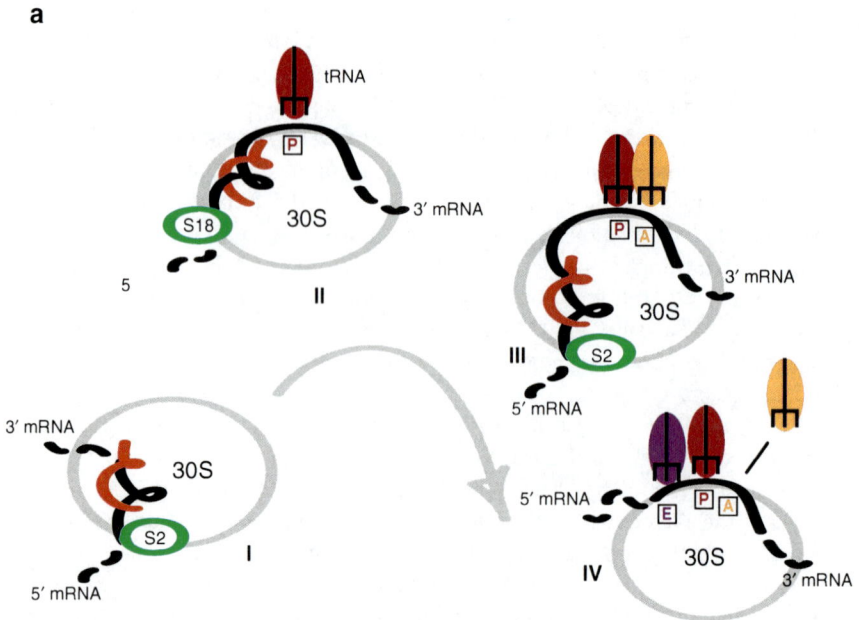

Fig. 1.7 Simplified scheme for messenger RNA motion on the ribosome

single-stranded 3' end of 16S rRNA. Subsequently, the SD duplex interacts with the 30S platform and is oriented towards ribosomal protein S2, while the rest of the mRNA molecule is still unbound (Fig. 1.7, state I). The following step is a simultaneous positional adjustment of the initiation codon in the P site, where it is stabilized by a codon–anticodon interaction with initiator tRNA[fMet], and a rearrangement of the SD duplex towards ribosomal protein S18 (Fig. 1.7, state II). At this stage of initiation, the mRNA is present in a tense conformation and is precisely positioned on the 30S subunit, so that the start codon will be read first and in the correct frame. Our findings of mRNA adjustments during the initiation process are in good agreement with earlier cross-linking results (Canonaco et al. 1989; Rinke-Appel et al. 1994; La Teana et al. 1995). Immediately after initiation, when one or several codons have already been translated by the ribosome but the SD interaction is still intact, a simultaneous movement of the complete mRNA in the 5' end direction and a lengthening of the SD helix take place (Fig. 1.7, state III). The SD helix is once again shifted towards ribosomal protein S2. Further translation of mRNA by the ribosome is accompanied by movement in the 3'–5' direction and leads to melting of the SD interaction. Eventually, at some stage during elongation, the 5' end of mRNA will no longer interact with the ribosome at all (Fig. 1.7, state IV) and become accessible for other ribosomes, which can lead to formation of a polysome.

1.2.8 Concluding Remarks

Over the course of past years we have shed light on the mechanism by which modifications on tRNAs stabilize tRNA–mRNA interaction on the ribosome. We showed how the ribosome, mRNA, and tRNA with modifications form contacts localized around the kink between the P and E codon in order to prevent P-tRNA from shifting frame. We described how contraction of the mRNA tunnel leads to formation of a new network of interactions between the mRNA downstream of the A-codon and the elongating ribosome. These interactions may safeguard against reading frame slippage during translocation and also align and prepare the mRNA prior to its movement into the A site for decoding. Furthermore, we have identified a novel intersubunit bridge in the form of ribosomal protein L31 that links the two subunits together and might serve as a safety belt during the large-scale conformational changes that accompanies translocation. These results represented a major step in our understanding of how the ribosome strongly and accurately safeguards the mRNA during translocation, and illuminates one of the fundamental questions in translation: how the ribosome reliably traverses the mRNA and ensures accurate reading of the genetic code.

1.3 Crystal Structure of the Eukaryotic 80S Ribosome

Although the eukaryotic ribosome was identified first, most structural information about the mechanism of protein biosynthesis has been obtained from prokaryotic systems.

The architecture of the universally conserved catalytic core where peptide bonds are formed and correct aminoacyl-tRNAs are selected to complement mRNA codons was revealed by X-ray structures of the full prokaryotic 70S ribosome and individual subunits (Jenner et al. 2010b; Yusupov et al. 2001, 2006; Selmer et al. 2006; Laurberg et al. 2008; Harms et al. 2001; Ban et al. 2000; Wimberly et al. 2000; Schuwirth et al. 2005). There are, however, many other processes also constituting integral steps in translation that differ greatly between prokaryotes and eukaryotes. For example, prokaryotes and eukaryotes employ fundamentally different mechanisms for initiation of protein synthesis (Jackson et al. 2010; Sonenberg and Hinnebusch 2009; Jackson 2005).

Efforts to crystallize eukaryotic ribosomes were until recently unsuccessful even though it represents a highly competitive field in structural biology. Most likely the difficulties stem from the larger size (minimal molecular weight of 3.3 MDa in yeast and plants) and more complex biochemistry. However, we obtained crystals of the full 80S ribosome from *Saccharomyces cerevisiae* and have determined the first X-ray structure of eukaryotic ribosome at 4.15 Å resolution (Ben-Shem et al. 2010). Shortly after, the Ban laboratory reported the determination of the X-ray structure of the isolated 40S subunit from *Tetrahymena thermophila* (Rabl et al. 2011).

1.3.1   Overall View of the 80S Yeast Ribosome

The crystals of the 80S ribosome from *S. cerevisiae* we obtained belong to space group P2$_1$ with cell parameters $a = 437$ Å, $b = 288$ Å, $c = 307$ Å, $\beta = 99°$, with two 80S molecules per asymmetric unit. The crystals diffract to higher than 3 Å resolution but in order to obtain a complete dataset from a single crystal to take full advantage of the anomalous signal from the bound osmium ions we collected a dataset at 4.15 Å resolution, with reflections useful for refinement to 4 Å (I/δI = 2.0 at 4.15 Å).

The final model of the 80S ribosome contains the entire rRNA moiety except for a single flexible expansion segment in 60S (ES27) and a small part of ES7 likewise from the 60S subunit. The model contains also the Cα backbone of all proteins with homologues in prokaryotic ribosome X-ray structures, including in most cases their eukaryote-specific additions. Three additional proteins whose location was identified by cryo-EM (RACK1, S19, L30e) (Halic et al. 2005; Taylor et al. 2009; Sengupta et al. 2004) were also resolved. The model also contains many additional α-helices and β-strands that belong to eukaryote-specific proteins which we chose not to assign to individual proteins since the biochemical data that may confirm their position are lacking. Figure 1.8 (a, b) shows the basic architecture of full 80S ribosome viewed from the exit and entrance sites.

Overall, the structure of the eukaryotic ribosome is considerably larger than its prokaryotic counterpart, but the core architecture is similar. Most of the rRNA expansion elements are located on the solvent-exposed sides of both the 60S and 40S subunits. The contacts between the two subunits of the yeast 80S ribosome were first visualized in low-resolution cryo-EM studies (Spahn et al. 2001a, 2004b). Our 4.15 Å model gives a more accurate and detailed view of the molecular components involved in these bridges between ribosomal subunits (Fig. 1.8d). The importance of the bridges is evident as they maintain communication pathways between the two subunits during protein synthesis. What is particularly noteworthy is the evolutionary conservation of intersubunit bridges at the core of the ribosome. All intersubunit bridges found in the crystal structure of the bacterial ribosome (Yusupov et al. 2001) have a corresponding bridge in the eukaryotic ribosome.

1.3.2   Ribosomal Domain Movements in the Ratcheted State

More than 40 years ago it was hypothesized that translocation of mRNA and tRNA during protein synthesis is coupled to intersubunit movements (Bretscher 1968; Spirin 1969). Recently, cryo-EM studies showed these movements to involve a rotation of the small subunit relative to the large subunit (Frank and Agrawal 2000; Gao et al. 2003; Valle et al. 2003) and that this ratchet-like intersubunit reorganization of the ribosome is essential for translocation (Horan and Noller 2007). Early X-ray structures of the *Escherichia coli* ribosome described intermediate states of ratcheting (Zhang et al. 2009); however, an X-ray structure of the ribosome in the fully ratcheted state was lacking. Fortunately, our 80S crystals capture the ribosome in

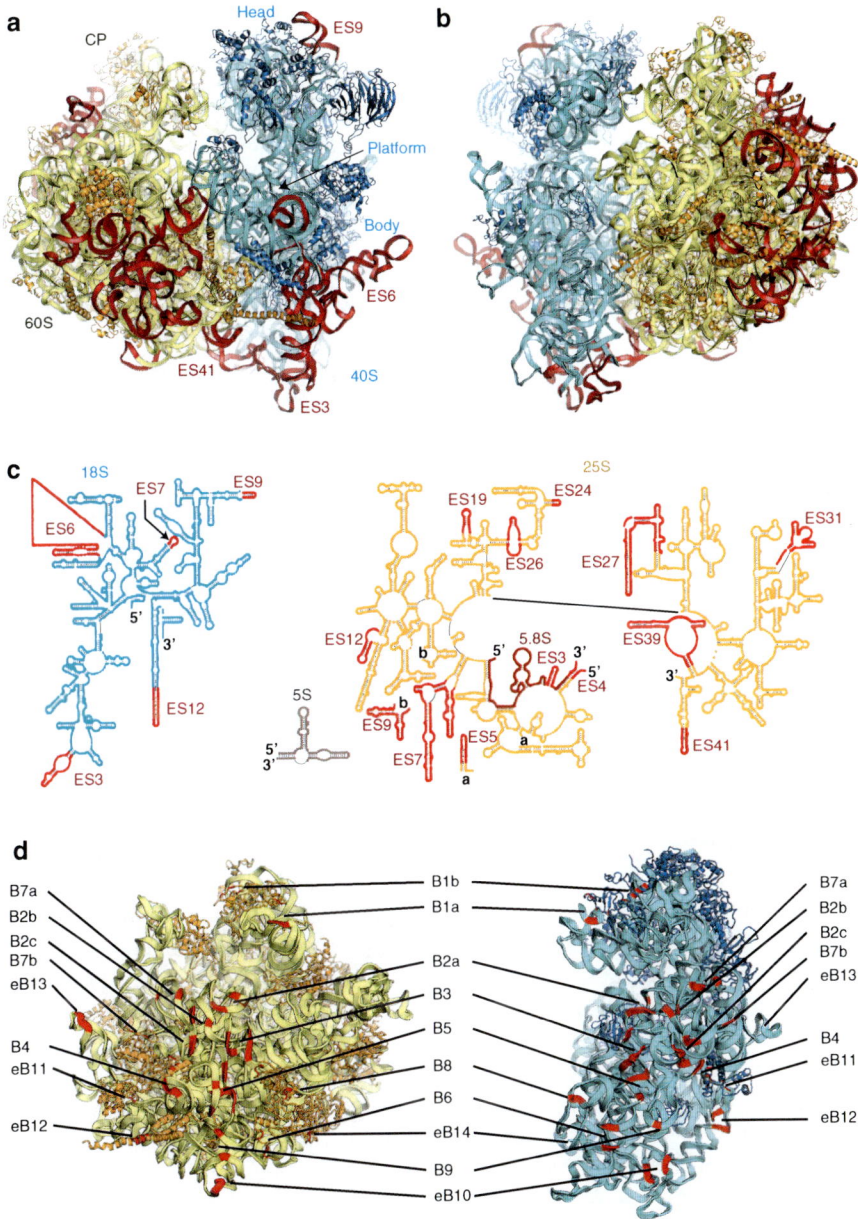

Fig. 1.8 X-ray structure of the 80S ribosome from *Saccharomyces cerevisiae*. (**a**) View from the E site. Proteins and rRNA in the 40S subunit are colored in *dark* and *light blue*, respectively, and in *dark* and *pale yellow*, respectively, in the 60S subunit (this color scheme will be maintained in all following figures unless otherwise indicated). Expansion segment are colored in *red*. (**b**) View from the A site. (**c**) Secondary structure diagram of 18S rRNA in *blue*, 5S rRNA in *brown*, 25S rRNA in *yellow*, and 5.8S in *dark red*, showing expansion segments in *red*. (**d**) Interface views of the 60S and 40S subunits with bridges numbered essentially as previously (Yusupov et al. 2001), and colored in *red*

the fully ratcheted conformation, which was also described in later X-ray studies of *E. coli* ribosome (Dunkle et al. 2011).

When compared to the structure of the unratcheted prokaryotic ribosome (Jenner et al. 2010b) our model of the eukaryotic ribosome shows a 5° counter-clockwise rotation of the 40S subunit body relative to the 60S subunit and a swiveling of the 40S head domain by 14° in the direction of the E-site tRNA (Fig. 1.9a, b). These characteristics are in agreement with cryo-EM observations demonstrating that vacant yeast ribosomes assume a ratcheted conformation similar to the one stabilized by the binding of eukaryotic elongation factor 2 (eEF2) (Spahn et al. 2004a). Indeed, our X-ray model fits well into the cryo-EM maps of the 80S-eEF2 complex (data not shown).

As a result of the large-scale movements implicated in ratcheting there are significant alterations in the bridges between the head domain of the small subunit and the large subunit, in comparison with the unratcheted prokaryotic ribosome (Fig. 1.9c–f). In the latter, the first bridge between the small subunit head domain and the large subunit, bridge B1a, is formed by the A-site finger (H38 of 23S) and protein S13 (Fig. 1.9d). Since head swiveling displaces components at the periphery of the head domain by as much as 25 Å, this bridge is rearranged in our model (Fig.1.9c, e). We find that ratcheting brings residues 1,239–1,241 at the tip of h33 (a component of the beak of 40S) as well as protein S15 (prokaryotic homolog, S19p), into proximity of the tip of H38 that bends significantly in order to form interactions with these partners. Conformational changes are also observed at the base of H38 where it contacts the central protuberance. In the second bridge between the head domain of 40S and the 60S subunit, B1b (Fig. 1.9c, f), the large shift in the position of protein S18 (S13p) places its largest helix, instead of the N-terminal loop, in contact with protein L11 (L5p) of the central protuberance. In addition, residues from loop 65–75 in S15 (S19p) may also interact with L11 (L5p) in the ratcheted state. The prokaryotic homologues of the two proteins, S15 (S19 p) and S1 (S13p), were shown to monitor the occupancy of the A and P sites in the non-ratcheted state (Jenner et al. 2010b). In eukaryotes these two proteins probably interact stronger than in prokaryotes because they have significant eukaryote-specific extensions in the interacting area.

The extent of the head's rotation may be limited or determined by numerous weak interactions between the head and the large subunit. Hence, multiple weak interactions facilitate a wide but precise ratcheting movement. The observed flexibility of the interacting partners is probably crucial for constantly adjusting the bridges as the ratcheting movement progresses.

Fig. 1.9 (continued) (**b**) View from the solvent side of 40S. (**c**) The bridge B1 in the ratcheted 80S. (**d**) Bridge 1 of the non-ratcheted prokaryotic ribosome. (**e**) *Close-up view* of bridge B1a. The tip of the A-site finger (ASF-H38) from 25S RNA forms interactions (colored in *red*) with the head of the 40S subunit including protein S15 (*magenta*). (**f**) View of bridge B1b formed between proteins S18 and protein L11. Residues thought to interact are indicated in *red*

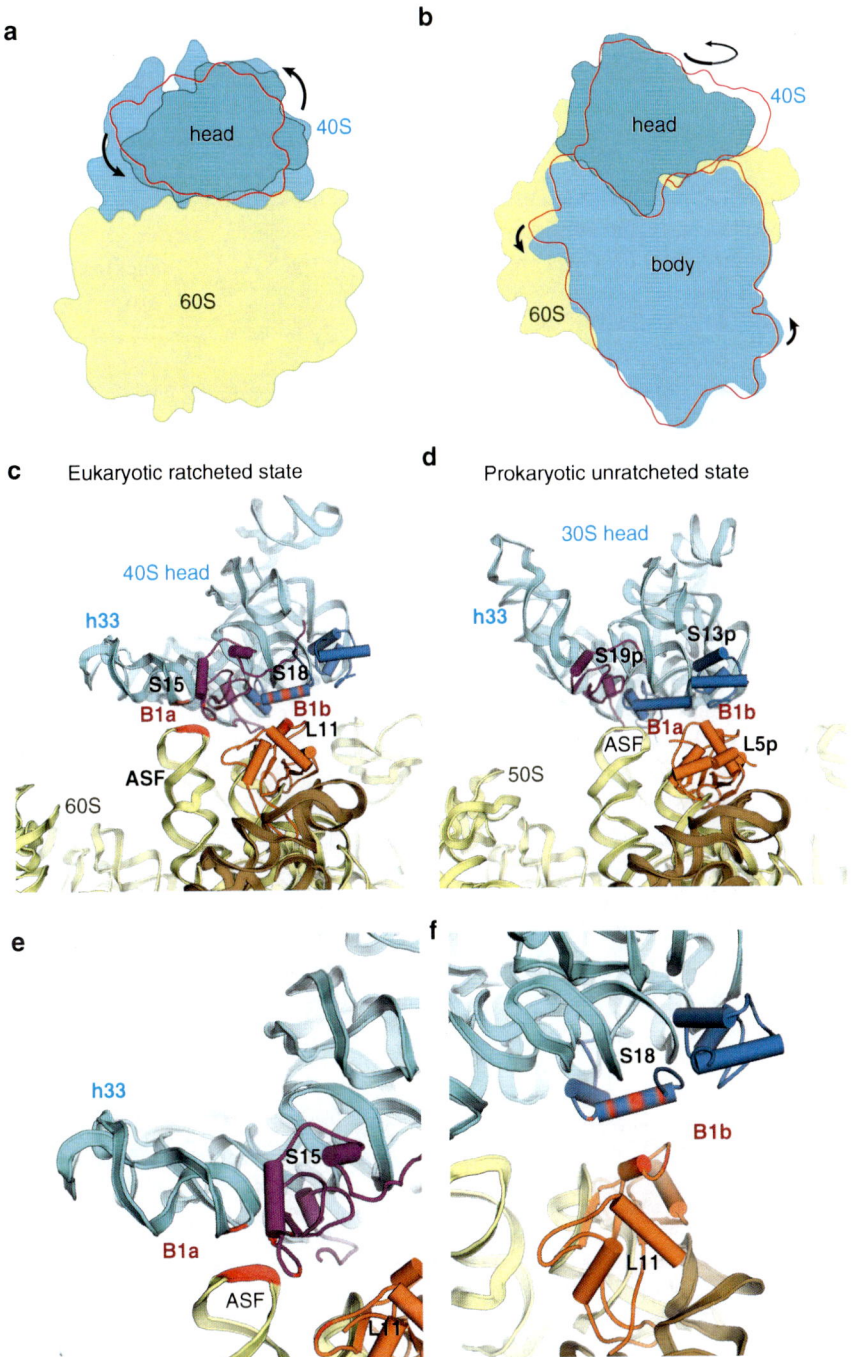

Fig. 1.9 Ratcheted state of the eukaryotic 80S ribosome. (**a, b**) Schematic representation of the motion from the unratcheted to the ratcheted state. The *red line* indicates the outline of the 40S in the unratcheted state, with *arrows* indicating the trajectory. (**a**) *Top view* of the yeast 80S ribosome.

1.3.3   Structure of the Entry and Exit Sites of mRNA: Implication to Initiation

Analysis of our 80S crystal structure at the entry and exit sites of the mRNA (Yusupova et al. 2001) on the small subunit reveals features that are unique to eukaryotes and may pertain to the ribosome interactions with mRNA and initiation factors. At the entry site in prokaryotes, h16 assumes a closed conformation where its tip is in proximity of S3p, a protein that forms the mRNA entry tunnel in collaboration with proteins S4p and S5p (Fig. 1.10a) (Jenner et al. 2010b; Yusupova et al. 2001). In contrast, this helix bends in eukaryotes to adopt a very different, open, orientation that extends away from the body (Fig. 1.10b). In prokaryotes a domain that belongs to S4p composed of two α-helices forms strong interactions with h16, virtually covering a large part of this RNA helix. In the eukaryotic homologue of S4p, protein S9, this domain does not exist, while a helix that could potentially interact with h16 tilts away. Thus, in yeast h16 is bare, with no rRNA–protein interactions, free to rotate around its base. These observations must be viewed in the context of the eukaryotic initiation step. Current models suggest that binding of factors eIF1 and eIF1A to 40S stimulates scanning by inducing h16 to adopt a closed orientation that stabilizes an opening of the mRNA entry tunnel latch (Fig. 1.10c, d). It was shown that binding of IRES to 80S ribosome also induces such changes (Spahn et al. 2001b, 2004c). The latch is formed by interactions between the beak of 40S and h18 at the body of the small subunit. Interestingly, the beak in eukaryotes has a considerably different structure and harbors an additional protein moiety, partially modeled here as the eukaryote-specific protein S17.

The exit site of the mRNA is more intricate in eukaryotes than in prokaryote and contains several additional components (Fig. 1.10e). In association with protein S5 (S7p), just above the mRNA path, a eukaryote-specific protein is poised to bind mRNA. This protein was modeled as S28e in accordance with biochemical data (Pisarev et al. 2008). We also located several protein secondary structure elements situated just below the proposed mRNA path, in a position similar to that occupied in prokaryotes by protein S18p that binds the Shine–Dalgarno sequence

--- →

Fig. 1.10 (continued) at the mRNA entry tunnel. (**c**) The latch is formed by interactions between h34 of the beak and h18 at the body of the small subunit. The superposition of the prokaryotic ribosome (16S rRNA colored *grey*) on the eukaryotic ratcheted ribosome (18S rRNA colored *blue*) shows that the beak in eukaryotes has a considerably different structure and harbors an additional protein moiety, partially modeled here as the eukaryote-specific protein S17 (colored in *magenta*). (**d**) Part of the secondary structure diagram of 18S rRNA from *S. cerevisiae* with expansion segment 9 (colored in *red*) (*upper panel*) and of 16S rRNA from *T. thermophilus*. The beak is formed by h33 and h34. (**e**) View of the back of the 40S showing the connection between the mRNA exit and entry sites in the yeast ribosome (prokaryotic mRNA shown in *red*). The eukaryote-specific additions to the N- and C-terminal of S2 (S5p) are shown in *magenta*. (**f**) The eukaryotic mRNA exit site (mRNA from the prokaryotic model in *red*). The eukaryote-specific bridge eB8 is marked with an *asterisk*

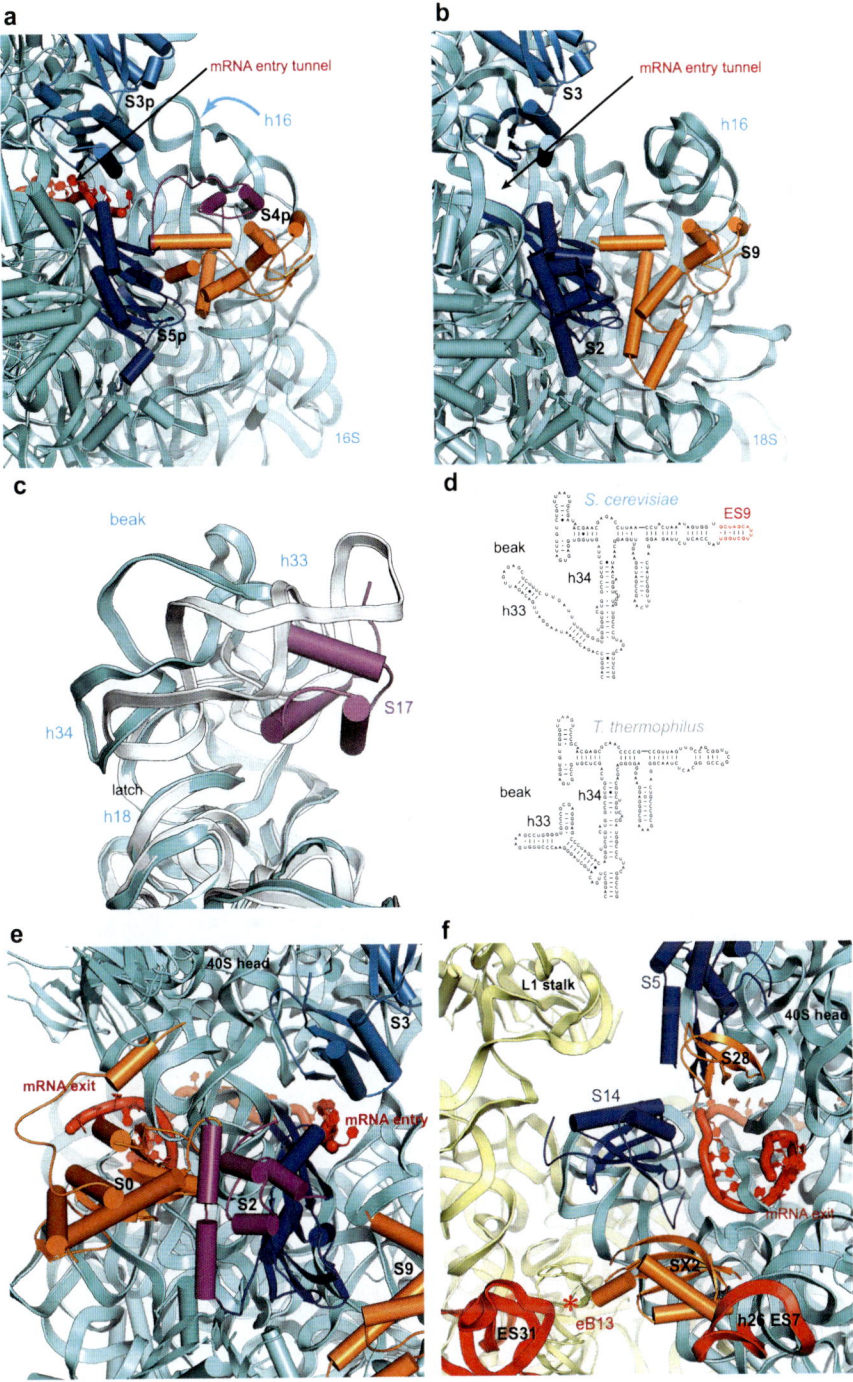

Fig. 1.10 Structure of the entry and exit sites of mRNA. Solvent side view of the mRNA entry site on the small subunit in the (**a**) prokaryote (mRNA shown in *red*) and in the (**b**) yeast ribosome. The additional prokaryotic domain of S4p is shown in magenta. (**c, d**) *Close-up view* of the latch

(indicated together in Fig. 1.10f as protein SX2) (Yusupova et al. 2006; Kaminishi et al. 2007; Korostelev et al. 2007). In addition, ES7 of 18S, an extension of h26, forms part of the mRNA exit site.

Another unique feature of the eukaryotic ribosome in this region is a direct contact between the mRNA entry and exit sites, which is established by a strong interaction between S0 (S2p), a part of the mRNA exit site, and the 80aa long extension of protein S2 (S5p), a component of the mRNA entry tunnel.

1.3.4   Concluding Remarks

The crystal structure of the 80S eukaryotic ribosome described here represents a breakthrough in structure/function studies of eukaryotic ribosomes that until recently only have been investigated by low-resolution cryo-EM methods. It allows rationalization, in structural terms, of existing biochemical and genetic information and will facilitate the design of future experimental models for investigating various aspects of protein synthesis. Further high resolution structures of the eukaryotic ribosome, from yeast as well as other species of eukaryotes, with its plethora of substrates, factors, and protein partners coupled with biochemical and biophysical studies will be needed to provide a molecular description of such complex phenomena as translation initiation, regulation, and ribosome assembly as well as for the development of drugs that will target the translational machinery.

References

Agrawal RK, Penczek P, Grassucci RA, Frank J (1998) Visualization of elongation factor G on the *Escherichia coli* 70S ribosome: the mechanism of translocation. Proc Natl Acad Sci USA 95:6134–6138

Ashe MP, De Long SK, Sachs AB (2000) Glucose depletion rapidly inhibits translation initiation in yeast. Mol Biol Cell 11:833–848

Ban N, Nissen P, Hansen J, Moore PB, Steitz TA (2000) The complete atomic structure of the large ribosomal subunit at 2.4 A resolution. Science 289:905–920

Ben-Shem A, Jenner L, Yusupova G, Yusupov M (2010) Crystal structure of the eukaryotic ribosome. Science 330:1203–1209

Bouadloun F, Srichaiyo T, Isaksson LA, Bjork GR (1986) Influence of modification next to the anticodon in tRNA on codon context sensitivity of translational suppression and accuracy. J Bacteriol 166:1022–1027

Bretscher MS (1968) Translocation in protein synthesis: a hybrid structure model. Nature 218:675–677

Brodersen DE, Clemons WM Jr, Carter AP, Wimberly BT, Ramakrishnan V (2002) Crystal structure of the 30S ribosomal subunit from *Thermus thermophilus*: structure of the proteins and their interactions with 16S RNA. J Mol Biol 316:725–768

Canonaco MA, Gualerzi CO, Pon CL (1989) Alternative occupancy of a dual ribosomal binding site by mRNA affected by translation initiation factors. Eur J Biochem 182:501–506

Dunkle JA, Wang L, Feldman MB, Pulk A, Chen VB et al (2011) Structures of the bacterial ribosome in classical and hybrid states of tRNA binding. Science 332:981–984

Frank J, Agrawal RK (2000) A ratchet-like inter-subunit reorganization of the ribosome during translocation. Nature 406:318–322

Frank J, Zhu J, Penczek P, Li Y, Srivastava S et al (1995) A model of protein synthesis based on cryo-electron microscopy of the *E. coli* ribosome. Nature 376:441–444

Gao H, Sengupta J, Valle M, Korostelev A, Eswar N et al (2003) Study of the structural dynamics of the *E coli* 70S ribosome using real-space refinement. Cell 113:789–801

Gustilo EM, Vendeix FA, Agris PF (2008) tRNA's modifications bring order to gene expression. Curr Opin Microbiol 11:134–140

Halic M, Becker T, Frank J, Spahn CM, Beckmann R (2005) Localization and dynamic behavior of ribosomal protein L30e. Nat Struct Mol Biol 12:467–468

Harms J, Schluenzen F, Zarivach R, Bashan A, Gat S et al (2001) High resolution structure of the large ribosomal subunit from a mesophilic eubacterium. Cell 107:679–688

Hoburg A, Aschhoff HJ, Kersten H, Manderschied U, Gassen HG (1979) Function of modified nucleosides 7-methylguanosine, ribothymidine, and 2-thiomethyl-N6-(isopentenyl)adenosine in procaryotic transfer ribonucleic acid. J Bacteriol 140:408–414

Horan LH, Noller HF (2007) Intersubunit movement is required for ribosomal translocation. Proc Natl Acad Sci USA 104:4881–4885

Jackson RJ (2005) Alternative mechanisms of initiating translation of mammalian mRNAs. Biochem Soc Trans 33:1231–1241

Jackson RJ, Hellen CU, Pestova TV (2010) The mechanism of eukaryotic translation initiation and principles of its regulation. Nat Rev Mol Cell Biol 11:113–127

Jenner L, Romby P, Rees B, Schulze-Briese C, Springer M et al (2005) Translational operator of mRNA on the ribosome: how repressor proteins exclude ribosome binding. Science 308:120–123

Jenner L, Rees B, Yusupov M, Yusupova G (2007) Messenger RNA conformations in the ribosomal E site revealed by X-ray crystallography. EMBO Rep 8:846–850

Jenner L, Demeshkina N, Yusupova G, Yusupov M (2010a) Structural rearrangements of the ribosome at the tRNA proofreading step. Nat Struct Mol Biol 17:1072–1078

Jenner LB, Demeshkina N, Yusupova G, Yusupov M (2010b) Structural aspects of messenger RNA reading frame maintenance by the ribosome. Nat Struct Mol Biol 17:555–560

Jorgensen F, Kurland CG (1990) Processivity errors of gene expression in *Escherichia coli*. J Mol Biol 215:511–521

Kaminishi T, Wilson DN, Takemoto C, Harms JM, Kawazoe M et al (2007) A snapshot of the 30S ribosomal subunit capturing mRNA via the Shine-Dalgarno interaction. Structure 15:289–297

Konevega AL, Soboleva NG, Makhno VI, Peshekhonov AV, Katunin VI (2006) The effect of modification of tRNA nucleotide-37 on the tRNA interaction with the P- and A-site of the 70S ribosome *Escherichia coli*. Mol Biol (Mosk) 40:669–683

Korostelev A, Trakhanov S, Laurberg M, Noller HF (2006) Crystal structure of a 70S ribosome-tRNA complex reveals functional interactions and rearrangements. Cell 126:1065–1077

Korostelev A, Trakhanov S, Asahara H, Laurberg M, Lancaster L et al (2007) Interactions and dynamics of the Shine Dalgarno helix in the 70S ribosome. Proc Natl Acad Sci USA 104:16840–16843

Korostelev A, Asahara H, Lancaster L, Laurberg M, Hirschi A et al (2008) Crystal structure of a translation termination complex formed with release factor RF2. Proc Natl Acad Sci USA 105:19684–19689

La Teana A, Gualerzi CO, Brimacombe R (1995) From stand-by to decoding site. Adjustment of the mRNA on the 30S ribosomal subunit under the influence of the initiation factors. RNA 1:772–782

Laurberg M, Asahara H, Korostelev A, Zhu J, Trakhanov S et al (2008) Structural basis for translation termination on the 70S ribosome. Nature 454:852–857

Ma J, Campbell A, Karlin S (2002) Correlations between Shine-Dalgarno sequences and gene features such as predicted expression levels and operon structures. J Bacteriol 184:5733–5745

Menichi B, Heyman T (1976) Study of tyrosine transfer ribonucleic acid modification in relation to sporulation in Bacillus subtilis. J Bacteriol 127:268–280

Ogle JM, Murphy FV, Tarry MJ, Ramakrishnan V (2002) Selection of tRNA by the ribosome requires a transition from an open to a closed form. Cell 111:721–732

Pisarev AV, Kolupaeva VG, Yusupov MM, Hellen CU, Pestova TV (2008) Ribosomal position and contacts of mRNA in eukaryotic translation initiation complexes. EMBO J 27:1609–1621

Rabl J, Leibundgut M, Ataide SF, Haag A, Ban N (2011) Crystal structure of the eukaryotic 40S ribosomal subunit in complex with initiation factor 1. Science 331:730–736

Rinke-Appel J, Junke N, Brimacombe R, Lavrik I, Dokudovskaya S et al (1994) Contacts between 16S ribosomal RNA and mRNA, within the spacer region separating the AUG initiator codon and the Shine-Dalgarno sequence; a site-directed cross-linking study. Nucleic Acids Res 22:3018–3025

Rozenski J, Crain PF, McCloskey JA (1999) The RNA modification database: 1999 update. Nucleic Acids Res 27:196–197

Schurr T, Nadir E, Margalit H (1993) Identification and characterization of E. coli ribosomal binding sites by free energy computation. Nucleic Acids Res 21:4019–4023

Schuwirth BS, Borovinskaya MA, Hau CW, Zhang W, Vila-Sanjurjo A et al (2005) Structures of the bacterial ribosome at 3.5 A resolution. Science 310:827–834

Selmer M, Dunham CM, Murphy FVt, Weixlbaumer A, Petry S et al (2006) Structure of the 70S ribosome complexed with mRNA and tRNA. Science 313:1935–1942

Sengupta J, Nilsson J, Gursky R, Spahn CM, Nissen P et al (2004) Identification of the versatile scaffold protein RACK1 on the eukaryotic ribosome by cryo-EM. Nat Struct Mol Biol 11:957–962

Shine J, Dalgarno L (1974) The 3′-terminal sequence of Escherichia coli 16S ribosomal RNA: complementarity to nonsense triplets and ribosome binding sites. Proc Natl Acad Sci USA 71:1342–1346

Sonenberg N, Hinnebusch AG (2009) Regulation of translation initiation in eukaryotes: mechanisms and biological targets. Cell 136:731–745

Spahn CM, Beckmann R, Eswar N, Penczek PA, Sali A et al (2001a) Structure of the 80S ribosome from Saccharomyces cerevisiae–tRNA-ribosome and subunit-subunit interactions. Cell 107:373–386

Spahn CM, Kieft JS, Grassucci RA, Penczek PA, Zhou K et al (2001b) Hepatitis C virus IRES RNA-induced changes in the conformation of the 40s ribosomal subunit. Science 291:1959–1962

Spahn CM, Gomez-Lorenzo MG, Grassucci RA, Jorgensen R, Andersen GR et al (2004a) Domain movements of elongation factor eEF2 and the eukaryotic 80S ribosome facilitate tRNA translocation. EMBO J 23:1008–1019

Spahn CM, Jan E, Mulder A, Grassucci RA, Sarnow P et al (2004b) Cryo-EM visualization of a viral internal ribosome entry site bound to human ribosomes: the IRES functions as an RNA-based translation factor. Cell 118:465–475

Spirin AS (1969) A model of the functioning ribosome: locking and unlocking of the ribosome subparticles. Cold Spring Harb Symp Quant Biol 34:197–207

Stark H, Orlova EV, Rinke-Appel J, Junke N, Mueller F et al (1997a) Arrangement of tRNAs in pre- and posttranslocational ribosomes revealed by electron cryomicroscopy. Cell 88:19–28

Stark H, Rodnina MV, Rinke-Appel J, Brimacombe R, Wintermeyer W et al (1997b) Visualization of elongation factor Tu on the Escherichia coli ribosome. Nature 389:403–406

Taylor DJ, Devkota B, Huang AD, Topf M, Narayanan E et al (2009) Comprehensive molecular structure of the eukaryotic ribosome. Structure 17:1591–1604

Urbonavicius J, Qian Q, Durand JM, Hagervall TG, Bjork GR (2001) Improvement of reading frame maintenance is a common function for several tRNA modifications. EMBO J 20:4863–4873

Urbonavicius J, Stahl G, Durand JM, Ben Salem SN, Qian Q et al (2003) Transfer RNA modifications that alter +1 frameshifting in general fail to affect −1 frameshifting. RNA 9:760–768

Valle M, Zavialov A, Sengupta J, Rawat U, Ehrenberg M et al (2003) Locking and unlocking of ribosomal motions. Cell 114:123–134

Weixlbaumer A, Murphy FV 4th, Dziergowska A, Malkiewicz A, Vendeix FA et al (2007a) Mechanism for expanding the decoding capacity of transfer RNAs by modification of uridines. Nat Struct Mol Biol 14:498–502

Weixlbaumer A, Petry S, Dunham CM, Selmer M, Kelley AC et al (2007b) Crystal structure of the ribosome recycling factor bound to the ribosome. Nat Struct Mol Biol 14:733–737

Wilson RK, Roe BA (1989) Presence of the hypermodified nucleotide N6-(delta 2-isopentenyl)-2-methylthioadenosine prevents codon misreading by *Escherichia coli* phenylalanyl-transfer RNA. Proc Natl Acad Sci USA 86:409–413

Wimberly BT, Brodersen DE, Clemons WM Jr, Morgan-Warren RJ, Carter AP et al (2000) Structure of the 30S ribosomal subunit. Nature 407:327–339

Yusupov MM, Yusupova GZ, Baucom A, Lieberman K, Earnest TN et al (2001) Crystal structure of the ribosome at 5.5 A resolution. Science 292:883–896

Yusupova GZ, Yusupov MM, Cate JH, Noller HF (2001) The path of messenger RNA through the ribosome. Cell 106:233–241

Yusupova G, Jenner L, Rees B, Moras D, Yusupov M (2006) Structural basis for messenger RNA movement on the ribosome. Nature 444:391–394

Zhang W, Dunkle JA, Cate JH (2009) Structures of the ribosome in intermediate states of ratcheting. Science 325:1014–1017

Chapter 2
A Passage Through the Ribosome by Cryo-EM

Partha P. Datta and Ananya Chatterjee

2.1   Introduction

From the first picture of the ribosome taken some 60 years ago by George E. Palade (1955), to the current near atomic level three-dimensional cryo-Electron Microscopic (3D cryo-EM) ribosomal structures (Fig. 2.1; Armache et al. 2010), electron micros-copy has always paved the way for the research in the translation field. Over the past several decades we have witnessed two beautiful examples where a significant portion of both the translation field and the single particle cryo-EM field has evolved together by complementing each other in Joachim Frank and Marin van Heel's labs Frank et al. (1981) and van Heel and Keegstra (1981). Due to the increasing sophis-tication of the cryo-EM field and its extensive application on the ribosome research, a plethora of knowledge has been achieved on the structure–function aspects of this magnificent macro-molecular machine. Thus, cryo-EM has established its firm place in the ribosome research alongside the X-ray crystallography and other biochemical methods. This chapter will focus on the conceptual advancements on our under-standing of the mechanism of translation that have been achieved largely due to the use of this cutting-edge biophysical technique.

P.P. Datta (✉) • A. Chatterjee
Department of Biological Sciences, Indian Institute of Science Education and Research-Kolkata,
P.O. BCKV campus main office, Mohanpur-741252, Nadia, West Bengal, India
e-mail: partha_datta@iiserkol.ac.in; partha_datta31@yahoo.com

J.D. Dinman (ed.), *Biophysical Approaches to Translational Control of Gene Expression*,
Biophysics for the Life Sciences 1, DOI 10.1007/978-1-4614-3991-2_2,
© Springer Science+Business Media New York 2013

T. aestivum 80S map

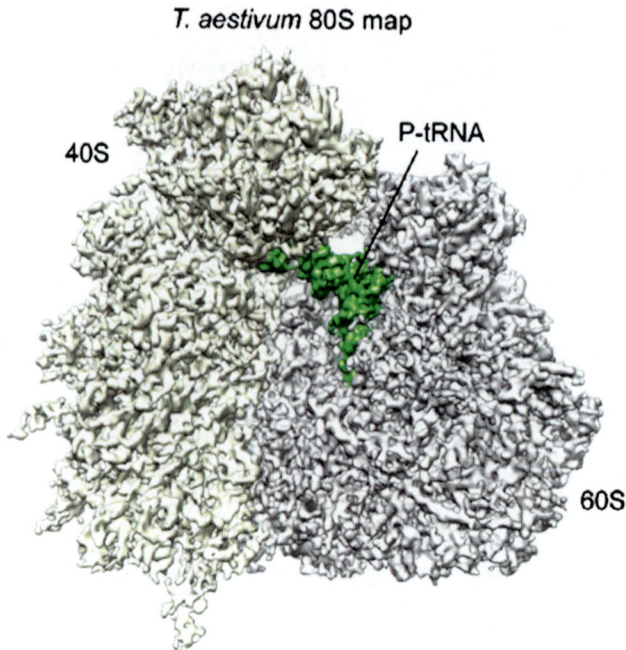

Fig. 2.1 A 5.5 Å resolution cryo-EM map of *Triticum asestivum* 80S ribosome. *Source*: From the "Fig 1, A" of the "Supporting Information" section of Armache et al. (2010)

2.1.1 Why Cryo-EM is a Sought After Technique in the Ribosome Field?

Cryo-electron microscopy is a powerful technique which is predominantly used in studying the 3D structure of large macromolecules in different functional states. Using liquid ethane, a dilute solution of a biological macromolecular complex is flash frozen in a thin layer of vitreous ice on an EM grid, and then single molecular entities are imaged under cryo-Transmission Electron Microscope (cryo-TEM) at low electron dose condition. Due to the flash freezing in noncrystalline ice, the randomly oriented individual macromolecules (particles) become trapped in their native forms. Two-dimensional (2D) projection images of those particles are collected under high (~50 K or more) magnification in the cryo-TEM. Since no heavy metal stains are used in the sample preparation, the signal to noise ratio of the images of the particles to the background ice becomes very low, resulting in low contrast images. To address this issue and to obtain a full spectrum of information, hundreds of thousands of individual images are collected at a series of under-defocus ranges in the cryo-TEM. From the digital images of those collected and subsequently selected good particles, contrast transfer functions (CTF) are calculated. Particles falling in the same orientations are classified, averaged, merged to other class averaged particles

Molecular Microscopy

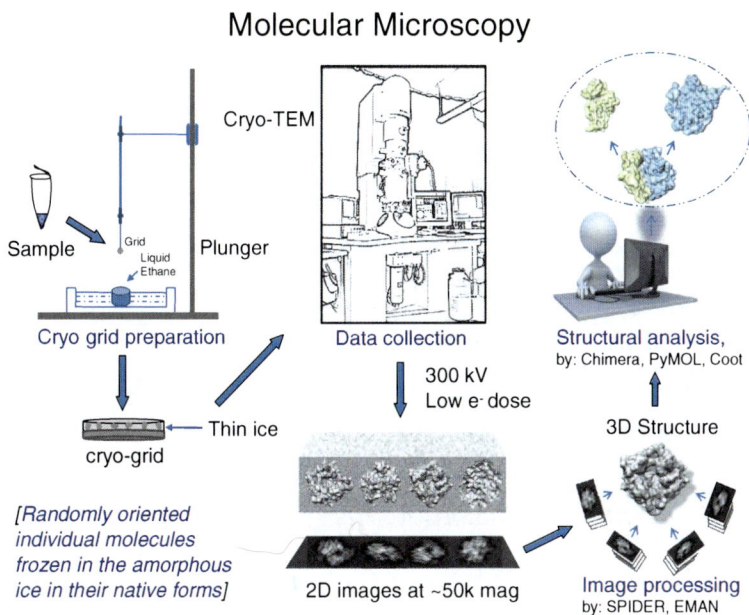

Fig. 2.2 Schematic showing the cryo sample preparation, cryo-Electron Microscopy, and image-processing methods

and then finally back projected using suitable image-processing software to generate a 3D cryo-EM density map (3D structure). To scoop out the functional information from the resulting cryo-EM structure, generally a compound approach is taken, where the entire or the portions of the 3D EM density map is interpreted at the atomic level through fitting or docking into its identifiable components with corresponding available atomic structure obtained from X-ray crystallography, NMR, or homology models. Thus, through this powerful method it is becoming increasingly possible to investigate the native forms and the dynamics of any molecular machines caught at any event of its biological actions. For a schematic of cryo-EM methodology see Fig. 2.2.

During protein synthesis, ribosome undergoes multiple dynamic conformational changes very rapidly which materialize not only from its inherent molecular character but also due to its interactions with numerous ligands. In order to grasp the detailed knowledge of the translation mechanisms one needs to trap these dynamic events in rapid successions and study the corresponding structures in great detail.

Due to the highly dynamic nature of the ribosome–ligand interactions, it is very challenging task for the X-ray crystallographers to obtain crystals from the series of structural forms during the translation events. On the other hand, since ribosome is a very large (~2.5 MDa) macromolecular machine it is impossible to study by NMR, which is otherwise a very useful technique for studying dynamic smaller molecules.

For studying the dynamic aspects of any macromolecule, FRET is a useful technique. However, here FRET has a limitation such that it can only be used primarily in the previously known dynamic regions of the ribosome thus any new dynamic information in a different region may get eluded from the study. While the X-ray crystallographic studies on ribosome had given us the atomic detailed information, which were honored through Nobel Prizes in 2009 (Nobel Prize in chemistry to Venkatraman Ramakrishnan, Thomas A. Steitz, and Ada E. Yonath), discovery of the intricate details of the many dynamic events of the translation process is being possible due to the single particle cryo-EM, 3D image-processing, and related molecular modeling methods. Thus for acquiring the missing knowledge to visualize the whole scenario of translation process, cryo-EM combined with time-resolved biochemical techniques and advanced image-processing methods can have a profound future.

2.1.2   Concept Developments to Meet the Challenges for Solving Various Ribosomal Complexes on Single Particle Cryo-EM

The cellular protein synthesizing machine ribosome is an asymmetric complex macromolecular assembly composed of two unequal ribo-nucleo-protein subunits. Not only the whole ribosome 70S and 80S in prokaryotes and eukaryotes respectively but also its individual small (30S and 40S in prokaryotes and eukaryotes respectively) and large (50S and 60S in prokaryotes and eukaryotes respectively) subunits are big enough to be visualized under electron microscope. Ribosome has been proved as a good object for developing the 3D image-processing tools based on the cryo-EM data. This is because of the large rRNA helices content in the ribosome, particularly due to its phosphate backbone; ribosome's contrast is naturally higher than other molecular machines made of protein only molecules. This property, along with the help of genetic tagging, was later exploited to separate rRNA densities from the rest of the densities comprising r-protein regions of the ribosome in a 3D cryo-EM map (Spahn et al. 1999). To reconstruct an initial 3D structure from the 2D EM projection images of such an asymmetric macromolecule, Random Conical Tilt (RCT) (Radermacher et al. 1987) and Common Line (CL) based methods (van Heel 1987 and Penczek et al. 1996) were developed. In RCT method, a pair of EM pictures of the same object area on the EM grid were taken in tilted (~45°) and untilted condition. Those provide two Eulerian angles of the projection images and the third angle was computed from the 2D alignment of the chosen particle images. That information helped to compute an initial 3D volume. In CL method, the common lines of any two 2D averaged classes were found as a resultant of the Fourier transform of the original 3D object to compute their relative angles. Information thus obtained was used to generate an ab initio model of the structure. In common practice, once an initial model is generated from a small set of particles a reference-based reconstruction method is used to generate a reasonably good 3D cryo-EM map from a fairly large number of dataset.

Using biochemical information and applying appropriate threshold level, the contours of the two ribosomal subunits were delineated within an 11.5Å resolution intact 70S ribosome (Gabashvili et al. 2000). Prior to this study a direct localization of multiple tRNA densities were made possible in a ribosome-bound tRNA complex by subtracting empty ribosome density from the complex (Agrawal et al. 1996). Those approaches became integral part of all the subsequent structural analyses of several functional complexes. To interpret moderate resolution EM densities into higher resolutions information, available atomic densities were fitted into corresponding EM densities and based on features and the contour of the EM map the atomic models were broken domain wise and fitted separately into the corresponding regions (Agrawal et al. 1998) as rigid body docking. Later on a flexible docking approach (Wriggers et al. 2000) was taken to obtain a better fitting of the atomic density into the corresponding EM map. Further improvements towards building quasi atomic models revealing the dynamic nature of the ribosome–ligand complexes into intricate details were based on normal mode analysis (Tama et al. 2003), real-space refinement (Gao et al. 2003), and molecular dynamics simulation (Li and Frank 2007; Trabuco et al. 2008) methods. In another line of approach, to track and localize one of the highly flexible components of the ribosome, like the C-terminal domain of the L7/L12 protein within the 3D EM volume, a heavy metal cluster, Nanogold, was used as a probe (Montesano-Roditis et al. 2001). Similarly to trace the dynamic behavior of some of the translation factors, like EF-G (Datta et al. 2005), and ribosome-recycling factor (RRF) (Barat et al. 2007), a relatively smaller heavy metal cluster, Undecagold, was used.

After a biochemical reaction among ribosome and ligands is done, possibilities are that there will be some heterogeneous complexes with partially bound ligands. Moreover, there will be conformational changes associated with the reaction condition. These lead to heterogeneity in the dataset and if minute attention is not given then appropriate structures may not be obtained. Most of the time it becomes impossible for the human eye to discriminate these heterogeneity. Thus, these different forms are to be separated in silico. To address those issues, successive separation techniques like supervised classification (Valle et al. 2002), unsupervised classification (Fu et al. 2007), maximum likelihood approach (Scheres et al. 2007), multiparticle cryo-EM method (Loerke et al. 2010) have been developed.

2.2   Contributions of Cryo-EM on the Dynamics of Ribosome Factors Interplay During Translation

The inter-subunit space in the 70S, i.e., the slit between the two associate subunits, 30S and 50S, provides the venue for the translation process where all the translation factors interact with the ribosome and to themselves in a dynamic and sequential manner to carry out the entire translation event. Extensive cryo-EM studies are revealing the dynamic nature of the translation process. Since bacterial and eukaryotic translation process have many similarities, except to some extent in the initiation and recycling process, and for the reason of simplicity, here we are describing mainly the concepts that were learnt from studying the bacterial translation system.

2.2.1 Initiation

In bacteria, with the help of initiation factors IF1, IF2, and IF3, an fMet-tRNA^{fMet} binds to the correct messenger RNA start codon on the 30S. Thus the 30S initiation complex (30SIC) is formed, which is then joined by the 50S to form the 70S initiation complex (70SIC) followed by subsequent release of the initiation factors. Following three-dimensional multivariate statistical analysis and three-dimensional classification into structural sets with low intraclass of three-dimensional reconstructions from randomly selected particle images, Simonetti et al. (2008) reported a detailed description of the 30SIC formation, containing mRNA, fMet-tRNA^{fMet}, and IF1 and IF2·GTP densities. The study revealed characteristic and precise positioning of the fMet-tRNA^{fMet} where it is held stable on the 30S. Its anticodon end is anchored to the decoding site of the 30S while its aminoacyl end is bound to the IF2. This observation thus helped explain the rate-limiting step in translation, i.e., the basis of the formation of a stable 30SIC.

First insights into the 70SIC also came from two cryo-EM studies by Allen et al. (2005) and Myasnikov et al. (2005). Together these studies shed lights on the mechanism of IF2-mediated fMet-tRNA^{fMet} positioning to the peptidyl/initiation (P/I) site of the ribosome. Analyses from the 70SICs complex comprising the mRNA, fMet-tRNA^{fMet}, and IF2 with either a non-hydrolyzable GTP analog or GDP show substantial conformational changes of the IF2 and of the entire ribosome. It also reveals that the GTP binding domain of IF2 interacts with the GTPase-associated center of the 50S subunit in a mode similar to Elongation Factor-G (EF-G) and Elongation Factor-Tu (EF-Tu). In the non-hydrolyzable GTP-bound state, IF2 interacts mostly with the small subunit and positions the fMet-tRNA^{fMet} to the P/I site, whereas in the GDP-bound state IF2 adopts a more compact conformation suitable for release from the ribosome. It also provided insights into the molecular mechanism for release of IF1 and IF3.

2.2.2 Elongation Cycle

Extensive cryo-EM studies have helped understand this process which is mainly the mechanism in which the polypeptide chain grows by one amino acid following each elongation cycle that consists of the following steps, like appropriate aminoacyl-tRNA selection and accommodation, peptidyl transferase reaction between the P-site tRNA$^{polypeptide-chain}$ and A-site aa-tRNA, followed by tRNA–mRNA translocation by one codon towards the 3′ mRNA direction, and subsequent release of an deacylated tRNA from the E site of the ribosome. Two elongation factors EF-Tu and

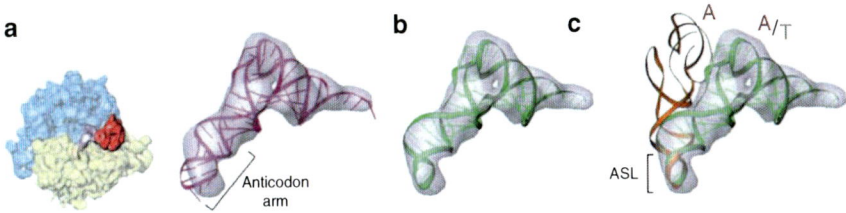

Fig. 2.3 Model of A/T-site tRNA and its interaction with ribosome. (**a**) Rendering of the fitted tRNA^{Phe} inside the cryo-EM density for the A/T-site tRNA. Evidently, the anticodon arm of the X-ray structure does not fit the features of the density for the A/T-site tRNA. (**b**) Modeled structure for the A/T-tRNA is represented in *green* ribbons, showing an evident improvement in the docking of the coordinates within the density. (**c**) Comparison of the modeled coordinates for the A/T-site tRNA (*green*) and the coordinates for the A-site tRNA (*gold*). (The docking of the A-site tRNA in this cryo-EM density did not require any modification in the anticodon arm of the X-ray structure.) The Anti-codon Stem Loops (ASLs) of both tRNAs are similarly oriented. *Source*: Fig 4 from the following reference Valle et al. (2003a,b)

EF-G play major role here. Structural and functional observations from relevant cryo-EM studies on bacteria are described below.

2.2.2.1 tRNA Selection and Accommodation

After the initiation step, an aminoacyl-tRNA is brought to the ribosome by EF-Tu for delivery of the next amino acid to the growing polypeptide chain. This is a crucial step where only that aa-tRNA which has the correct nucleotides at its anticodon end against the specific codon sequence in the translating mRNA will be selected and then accommodated into the ribosome for the subsequent peptidyl transferase and translocation process. Fairly detailed mechanisms underlying this important aspect were first revealed by cryo-EM by studying various 70S-Ternary Complexes (70S·TC), [70S·EF-Tu·aa-tRNA·GTP+Kirromycin; and its derivative complexes]. Results from several of those studies (Stark et al. 2002; Valle et al. 2002; 2003a,b; Li et al. 2008; Schuette et al. 2009; Villa et al. 2009) depicted that the EF-Tu-bound incoming cognate aa-tRNA (the CCA end of the aa-tRNA is attached to the EF-Tu, whereas anticodon end is free) possess a partially bended conformation at its anticodon end, compared to the X-ray structure of free-standing tRNA, (see Fig. 2.3) during the initial steps of aa-tRNA selection prior to the EF-Tu·GDP is released. This partial bending in a suitable angle allows interaction between the anticodon of the tRNA and the nucleotides of the translating mRNA in the decoding center while that particular aa-tRNA is still attached to the EF-Tu; thus screening for the cognate tRNAs are facilitated. After the cognate tRNA is recognized, a series of conformational changes take place on the associated ribosomal components and the EF-Tu, GTP hydrolysis occurs, and the EF-Tu releases the CCA end of the aa-tRNA and goes out of the ribosome. Soon after the cognate tRNA firmly establishes its interaction with the codon (referred as the A/T state), and gets released from the EF-Tu, the deformity in the anticodon arm is relieved. Mechanical energy released from this, in turn, paves the way for the subsequent positioning of the A/T tRNA into the A/A state. Thus a molecular spring like mechanism is deciphered by the cryo-EM study during the tRNA selection and accommodation process. Furthermore, a very recent cryo-EM study (Agirrezabala et al. 2011) based on the analyses of the 70S·TC complexes containing a near-cognate aa-tRNA, in addition to its cognate counterpart, observed distinct structural changes, which indicates an induced fit mechanism during aa-tRNA incorporation. The atomic details of some of the above findings have been confirmed by the crystal structures of different 70S·TC complexes (Schmeing et al. 2009; Voorhees et al. 2010).

2.2.2.2   Translocation

EF-G, a GTPase enzyme, binds to the translating ribosome and promotes the step of translocation, a process in which tRNA moves from the A to the P site after the peptide bond formation, and the mRNA advances by one codon towards the 3′ end. Agrawal et al. (1998) first visualized EF-G density inside a 70S. Molecular fitting of the crystal structure of EF-G into the corresponding cryo-EM density map revealed a large conformational change that was mainly associated with its domain IV. The domain IV density showed a molecular mimicry against the anticodon arm of the tRNA in the structurally homologous ternary complex of Phe·tRNA[Phe]·EF-Tu·GTP analog. The apex of the domain IV was found to be mutually exclusive to the positioning of the anticodon arm of the A-site tRNA on the translating ribosome. This implied that the EF-G displaced the A-site tRNA to the P site by physical interaction with the anticodon arm. Strong support to this phenomenon also came from a recent cryo-EM study (Ratje et al. 2010). By analyzing the 3D cryo-EM maps of two complexes, where EF-G was bound to the 70S in GTP and GDP form (70S·EF-G·non-hydrolyzable GTP analog complex and 70S·EF-G·GDP·fusidic acid complex respectively), Agrawal et al. (1999) observed large conformational changes associated with the EF-G and the ribosome as well. Further analysis comprising a movie from those complexes revealed a surprising result whereby a ratchet-like rotation of the 30S subunit relative to the 50S subunit in the direction of the mRNA movement was observed (Frank and Agrawal 2000) in the presence of EF-G in GDP state. Subsequent cryo-EM study by Valle et al. (2003a,b) at better resolutions revealed that the ratchet-like rotation was accompanied by ~20 Å movement of the L1 stalk of the 50S subunit opposite to the direction of the mRNA movement. The study implied that the L1 stalk was involved in the translocation of deacylated tRNAs from the P to the E site and the associated ribosomal motions could only occur when the P-site tRNA was deacylated.

Further insights into the dynamic nature of the translocation processes were followed by cryo-EM using heavy metal cluster probing in 70S·EF-G·GMPPCP and 70S·EF-G·GDP·fusidic acid complexes. By labeling an amino acid near the tip of the G′ domain of EF-G with maleimide undecagold cluster, Datta et al. (2005) traced a large movement of the G′ domain of the EF-G associated with the GTP hydrolysis. Most importantly this study for the first time revealed structural evidence of the involvement of the highly mobile C-terminal domain of ribosomal protein L7/L12 in the translation process by revealing molecular interaction of the CTD of L7/L12 with the G′ domain of the EF-G in GDP state. This observation was later supported by other combined X-ray crystallographic and cryo-EM (Harms et al. 2008) and X-ray crystallographic studies (Gao et al. 2009). Using combined X-ray crystallographic and cryo-EM approach Connell et al. (2007) revealed a detailed insight into the ribosome-dependent GTPase activity of EF-G during the translocation process. It showed that the dynamic reorganization of the functionally important, P loop, switch I and II regions of EF-G which interact with the bound GTP, occurred upon EF-G's interactions with the ribosome during the ratcheting motions.

Our understanding on the pathways of tRNA translocation on the ribosome has been advanced further by finding of tRNAs in distinct hybrid states (A/P and P/E) in spontaneously ratcheted 70S-tRNA complexes (Agirrezabala et al. 2008). Most recently, using multiparticle cryo-EM analysis Ratje et al. (2010) reported direct structural and mechanistic insight into the previously unseen tRNA intermediates involved in the universally conserved translocation process. They found two previously unseen subpopulations of translocation complexes having a novel intra-subunit tRNA in pe/E hybrid state that is stabilized by domain IV of EF-G, which interacts with the swiveled 30S-head conformation thus advancing the field of research further.

2.2.3   Termination

When the translating ribosome reaches to a stop codon UAA, UAG, and UGA moving into the ribosomal A site, the elongation cycle stops due to the nonavailability of a corresponding tRNA. Instead, depending on the nature of the stop codons, either one of the class I release factors, RF1 or RF2, binds to the decoding center of the ribosome, and triggers the hydrolysis of the ester bond that links the polypeptide with the P-site tRNA, thereby facilitating the release of the polypeptide from the ribosome. Much of the structural and functional insights of the above phenomenon came from several cryo-EM studies. Rawat et al. (2006) visualized ribosome-bound RF1 is in an open conformation, unlike the closed conformation observed in the crystal structure of the free factor. Similar results were obtained previously for RF2 study (Rawat et al. 2003; Klaholz et al. 2003). The open form of RF1 and RF2 allowed their simultaneous access to both the decoding center and the peptidyl-transferase center (PTC) on the ribosome. By connecting the ribosomal decoding center with the PTC through two conserved motifs, RF1 and RF2 functionally mimic a tRNA molecule in the A site; this showed another example of molecular mimicry as was seen previously between the EF-G and the TC during the elongation event. Finally RF3, a GTPase enzyme, binds to the termination complex and upon GTP hydrolysis induces large conformational changes in the ribosome that break the interactions of the class I RF with both the decoding center and the GTPase-associated center of the ribosome. This releases the class I RFs from the ribosome. It also helps to move the deacylated tRNA from the P site to the E site of the ribosome and thus completes the termination process (Klaholz et al. 2004; Gao et al. 2007a,b).

2.2.4   Recycling

Even after the actions of RFs at the end of the termination step a deacylated tRNA and mRNA remain associated with the ribosome. In order to reuse the ribosome, the RRF, together with EF-G, disassembles this post-termination (PoTC) complex into mRNA, tRNA, and the ribosomal subunits. The first report of the visualization of the RRF density in a cryo-EM map of a 70S·RRF complex came from Agrawal et al. (2004). Comparative study of PoTC·RRF with an empty 70S revealed that the RRF induced considerable conformational changes in the ribosome. The domain II of the RRF was facing towards the 30S. RRF interacted mainly with those rRNA helices of the 50S that were involved in the formation of two central inter-subunit bridges, B2a and B3. This gave the idea that the movement of RRF would disrupt those bridges and thus dissociate the subunits. Further insights were needed to identify the sequential steps of dissociation of the ribosomal subunits, and release of mRNA and deacylated tRNA from the PoTC. Using Undecagold-labeled RRF in similar PoTC·RRF complexes and applying sequential supervised classification from a heterogeneous data set, Barat et al. (2007), showed that RRF was capable of spontaneously moving from its initial binding site on the 70S to a site exclusively on the large 50S ribosomal subunit and on the way disrupted crucial inter-subunit bridges and thereby led to the dissociation of the two ribosomal subunits. The study also reported a substantial amount of intra-domain movement of the domain II relative to the domain I of the RRF. Inside the cell RRF's action is augmented by the assistance of EF-G. Using 50S·EF-G·RRF and 50S·EF-G·GMPPNP·RRF complexes Gao et al. (2005, 2007a,b) reported the nature of interaction of EF-G with RRF. The studies revealed that upon GTP hydrolysis the domain IV of EF-G pushes the domain II of RRF in such a way that ultimately leads to the disruption of the inter-subunit bridges, B2a and B3, thereby dissociates the two subunits. Further insights into the recycling event came from a very recent study by Yokoyama et al. (2012) which reported intermediate steps of the recycling process, where domain II of RRF and domain IV of EF-G adopted novel conformations. They showed that binding of EF-G to the PoTC·RRF complex reverted the ribosome from the ratcheted to unratcheted state. Due to EF-G binding on the PoTC·RRF complex, both RRF and the deacylated P-site tRNA were pushed towards the E site, with a large rotational movement of domain II of RRF towards the 30S ribosomal subunit.

2.3 Contributions of Cryo-EM on Understanding of Cotranslational Translocation of Nascent Polypeptide Chain

In cotranslational translocation process the signal sequence for translocation of the emerging polypeptide from the exit tunnel to endoplasmic reticulum (ER, for eukaryotes) or in plasma membrane (for prokaryotes) is recognized by signal recognition particle (SRP). SRPs recognize this signal and bind to it as well as it also stall the elongation process until the translating ribosome is docked to the translocon (Sec61 and SecYEG in eukaryotes and prokaryotes respectively) on the ER or plasma membrane. To know how this SRP molecule recognize signal sequence as well as stall the elongation process simultaneously, cryo-EM studies of complexes of elongation stalled ribosome with SRP were studied by Halic et al. (2004) and then it was known that the SRP molecule mainly contains two domains, S domain which binds to the ribosome near the exit channel and scan the emerging nascent polypeptide for the signal sequence and as soon as it recognize the signal sequence it undergoes a conformation change which increases its ribosome binding affinity. Simultaneously another domain of SRP Alu domain (separated by S domain by two hinges) binds to the elongation factor binding site, thus directly compensating the binding of elongation factor and stalling elongation process. This SRP molecule then binds to the SRP receptor in the ER or plasma membrane and the S domain of SRP undergoes a conformational change which exposes the translocon binding site in ribosome and the elongation stalled ribosome complex is docked into the translocon (Halic et al. 2006). The movement of polypeptide chain through the protein conducting channel (PCC) or translocon and the structure of PCC were first studied by Beckmann et al. (1997) by cryo-EM of translating ribosome with Sec61 complex. It shows that the PCC has a very compact structure and it opens only in presence of signal sequence to diameter of 15 Å through which only the nascent protein having α-helices can pass through, preventing the conductance of ions through it (Beckmann et al. 1997; Beckmann et al. 2001). The structure of this complex (in bacteria Sec YEG • Ribosome) in lipid membrane environment was also studied by cryo-EM by Frauenfeld et al. (2011) which was quite difficult to see by crystallography. The initiation of protein folding, i.e., the α-helices formation in the nascent polypeptide has been shown to take place within the exit tunnel which leads to many hypothesis like (i) protein folding occurring in hierarchy with α-helix formation taking place first, (ii) the α-helix formation preventing many nonspecific interaction of the nascent peptide with the exit tunnel. The compactness in peptide helps to identify many patterns required for regulatory events like signal sequence recognition by SRP (signal recognition particle) and also insertion of the membrane proteins into the membrane (Bhushan et al. 2010). Many ribosome-associated chaperons in eukaryotes help the folding of the nascent polypeptide chain emerging from the exit tunnel. But only one such chaperon is known in case of bacteria which is trigger factor. This trigger factor has been found to bind on mouth of the exit tunnel forming an arc kind of structure and interacting with the nascent polypeptide chain, assisting it to take proper folding (Merz et al. 2008). The SRP and the trigger factor together have been found to decide whether the nascent polypeptide will be exported or it will be folded in the cytosol (Kramer et al. 2009).

2.4 Insights into the Structural and Functional Implications of Ribosomal Maturation Factors on the Ribosome Assembly by Cryo-EM

During the structural assembly of ribosomal components into a functionally mature ribosomal subunits several protein factors play crucial roles. Cryo-EM studies from three such maturation factors, namely Era, RbfA, and RsgA which are having overlapping functions, are now facilitating our understanding of the 30S maturation. A brief description is given below.

2.4.1 Era

Era (*Escherichia coli* Ras-like protein) is an essential GTPase in bacteria. It is highly conserved and binds to the 16S rRNA of the 30S. Its depletion results into accumulation of an unprocessed precursor of the 16S rRNA. A comparative cryo-EM study by Sharma et al. (2005) on a 30S vs. 30S·Era complex localized Era's density in the cleft between the head and the platform of the 30S subunit. After fitting of the individual domains of an X-ray crystallographic structure of Era into the corresponding cryo-EM density as well as fitting of the X-ray crystallographic structure of the ribosomal components into the rest of the 30S revealed that the RNA binding KH motif of the C-terminal domain of Era interacts with the conserved nucleotides in the 3' region of the 16S rRNA. Further analysis showed that Era makes contact with several assembly elements of the 30S subunit. The study revealed that the positioning of Era would lock the 30S in a conformation that is not favorable for association with the 50S, thus until the maturation process is complete the 30S will not be involved in the translation process.

2.4.2 RbfA

RbfA (Ribosome binding factor A) is required for processing of the 5' end of the 16S rRNA during assembly of the 30S. RbfA prevents formation of an altered 5' helix of the 17S rRNA during cold-shock; otherwise 17S rRNA will not be

Fig. 2.4 Comparison of the
binding positions of RbfA
and Era on the 30S Subunit:
(**a**) Binding position of RbfA
(*red*) and Era (*magenta*;
Sharma et al. 2005) on the
30S subunit. (**b**) RbfA (*red*)
and Era (*magenta*) interact
with a common structural
element, h28, of the 16S
rRNA (*cyan*). The thumbnail
to the *left* depicts the
orientation of the 30S
subunit. *Source*: Fig "No. 7"
from the following reference
Datta et al. (2007)

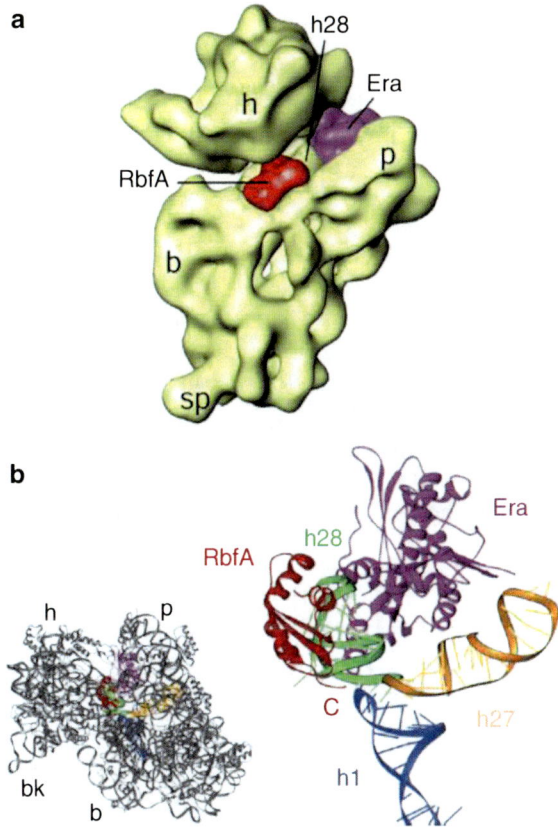

processed efficiently to 16S rRNA. Thus it was important to investigate how RbfA
was working. Even after several try crystallization of 30S·RbfA was not possible by
a Japanese research group (personal communication) which led Datta et al. (2007)
to study the 30S·RbfA complex under cryo-EM. Analyses of the 30S·RbfA cryo-
EM map through molecular docking showed that RbfA bound to the 30S subunit in
a position overlapping to the binding sites of the A- and P-site tRNAs, and RbfA's
functionally important C terminus extended toward the 5' end of the 16S rRNA,
which could provide stability to form the canonical helix 1. It also revealed that
binding of RbfA displaced a portion of the 16S rRNA encompassing helix 44 and 45
from its consensus position of a matured 30S. Those portions of 16S rRNA are
known to be directly involved in mRNA decoding and tRNA binding. From biochemical study it was known that over-expression of Era could functionally complement the function of RbfA (i.e., providing stability to the h1) in a ΔRbfA strain
during cold-shock. Datta et al. (2007) found that portions of RbfA and Era were able
to interact to the either side of neck region of the 30S that was composed of h28. The
h28 in turn interacted with the h1, thus in the absence of RbfA the Era could provide
stability to the h1 indirectly by interacting with the h28 (see Fig. 2.4).

2.4.3   RsgA

RsgA (Ribosome biogenesis GTPase) is another factor involved in the late-stage ribosome maturation of the 30S subunit. The role of RsgA is to release RbfA, from the mature 30S subunit in a GTP-dependent manner. Through a very recent cryo-EM study, Guo et al. (2011) have determined the structure of the 30S·RsgA in the presence of GMPPNP. They showed that RsgA was bound to the central part of the 30S subunit, close to the decoding center. Like RbfA, RsgA also interacted with the h44, but in the opposite side, closer towards the inter-subunit side. Careful observation by Guo et al. (2011) in their study provided strong supporting evidence of the large displacement of the h44 by RbfA and they subsequently proposed a release mechanism of the RbfA from the 30S by RsgA. Furthermore, RsgA's presence would prevent binding of all the initiation factors, as well as A-, P-site tRNAs and the 50S subunit to the 30S. Together with available biochemical and genetic data, Guo et al. (2011) further suggested that RsgA might be a general checkpoint protein in the late stage of the 30S subunit biogenesis process.

2.5   Insights into the Organeller Ribosomes as Revealed by Cryo-EM

Eukaryotic organelles like Mitochondria and Chloroplast contain their own translating system. These organelles are believed to be evolved from endosymbiosis between prokaryotic Gray et al. (2001) and eukaryotic cells and so it was expected that their ribosome to be functionally and structurally similar to prokaryotic ribosome. But it was seen that both mitoribosome and chloro-ribosome are quite different from bacterial ribosome.

2.5.1   Mitochondrial Ribosome

Through comparative analysis of the secondary structure of the bacterial and the mammalian mitochondrial ribosome's rRNAs, significant deleted portions were observed in the later case, indicating a less number of rRNA helices in the 55S mitoribosome. Though the mitoribosomes are generally heavier and larger in size than that of bacterial ribosome, its sedimentation coefficient is quite less compared to 70S in bacteria, suggesting that mitoribosome is porous in nature. It was found that its molecular composition is also different compared to bacterial ribosome. The bovine 55S mitoribosome is comprised of two asymmetric subunits, a small (28S) subunit and large (39S) subunit. The small subunit contains a 12S rRNA (955 nucleotides) with 29 proteins and the large subunit contains a 16S rRNA (1,571 nucleotides) with 48 proteins (O'Brien 1971, 2002; Koc et al. 2000). When compared to its bacterial counterpart, small subunit (30S) is composed of 16S rRNA (1,500 nucleotides, on average) with roughly 20 proteins and the large (50S) subunit contains two ribosomal RNA components, 5S rRNA (120 nucleotides, on average) and 23S rRNA (3,000 nucleotides, on average), with more than 30 proteins. The RNA:protein is completely reversed in mitoribosome (1:2) compared to prokaryotic ribosome (2:1). The cryo-EM study of the structural organization of mitoribosome (Sharma et al. 2003) has shown that the structural organization in both subunits is markedly different from that of the bacterial ribosome. Major portion of the rRNA component of the mitoribosome is heavily shielded by peripheral protein masses whereas in bacterial ribosome proteins are present only in discrete patches in the periphery. The inter-subunit bridges which are mostly RNA–RNA bridges in bacterial ribosome are dominated by protein–protein bridges in mitoribosome. In both the organellar ribosome rRNA is significantly shortened or rRNA helices are missing compared to bacterial rRNA because of which many structural differences are seen like in mitoribosome beak structure is absent in the head region, spur is absent, the shoulder is narrower and the factor binding region is more opened, but the basic structural and functional domains were unaltered .The rRNA segment interacting with the tRNA at P site and A site is highly conserved showing the importance of the positioning of the A-site and P-site tRNAs during peptidyl transferase reaction. Moreover, the ribosomal proteins which are homologous to bacterial ribosomal proteins are comparatively larger in size by N-terminal or C-terminal extension. It was believed that this increase in the protein size and number will structurally and functionally compensate the reduced or truncated rRNA. In contrast to the earlier belief, only after this study it was understood that the enlarged and additional protein present in the mitoribosome do not necessarily compensate for the majority of the missing rRNA segment. Through this cryo-EM studies the topological differences in the mRNA entry and polypeptide exit site was seen, suggesting the mechanistic divergence of the protein synthesis on the ribosome (Sharma et al. 2003). Many new features like another opening in the polypeptide exit tunnel at 63 Å from pedtidyl transferase center known as polypeptide-accessible site (PAS) were located. The solvent site of the polypeptide exit tunnel was found to be rich in protein in all organellar ribosomes suggesting that this kind of features may be necessary for cotranslational insertion of the polypeptide into the membrane (Sharma et al. 2007). The mitoribosomal E site is markedly different due to the loss of rRNA segments. The weak interaction of the tRNA in E site proposes that E site is either weak or nonexistent (Mears et al. 2006). The most recently studied mitoribosome is from a protozoan *Leishmania tarentolae* having sedimentation coefficient of 50S. The cryo-EM study of (Lmr) showed that most of the landmark features of a typical ribosome are present in the Lmr (Sharma et al. 2009). The difference found was the inter-subunit space is larger and is connected to the solvent through several newly discovered tunnels. The 9S Lm-rRNA lacks 24 of the 45 bacterial rRNA helices (Maslov et al. 2006). Furthermore, a significant amount of additional protein mass is found in the base portion of the Lmr SSU. Similar to the extra protein mass present in mammalian mitochondrial (Sharma et al. 2003) and chloroplast (Sharma et al. 2007) ribosomes (see below), a protein-rich base thus seems to be a characteristic feature of organellar SSUs. So, through the cryo-EM study of Lmr it was seen the key architectural elements are present in Lmr and the basic functioning of the ribosome is conserved. Through Cryo-EM and 3D reconstruction of organellar ribosome many of the unknown structural and functional features were known compared to the bacterial ribosome. Moreover, the function of the organellar-specific ribosomal proteins was determined.

As we know the first step of protein synthesis is formation of initiation complex which is composed of smaller subunit, mRNA, initiator tRNA, and initiation factors (IF1, IF2, and IF3) in bacteria. In mitochondrial translation initiation only two initiation factors are required $IF2_{mt}$ and $IF3_{mt}$. IF1 is missing in the mitochondrial translation initiation; the role of IF1 in bacterial translation initiation is to bind to the initiation complex such that A site is blocked so that the initiator tRNA can only bind to the P site. In mitochondrial translation $IF2_{mt}$ can perform both the function of IF1 and IF2 (Gaur et al. 2008). On comparing the IF2 and $IF2_{mt}$ it was found that $IF2_{mt}$ is composed of four domains (III–VI) which are homologous to IF2 except it has a 37 amino acid insert between domain V and VI (Spencer and Spremulli 2005). And it was found that any mutation or deletion in this 37 amino acid insert necessitates the presences of IF1 (Spencer and Spremulli 2005). The Cryo-EM study and 3D structure reconstruction have shown that 37 amino acids insert in $IF2_{mt}$ occupies the same binding site in smaller subunit as done by IF1. So insertion domain of $IF2_{mt}$ mimics the function of the IF1 by structurally occupying the same binding site (Yassin et al. 2011, 2011a).

2.5.2   Chloroplast Ribosome

In case of chloro-ribosome the RNA:protein is 2:3. The smaller subunit of chloro-ribosome is composed of 16S rRNA (1,491 nucleotides) with 25 proteins and larger subunit is composed of three rRNAs instead of two with 33 proteins (Yamaguchi and Subramanian 2000; 2003). It also contains six plastid-specific ribosomal proteins (PSRPs) (Yamaguchi and Subramanian 2000; 2003). Through the cryo-EM studies many of the structural features of chloro-ribosome were determined like the presence of an extra mass of density at the spur of the 30S subunit and much larger densities for protein L1 protuberance and protein L7/L12 stalk base in the 50S subunit. And also the location and function of this PSRPs has been determined like the PSRP1 was found to bind near the neck region of 30S, PSRP4 in head region and PSRP2-3 is located in the bottom of 30S. The 50S subunit possesses PSRP5 and PSRP6, the tentative position of PSRP5 was determined to be located near the E site. This position of PSRP5 may reflect a possible role of PSRP5 in ejection of the deacylated tRNAs from the chloro-ribosome. PSRP1 has sequence homology with pY protein (cold stress protein) in *Escherichia coli* and CTD has sequence homology with light-regulated transcript A(lrtA) present in Cynobacterium *Synechococcus* sp., that is why it is believed its expression is also regulated in light dependent manner. The binding site of PSRP1 to chloro-ribosome is found to be same as that of binding site of protein pY in *E. coli* 70S ribosome and experimentally shown that PSRP1 stabilizes

70S formation like protein pY. Thus, proposing that PSRP1 also stabilizes the formation 70S in dark condition. PSRP1 CTD also found to interact with the pRRF bound to the chloro-50S; they together block the P-site and A-site tRNA binding site supporting the role of PSRP1 as stress factor, and PSRP1 along with the pRRF inhibits protein synthesis during stress. As soon as the optimal condition returns EF-G and RRF help in removing this stress factor from ribosome, and also cause recycling of 30S and 50S for initiating new protein synthesis (Sharma et al. 2010).

2.6   Insights into the Eukaryotic Translation Initiation Through Cryo-EM

Since mechanisms of eukaryotic translation initiation process are significantly different than that of the bacterial system, and significant cryo-EM studies have revealed some of those important information, we are discussing this topic here and omitting other eukaryotic translation events.

Initiation in eukaryotes can take place by two means, one is for 5′ capped mRNAs which involves 12 initiation factors and is a very complex process, and another is by internal ribosome entry site (IRES). IRES are sequences which are found in the upstream of few cellular mRNAs and viral RNAs.

For initiating the translation of 5′ capped mRNAs an initiation complex of 43S is formed involving initiation factors eIF1A, eIF1, eIF3, eIF5, eIF2·GTP·Met-tRNA and 40S subunit, thereby forming 43S complex. On binding of eIF4F to the 5′ cap of the mRNA this 43S complex become capable for scanning the unstructured mRNA for the initiation codon AUG. Once the initiation codon is recognized the eIF1 dissociates and P site becomes accessible, eIF5B then binds to the initiation complex. Binding of eIF5B increases the binding affinity of Met-tRNA to P site, finally forming a 48S complex. Binding of eIF5B also catalyzes the association of 60S to 48S initiation complex forming 80S initiation complex. On binding of 60S eIF2, eIF3 and eIF5 are released and thus the complex is ready for elongation (review by Allen and Frank 2007).

The mostly studied initiation complex is 40S·eIF1·eIF1A; eIF1 and eIF1A are highly conserved initiation Factors (12 and 17 kDa respectively). They cooperatively bind to the 40S subunit with high affinity forming a 43S mRNA complex along with eIF3 and eIF2, that is competent for scanning and recognizing the initiation codon (AUG). The cryo-EM reconstruction of 40S·eIF1·eIF1A shows a striking difference in the conformation of the 40S subunit like a new connection was observed between the head and shoulder on the solvent side (between 18S rRNA helix and the ribosomal protein rpS3), the beak and the platform exhibit a slightly altered conformation. Additionally the "latch" (latch is formed by non-covalent interaction between 18S rRNA helices 18 and 34) of the mRNA entry site was closed in empty 40S but in 40S·eIF1·eIF1A structure it was not visible. The latch is open in 40S·eIF1·eIF1A complex making the mRNA entry channel more accessible, which allows mRNA to dock into the mRNA binding channel directly into the initiation complex, instead of small tunnel. The latch must open to allow mRNA binding, because in *in vivo*, a large protein complex (eIF4F) is bound to the 5' end of the mRNA. The latch is proposed to clamp around the mRNA, either after initiation codon recognition or after subunit joining, to trap the mRNA on the ribosome, preventing mRNA dissociation. eIF1 is hypothesized to indirectly monitor initiation codon recognition by influencing the conformation of the platform and the positions of mRNA and tRNA. eIF1A is expected to bind in a similar position to IF1. Both the eIF1 and eIF1A are required for the full conformational change, and the change is fundamentally required for translation initiation (Passmore et al. 2007). Similar kind of conformational change in 40S subunit is also shown in a 43S complex of 40S and multifactor complex (eIF1, eIF1A, eIF3, and eIF5) in budding yeast saying that this conformational change in 40S subunit leads to opening of head/body/platform junction, thus opening the mRNA entry channel (Gilbert et al. 2007).

The second mode of translation initiation is by IRES, present in the upstream of few cellular mRNA and in viral genomic RNAs. They manipulate the translation system in such a manner that even in the absence of translation initiation factors they begin the translation of mRNA containing IRES in their upstream region. This kind of translation initiation mechanism is seen in viral mRNA and also in few mRNA expressing during stress conditions when the normal cellular translation is stopped. IRES can bind directly to the 40S subunit and can bypass the initiation process directly proceeding to the elongation step. Some of the IRES of viruses do not require any translation initiation factor and not even initiator tRNA like Cricket paralysis virus (CrPV). But few of the viral IRES require some of the initiation factors like Hepatitis C virus (HCV) requires eIF2, eIF3, and Met-tRNA and polio virus requires eIF4E (Ji et al. 2004). Through cryo-EM, translation initiation by two IRES of HCV and CrPV was studied. In HCV, IRES consist of three domains; domain II, domain III, and domain IV containing the AUG start codon. The HCV IRES binds to the solvent side of the 40S subunit, where domain II interacting with S5 which forms the mRNA exit channel along with S14 and the apex of the domain II binds close to the tRNA E site. Domain III is mainly involved in binding of 40S subunit and eIF3 (Berry et al. 2011). These interactions induce a conformational change in the 40S subunit which is required for placing the start codon at appropriate position in P site to initiate translation and also 60S subunit binding. The conformational changes induced in the 40S are similar to the conformation changes which take place upon binding of eIFs to the 40S in normal translation initiation. To precede to the elongation step, the domain II should be removed from the E site during the first translocation step, that is the reason why IRES translation initiation is more sensitive to cyclohexamide than normal translation initiation as they bind to the E site and prevent the removal of domain II from the E site (Boehringer et al. 2005). The mechanism by which the IRES of CrPV initiate translation is a bit different from that of HCV. CrPV IRES contains three psuedoknots; PK1 which binds to the P site, PK2 and PK3 which binds to the E site. PK1 mimics the initiator tRNA, that is the reason why CrPV IRES do not require even initiator tRNA to initiate translation and the binding of PK1 to the P site positions the first codon to the A site. As it was seen in the HCV the binding of IRES of CrPV also induces a conformational change in the 40S subunit, thus preparing the ribosome to proceed for the further steps.

The binding of the 60S subunit to the initiation complex also induces a conformational change in the IRES of CrPV, increasing the A-site accessibility and proceeding for elongation. The basic principle by which this IRES initiate translation is that they mimic the eIFs and tRNA and induce conformational changes as done by the eIFs, thus bypassing the initiation process and proceeding directly to the elongation step (Spahn et al. 2004; Schuler et al. 2006; Hellen 2007).

2.7   Conclusions and Future Directions

It is no doubt from the above studies that a significant amount of knowledge has been gained on the mechanisms of translation process through the use of the cryo-EM techniques. At the same time cryo-EM techniques have advanced significantly because the need to visualize the translation events in better resolution has always been the urge in the cryo-EM and the ribosome community. Due to the concerted contribution of several gifted mathematicians, physicists, and biochemists, tremendous amounts of advancement have been achieved in the image processing and rendering and analysis software. At the same time increased sophistication on the electron microscopy hardware, automation, usage of high resolution digital camera, and powerful computer hardware have become instrumental in pushing both the cryo-EM and the ribosome fields to their boundaries and beyond the new horizons. It is necessary to continue the above practice in the future as well.

The challenging problems that lie ahead are to achieve atomic resolution of ribosome complexes through cryo-EM, clearly seeing the translating ribosomes in their native organization inside the cell and finding out causes of aberrations in translation process that may lead to disease manifestation in humans.

References

Agirrezabala X, Lei J, Brunelle JL, Ortiz-Meoz RF, Green R, Frank J (2008) Visualization of the hybrid state of tRNA binding promoted by spontaneous ratcheting of the ribosome. Mol Cell 32(2):190–197

Agirrezabala X, Schreiner E, Trabuco LG, Lei J, Ortiz-Meoz RF, Schulten K, Green R, Frank J (2011) Structural insights into cognate versus near-cognate discrimination during decoding. EMBO J 30(8):1497–1507

Agrawal RK, Penczek P, Grassucci RA, Li Y, Leith A, Nierhaus KH, Frank J (1996) Direct visualization of A-, P-, and E-site transfer RNAs in the *Escherichia coli* ribosome. Science 271(5251):1000–1002

Agrawal RK, Penczek P, Grassucci RA, Frank J (1998) Visualization of elongation factor G on the Escherichia coli 70S ribosome: the mechanism of translocation. Proc Natl Acad Sci U S A 95:6134–6138

Agrawal RK, Heagle AB, Penczek P, Grassucci RA, Frank J (1999) EF-G-dependent GTP hydrolysis induces translocation accompanied by large conformational changes in the 70S ribosome. Nat Struct Biol 6:643–647

Agrawal RK, Sharma MR, Kiel MC, Hirokawa G, Booth TM, Spahn CM, Grassucci RA, Kaji A, Frank J (2004) Visualization of ribosome-recycling factor on the Escherichia coli 70S ribosome: functional implications. Proc Natl Acad Sci U S A 101(24):8900–8905

Allen GS, Frank J (2007) Structural insights on the translation initiation complex: ghosts of a universal initiation complex. Mol Microbiol 63(4):941–950

Allen GS, Zavialov A, Gursky R, Ehrenberg M, Frank J (2005) The cryo-EM structure of a translation initiation complex from Escherichia coli. Cell 121(5):703–712

Armache JP, Jarasch A, Anger AM, Villa E, Becker T, Bhushan S, Jossinet F, Habeck M, Dindar G, Franckenberg S, Marquez V, Mielke T, Thomm M, Berninghausen O, Beatrix B, Soding J, Westhof E, Wilson DN, Beckmann R (2010) Localization of eukaryote-specific ribosomal proteins in a 5.5-A cryo-EM map of the 80S eukaryotic ribosome. Proc Natl Acad Sci U S A 107:19754–19759

Barat C, Datta PP, Raj VS, Sharma MR, Kaji H, Kaji A, Agrawal RK (2007) Progression of the ribosome recycling factor through the ribosome dissociates the two ribosomal subunits. Mol Cell 27(2):250–261

Beckmann R, Bubeck D, Grassucci R, Penczek P, Verschoor A, Blobel G, Frank J (1997) Alignment of conduits for the nascent polypeptide chain in the ribosome-Sec61 complex. Science 278(5346):2123–2126

Beckmann R, Spahn CM, Eswar N, Helmers J, Penczek PA, Sali A, Frank J, Blobel G (2001) Architecture of the protein-conducting channel associated with the translating 80S ribosome. Cell 107:361–372

Berry KE, Waghray S, Mortimer SA, Bai Y, Doudna JA (2011) Crystal structure of the HCV IRES central domain reveals strategy for start-codon positioning. Structure 19:1456–1466

Bhushan S, Gartmann M, Halic M, Armache JP, Jarasch A, Mielke T, Berninghausen O, Wilson DN, Beckmann R (2010) α-Helical nascent polypeptide chains visualized within distinct regions of the ribosomal exit tunnel. Nat Struct Mol Biol 17:3

Boehringer D, Thermann R, Ostareck-Lederer A, Lewis JD, Stark H (2005) Structure of the hepatitis C virus IRES bound to the human 80S ribosome: remodeling of the HCV IRES. Structure 13:1695–1706

Connell SR, Takemoto C, Wilson DN, Wang H, Murayama K, Terada T, Shirouzu M, Rost M, Schüler M, Giesebrecht J, Dabrowski M, Mielke T, Fucini P, Yokoyama S, Spahn CM (2007) Structural basis for interaction of the ribosome with the switch regions of GTP-bound elongation factors. Mol Cell 25(5):751–764

Datta PP, Sharma MR, Qi L, Frank J, Agrawal RK (2005) Interaction of the G' domain of elongation factor G and the C-terminal domain of ribosomal protein L7/L12 during translocation as revealed by Cryo-EM. Mol Cell 20:723–731

Datta PP, Wilson DN, Kawazoe M, Swami NK, Kaminishi T, Sharma MR, Booth TM, Takemoto C, Fucini P, Yokoyama S, Agrawal RK (2007) Structural aspects of RbfA action during small ribosomal subunit assembly. Mol Cell 28(3):434–445

Frank J, Agrawal RK (2000) A ratchet-like inter-subunit reorganization of the ribosome during translocation. Nature 406(6793):318–322

Frank J, Shimkin B, Dowse H (1981) SPIDER—a modular software system for electron image processing. Ultramicroscopy 6:343–358

Frauenfeld J, Gumbart J, van der Sluis EO, Funes S, Gartmann M, Beatrix B, Mielke T, Berninghausen O, Becker T, Schulten K, Beckmann R (2011) Cryo-EM structure of the ribosome–SecYE complex in the membrane environment. Nat Struct Mol Biol 18:5

Fu J, Gao H, Frank J (2007) Unsupervised classification of single particles by cluster tracking in multi-dimensional space. J Struct Biol 157(1):226–239

Gabashvili IS, Agrawal RK, Spahn CM, Grassucci RA, Svergun DI, Frank J, Penczek P (2000) Solution structure of the E. coli 70S ribosome at 11.5 Å resolution. Cell 100(5):537–549

Gao H, Sengupta J, Valle M, Korostelev A, Eswar N, Stagg SM, Van Roey P, Agrawal RK, Harvey SC, Sali A, Chapman MS, Frank J (2003) Study of the structural dynamics of the E. coli 70S ribosome using real-space refinement. Cell 113(6):789–801

Gao N, Zavialov AV, Li W, Sengupta J, Valle M, Gursky RP, Ehrenberg M, Frank J (2005) Mechanism for the disassembly of the posttermination complex inferred from cryo-EM studies. Mol Cell 18(6):663–674

Gao H, Zhou Z, Rawat U, Huang C, Bouakaz L, Wang C, Cheng Z, Liu Y, Zavialov A, Gursky R, Sanyal S, Ehrenberg M, Frank J, Song H (2007a) RF3 induces ribosomal conformational changes responsible for dissociation of class I release factors. Cell 129(5):929–941

Gao N, Zavialov AV, Ehrenberg M, Frank J (2007b) Specific interaction between EF-G and RRF and its implication for GTP-dependent ribosome splitting into subunits. J Mol Biol 374(5):1345–1358

Gaur R, Grasso D, Datta PP, Krishna PD, Das G, Spencer A, Agrawal RK, Spremulli L, Varshney U (2008) A single mammalian mitochondrial translation initiation factor functionally replaces two bacterial factors. Mol Cell 29(2):180–190

Gao YG, Selmer M, Dunham CM, Weixlbaumer A, Kelley AC, Ramakrishnan V (2009) The structure of the ribosome with elongation factor G trapped in the posttranslocational state. Science 326(5953):694–699

Gilbert R, Gordiyenko Y, Haar T, Sonnen A, Hofmann G, Nardelli M, Stuart D, McCarthy J (2007) Reconfiguration of yeast 40S ribosomal subunit domains by the translation initiation multifactor complex. Proc Natl Acad Sci U S A 104(14):5788–5793

Gray MW, Burger G, Lang BF (2001) The origin and early evolution of mitochondria. Genome Biol 2:1018.1–1018.5

Guo Q, Yuan Y, Xu Y, Feng B, Liu L, Chen K, Sun M, Yang Z, Lei J, Gao N (2011) Structural basis for the function of a small GTPase RsgA on the 30S ribosomal subunit maturation revealed by cryoelectron microscopy. Proc Natl Acad Sci U S A 108(32):13100–13105

Halic M, Becker T, Pool MR, Spahn CM, Grassucci RA, Frank J, Beckmann R (2004) Structure of the signal recognition particle interacting with the elongation-arrested ribosome. Nature 427(6977):808–814

Halic M, Gartmann M, Schlenker O, Mielke T, Pool MR, Sinning I, Beckmann R (2006) Signal recognition particle receptor exposes the ribosomal translocon binding site. Science 312(5774): 745–747

Harms JM, Wilson DN, Schluenzen F, Connell SR, Stachelhaus T, Zaborowska Z, Spahn CM, Fucini P (2008) Translational regulation via L11: molecular switches on the ribosome turned on and off by thiostrepton and micrococcin. Mol Cell 30(1):26–38

Heel MV, Keegstra W (1981) IMAGIC: a fast, flexible and friendly image analysis software system. Ultramicroscopy 7:113–130

Hellen C (2007) Bypassing translation initiation. Structure 15:4–6

Ji H, Fraser CS, Yu Y, Leary J, Doudna JA (2004) Coordinated assembly of human translation initiation complexes by the hepatitis C virus internal ribosome entry site RNA. Proc Natl Acad Sci U S A 101:49

Klaholz BP, Pape T, Zavialov AV, Myasnikov AG, Orlova EV, Vestergaard B, Ehrenberg M, van Heel M (2003) Structure of the Escherichia coli ribosomal termination complex with release factor 2. Nature 421(6918):90–94

Klaholz BP, Myasnikov AG, Van Heel M (2004) Visualization of release factor 3 on the ribosome during termination of protein synthesis. Nature 427(6977):862–865

Koc EC, Burkhart W, Blackburn K, Moseley A, Koc H, Spremulli LL (2000) A proteomics approach to the identification of mammalian mitochondrial small subunit ribosomal proteins. J Biol Chem 275:32585–32591

Kramer G, Boehringer D, Ban N, Bukau B (2009) The ribosome as a platform for co-translational processing, folding and targeting of newly synthesized proteins. Nat Struct Mol Biol 16:6

Li W, Frank J (2007) Transfer RNA in the hybrid P/E state: correlating molecular dynamics simulations with cryo-EM data. Proc Natl Acad Sci U S A 104(42):16540–16545

Li W, Agirrezabala X, Lei J, Bouakaz L, Brunelle JL, Ortiz-Meoz RF, Green R, Sanyal S, Ehrenberg M, Frank J (2008) Recognition of aminoacyl-tRNA: a common molecular mechanism revealed by cryo-EM. EMBO J 27(24):3322–3331

Loerke J, Giesebrecht J, Spahn CM (2010) Multiparticle cryo-EM of ribosomes. Methods Enzymol 483:161–177

Maslov DA, Sharma MR, Butler E, Falick AM, Gingery M, Agrawal RK, Spremulli LL, Larry SL (2006) Isolation and characterization of mitochondrial ribosomes and ribosomal subunits from Leishmania tarentolae. Mol Biochem Parasitol 148:69–78

Mears JA, Sharma MR, Gutell RR, McCook AS, Richardson PE, Caulfield TR, Agrawal RK, Harvey SC (2006) A structural model for the large subunit of the mammalian mitochondrial ribosome. J Mol Biol 358:193–212

Merz F, Boehringer D, Schaffitzel C, Preissler S, Hoffmann A, Maier T, Rutkowska A, Lozza J, Ban N, Bukau B, Deuerling E (2008) Molecular mechanism and structure of Trigger Factor bound to the translating ribosome. EMBO J 27:1622–1632

Montesano-Roditis L, Glitz DG, Traut RR, Stewart PL (2001) Cryo-electron microscopic localization of protein L7/L12 within the *Escherichia coli* 70 S ribosome by difference mapping and Nanogold labeling. J Biol Chem 276(17):14117–14123

Myasnikov AG, Marzi S, Simonetti A, Giuliodori AM, Gualerzi CO, Yusupova G, Yusupov M, Klaholz BP (2005) Conformational transition of initiation factor 2 from the GTP- to GDP-bound state visualized on the ribosome. Nat Struct Mol Biol 12(12):1145–1149

O'Brien TW (1971) The general occurrence of 55S ribosomes in mammalian liver mitochondria. J Biol Chem 246:3409–3417

O'Brien TW (2002) Evolution of a protein-rich mitochondrial ribosome: implications for human genetic disease. Gene 286:73–79

Palade GE (1955) A small particulate component of the cytoplasm. J Biophys Biochem Cytol 1(1)

Passmore LA, Schmeing TM, Maag D, Applefield DJ, Acker MG, Algire MA, Lorsch JR, Ramakrishnan V (2007) The eukaryotic translation initiation factors eIF1 and eIF1A induce an open conformation of the 40S ribosome. Mol Cell 26:41–50

Penczek PA, Zhu J, Frank J (1996) A common-lines based method for determining orientations for N > 3 particle projections simultaneously. Ultramicroscopy 63:205–218

Radermacher M, Wagenknecht T, Verschoor A, Frank J (1987) Three-dimensional reconstruction from a single-exposure, random conical tilt series applied to the 50S ribosomal subunit of *Escherichia coli*. J Microsc 146:113–136

Ratje AH, Loerke J, Mikolajka A, Brünner M, Hildebrand PW, Starosta AL, Dönhöfer A, Connell SR, Fucini P, Mielke T, Whitford PC, Onuchic JN, Yu Y, Sanbonmatsu KY, Hartmann RK, Penczek PA, Wilson DN, Spahn CM (2010) Head swivel on the ribosome facilitates translocation by means of intra-subunit tRNA hybrid sites. Nature 468(7324):713–716

Rawat UB, Zavialov AV, Sengupta J, Valle M, Grassucci RA, Linde J, Vestergaard B, Ehrenberg M, Frank J (2003) A cryo-electron microscopic study of ribosome-bound termination factor RF2. Nature 421(6918):87–90

Rawat U, Gao H, Zavialov A, Gursky R, Ehrenberg M, Frank J (2006) Interactions of the release factor RF1 with the ribosome as revealed by cryo-EM. J Mol Biol 357(4):1144–1153

Scheres SH, Gao H, Valle M, Herman GT, Eggermont PP, Frank J, Carazo JM (2007) Disentangling conformational states of macromolecules in 3D-EM through likelihood optimization. Nat Methods 4(1):27–29. Epub 2006 Dec 10

Schmeing TM, Voorhees RM, Kelley AC, Gao YG, Murphy FV IV, Weir JR, Ramakrishnan V (2009) The crystal structure of the ribosome bound to EF-Tu and aminoacyl-tRNA. Science 326(5953):688–694

Schuette JC, Murphy FV IV, Kelley AC, Weir JR, Giesebrecht J, Connell SR, Loerke J, Mielke T, Zhang W, Penczek PA, Ramakrishnan V, Spahn CM (2009) GTPase activation of elongation factor EF-Tu by the ribosome during decoding. EMBO J 28(6):755–765

Schuler M, Connell SR, Lescoute A, Giesebrecht J, Dabrowski M, Schroeer B, Mielke T, Penczek PA, Westhof E, Spahn C (2006) Structure of the ribosome-bound cricket paralysis virus IRES RNA. Nat Struct Mol Biol 13:12

Sharma MR, Koc EC, Datta PP, Booth TM, Spremulli LL, Agrawal RK (2003) Structure of the mammalian mitochondrial ribosome reveals an expanded functional role for its component proteins. Cell 115:97–108

Sharma MR, Barat C, Wilson DN, Booth TM, Kawazoe M, Hori-Takemoto C, Shirouzu M, Yokoyama S, Fucini P, Agrawal RK (2005) Interaction of Era with the 30S ribosomal subunit implications for 30S subunit assembly. Mol Cell 18(3):319–329

Sharma MR, Wilson DN, Datta PP, Barat C, Schluenzen F, Fucini P, Agrawal RK (2007) Cryo-EM study of the spinach chloroplast ribosome reveals the structural and functional roles of plastid-specific ribosomal proteins. Proc Natl Acad Sci U S A 104(49):19315–19320

Sharma MR, Bootha TM, Simpsonb L, Maslovc DA, Agrawal RK (2009) Structure of a mitochondrial ribosome with minimal RNA. Proc Natl Acad Sci U S A 106(24):9637–9642

Sharma MR, Donhofer A, Barat C, Marquez V, Datta PP, Fucini P, Wilson DN, Agrawal RK (2010) PSRP1 is not a ribosomal protein, but a ribosome-binding factor that is recycled by the ribosome-recycling factor (RRF) and elongation factor G (EF-G). J Biol Chem 285(6): 4006–4014

Simonetti A, Marzi S, Myasnikov AG, Fabbretti A, Yusupov M, Gualerzi CO, Klaholz BP (2008) Structure of the 30S translation initiation complex. Nature 455(7211):416–420

Spahn CM, Grassucci RA, Penczek P, Frank J (1999) Direct three-dimensional localization and positive identification of RNA helices within the ribosome by means of genetic tagging and cryo-electron microscopy. Structure 15:1567–1573

Spahn C, Jan E, Mulder A, Grassucci RA, Sarnow P, Frank J (2004) Cryo-EM visualization of a viral internal ribosome entry site bound to human ribosomes: the IRES functions as an RNA-based translation factor. Cell 118:465–475

Spencer AC, Spremulli LL (2005) The interaction of mitochondrial translational initiation factor 2 with the small ribosomal subunit. Biochim Biophys Acta 1750:69–81

Stark H, Rodnina MV, Wieden HJ, Zemlin F, Wintermeyer W, Van Heel M (2002) Ribosome interactions of aminoacyl-tRNA and elongation factor Tu in the codon-recognition complex. Nat Struct Biol 9:849–854

Tama F, Valle M, Frank J, Brooks CL 3rd (2003) Dynamic reorganization of the functionally active ribosome explored by normal mode analysis and cryo-electron microscopy. Proc Natl Acad Sci U S A 100(16):9319–9323

Trabuco LG, Villa E, Mitra K, Frank J, Schulten K (2008) Flexible fitting of atomic structures into electron microscopy maps using molecular dynamics. Structure 16(5):673–683

Valle M, Sengupta J, Swami NK, Grassucci RA, Burkhardt N, Nierhaus KH, Agrawal RK, Frank J (2002) Cryo-EM reveals an active role for aminoacyl-tRNA in the accommodation process. EMBO J 21:3557–3567

Valle M, Zavialov A, Li W, Stagg SM, Sengupta J, Nielsen RC, Nissen P, Harvey SC, Ehrenberg M, Frank J (2003a) Incorporation of aminoacyl-tRNA into the ribosome as seen by cryo-electron microscopy. Nat Struct Biol 10:899–906

Valle M, Zavialov A, Sengupta J, Rawat U, Ehrenberg M, Frank J (2003b) Locking and unlocking of ribosomal motions. Cell 114:123–134

Van Heel M (1987) Angular reconstitution: a posteriori assignment of projection directions for 3D reconstruction. Ultramicroscopy 21:111–124

Villa E, Sengupta J, Trabuco LG, LeBarron J, Baxter WT, Shaikh TR, Grassucci RA, Nissen P, Ehrenberg M, Schulten K, Frank J (2009) Ribosome-induced changes in elongation factor Tu conformation control GTP hydrolysis. Proc Natl Acad Sci U S A 106(4):1063–1068

Voorhees RM, Schmeing TM, Kelley AC, Ramakrishnan V (2010) The mechanism for activation of GTP hydrolysis on the ribosome. Science 330(6005):835–838

Wriggers W, Agrawal RK, Drew DL, McCammon A, Frank J (2000) Domain motions of EF-G bound to the 70S ribosome: insights from a hand-shaking between multi-resolution structures. Biophys J 79(3):1670–1678

Yamaguchi K, Subramanian AR (2000) The plastid ribosomal proteins. J Biol Chem 275(37): 28466–28482

Yamaguchi K, Subramanian AR (2003) Proteomic identification of all plastid-specific ribosomal proteins in higher plant chloroplast 30S ribosomal subunit. Eur J Biochem 270:190–205

Yassin AS, Agrawal RK, Banavali NK (2011) Computational exploration of structural hypotheses for an additional sequence in a mammalian mitochondrial protein. PLoS One. 6(7):e21871. Epub 2011 Jul 11

Yassin AS, Haqueb Md E, Datta PP, Elmoreb K, Banavalia NK, Spremullib LL, Agrawal RK (2011a) Insertion domain within mammalian mitochondrial translation initiation factor 2 serves the role of eubacterial initiation factor 1. Proc Natl Acad Sci U S A 108(10):3918–3923

Yokoyama T, Shaikh TR, Iwakura N, Kaji H, Kaji A, Agrawal RK (2012) Structural insights into initial and intermediate steps of the ribosome-recycling process. EMBO J 31(7):1836–1846. doi:10.1038/emboj.2012.22

Chapter 3
Molecular Dynamics Simulations of the Ribosome

Karissa Y. Sanbonmatsu, Scott C. Blanchard, and Paul C. Whitford

3.1 Introduction

To manufacture proteins encoded in the mRNA, the ribosome must convert the sequence of nucleotides inscribed on the mRNA to the sequence of amino acids that comprise the corresponding protein (Fig. 3.1). That is, the ribosome must translate the four-letter language of nucleic acids into the 20-letter language of proteins. The ribosome uses a suite of translating molecules (tRNAs) as a molecular look-up table, allowing it to convert from mRNA to protein. Each tRNA carries an amino acid corresponding to a three-nucleotide codon, which may appear in the sequence on the mRNA. In addition, each tRNA has a three-nucleotide anticodon, cognate (through Watson–Crick base pairing) to the corresponding codon in the mRNA. At any given time during elongation, the ribosome exposes one codon in the message to solution. The ribosome must then select only tRNAs with amino acids that correspond to that codon, while rejecting all other tRNAs. Subsequently, the ribosome must move itself along the message by exactly three nucleotides to

K.Y. Sanbonmatsu (✉)
Theoretical Division, Los Alamos National Laboratory, Theoretical Biology
and Biophysics Group, MS K710, Los Alamos, NM 87545, USA
e-mail: kys@lanl.gov

S.C. Blanchard (✉)
Department of Physiology and Biophysics, Weill Cornell Medical College,
1300 York Avenue, BX 75, New York, NY 10065, USA
e-mail: scb2005@med.cornell.edu

P.C. Whitford
Center for Theoretical Biological Physics and Department of Physics,
University of California at San Diego, San Diego, CA 92093, USA
e-mail: pwhitfor@ctbp.ucsd.edu

J.D. Dinman (ed.), *Biophysical Approaches to Translational Control of Gene Expression*,
Biophysics for the Life Sciences 1, DOI 10.1007/978-1-4614-3991-2_3,
© Springer Science+Business Media New York 2013

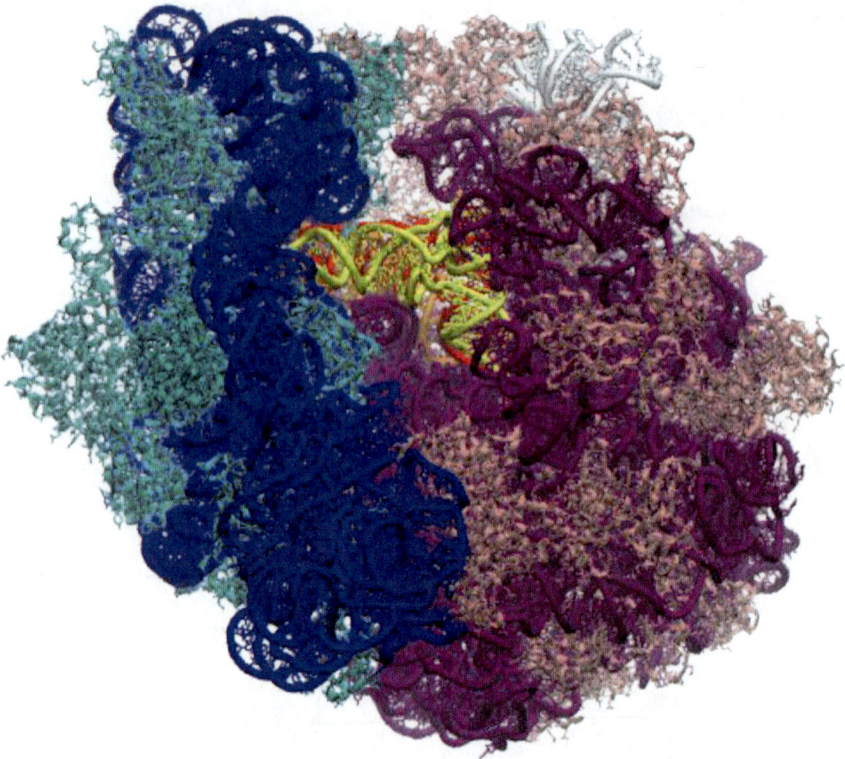

Fig. 3.1 Overview of ribosome structure determined by Yusupov and coworkers. *Blue*, small subunit 16S rRNA. *Cyan*, small subunit proteins. *Magenta*, large subunit 23S rRNA. *Pink*, large subunit proteins. *White*, large subunit 5S rRNA. *Yellow*, aminoacyl-tRNA. *Red*, peptidyl-tRNA

prepare for the next tRNA. The feat of discriminating between correct and incorrect amino acids and moving large ligands (~60 Å in length from anticodon to elbow) through the ribosome, from site to site with exquisite precision, has perplexed researchers for almost half a century. Recent studies have been able to quantify the stochastic nature of these processes (Feldman et al. 2010; Munro et al. 2007; Blanchard et al. 2004; Blanchard 2009) and conclude that the ribosome is highly dynamic in nature.

For elongation to occur, tRNAs must bind to the ribosome, move into the A site, proceed to the P site, continue to the E site, and dissociate from the ribosome (Moazed and Noller 1989). The inner structure of the ribosome produces precise alignment of the tRNA with the mRNA codon of the 30S (Fig. 3.2) and the peptidyl transferase center on the 50S (Fig. 3.3) (Demeshkina et al. 2010; Yusupova et al. 2006; Ogle et al. 2001; Voorhees et al. 2009). On the 50S in particular, the 3'-CCA end of the tRNA is surrounded by a dense thicket of ribosomal RNA in the A site and the P site (Fig. 3.4). In the E site, nucleotides of the 3'-CCA end are interca-lated between two ribosomal nucleotides (Selmer et al. 2006; Gao et al. 2009).

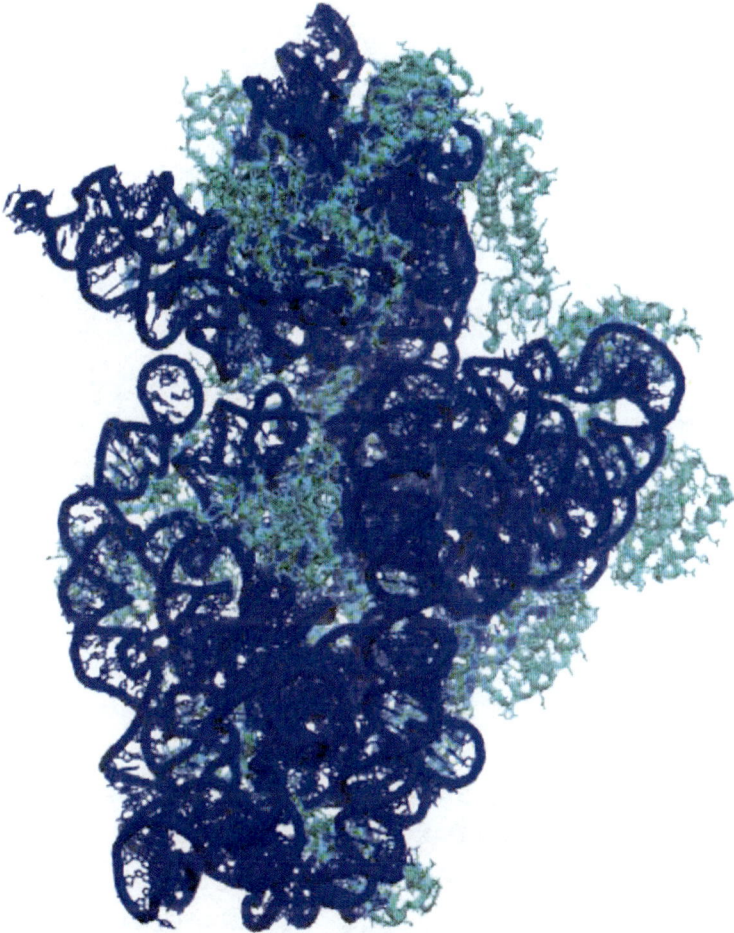

Fig. 3.2 30S small ribosomal subunit

The precise alignment of tRNA in each state presents a paradox: how can the tRNA be surrounded by many specific ribosome interactions, yet move to two other completely different states with equally complex interactions? The solution to this paradox is that both the ribosome and the tRNA are extremely dynamic, as evidenced by single-molecule FRET (smFRET), cryo electron microscopy (cryo-EM), and X-ray crystallography data (Munro et al. 2007; Blanchard et al. 2004; Blanchard 2009; Zhang et al. 2008, 2009; Schuwirth et al. 2005; Fischer et al. 2010). While smFRET provides excellent information on the dynamics of the ribosome, it has low spatial resolution. Cryo-EM and crystallographic studies provide higher resolution spatial information but provide little information about dynamics. Molecular dynamics simulations help to bridge the gap between single molecule dynamics experiments and structural biology studies.

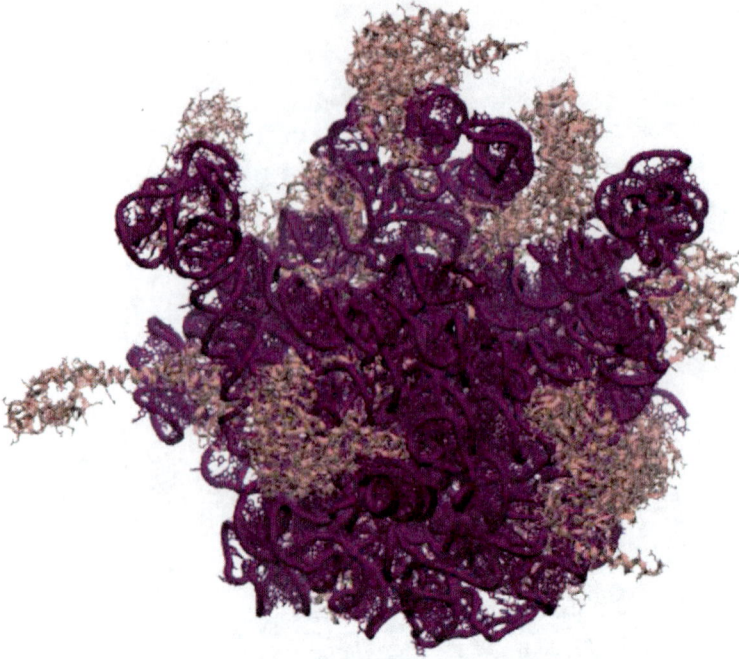

Fig. 3.3 50S large ribosomal subunit

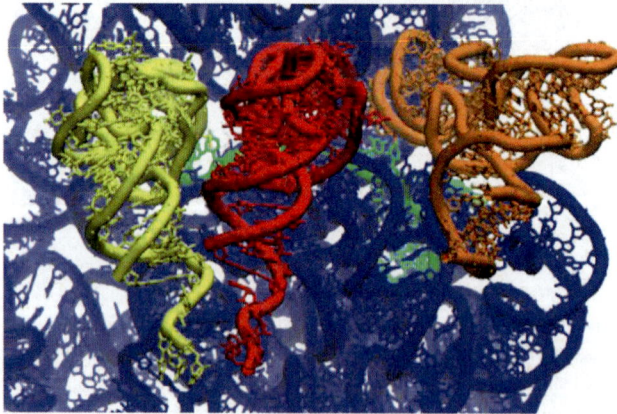

Fig. 3.4 tRNA binding sites on small subunit. Aminoacyl tRNA (*yellow*) bound to A site. Peptidyl tRNA (*red*) bound to P site. Exit tRNA (*orange*) bound to E site

With regard to medical applications, the bacterial ribosome is a major target of antibiotics. Approximately 50 % of antibiotics used in US hospitals target the ribosome (Mankin 2006). Although the molecular mechanism of several new antibiotic targets remains obscure, the major classes affect elongation events including decoding, peptide bond formation, and/or translocation (Blanchard et al. 2010; Carter et al. 2000; Schlunzen et al. 2001; Pioletti et al. 2001). A major problem for national health is multi-drug resistant bacteria. Mechanistic studies of the ribosome will help to find new combination therapies that can be used to combat resistant bacteria. Specifically, molecular dynamics simulations shed new light on the conformational transitions between states that occur during protein synthesis. A future class of antibiotics could be those that target transitions, rather than the major states of the ribosome.

3.2   tRNA Dynamics

The tRNA itself shows large fluctuations during translation. It binds to the ribosome in a highly kinked configuration called the A/T state where the "A" represents the anticodon interacting with the 30S subunit A site and the "T" represents the ternary complex interacting with the 50S subunit at the GTPase activity center (Fig. 3.5) (Valle et al. 2002, 2003a). Single molecule FRET studies also show rapid tRNA fluctuations between several intermediates occurring before accommodation, where the tRNA moves into the fully bound "A/A" position (anticodon interacting with the 30S A site and 3′-CCA end interacting with the 50S A site). These include transitions between the initial binding configuration, the codon–anticodon recognition configuration, the GTPase activation configuration, and the A/T state. This led to the conclusion that the aminoacyl-tRNA samples a broad ensemble of configurations prior to accommodation (Geggier et al. 2010).

In addition to tRNA selection, cryo-EM reconstructions of intermediates during translocation demonstrate tRNA flexibility. After accommodation and peptidyl transferase, the tRNA in the A/A position moves to the A/P position (anticodon in 30S A site and 3′-CCA end in 50S P site), while the tRNA in the P/P position moves to the P/E position (anticodon in the 30S P site and 3′-CCA end in the 50S E site). Cryo-EM reconstructions show the tRNA to be in a curved configuration, albeit a curved configuration which differs from the curved configuration required for the A/T position (Connell et al. 2008). These two curved configurations are suggestive of an energy storage-release mechanism that allows the tRNA to propel itself through the ribosome (Frank et al. 2005). This enthalpic energy source would of course be combined with the source of gated thermal fluctuations resulting from the thermal bath (Whitford et al. 2010a).

The motions of tRNA within the ribosome have been the subject of intense investigation using both bulk- and single-molecule techniques. smFRET studies measuring time-dependent changes in the intermolecular distance between A- and P-site tRNAs have proven particularly revealing (Munro et al. 2007). During tRNA selection aa-tRNA was shown to enter the A site through reversible excursions between multiple intermediate states. By analyzing the nature of these dynamics for both correct (cognate) and incorrect substates (near-cognate), these dynamics were shown to be directly related to the fidelity mechanism. After peptide-bond formation, but

a

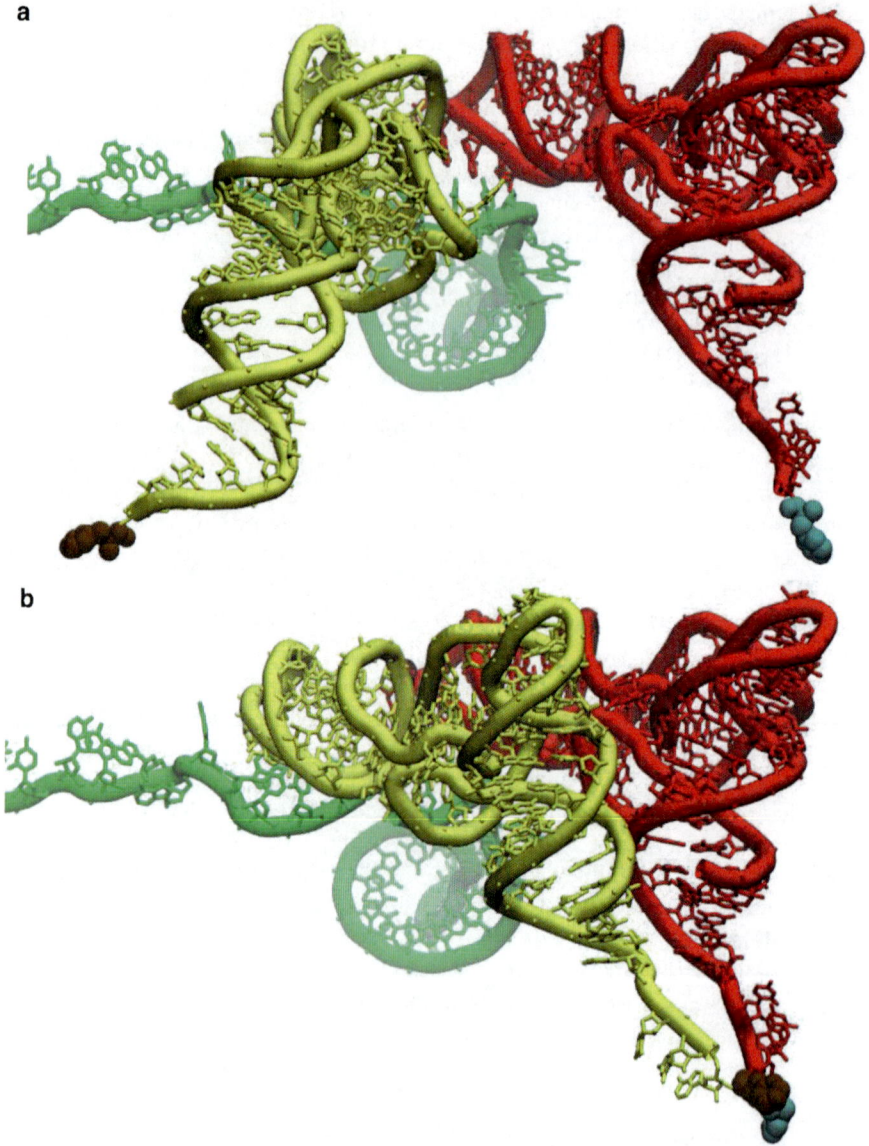

b

Fig. 3.5 Dynamics of the aminoacyl-tRNA during tRNA selection (accommodation step) in all-atom structure-based simulations. (**a**) Before accommodation. (**b**) After accommodation

prior to translocation, A- and P-site tRNAs within the pre-translocation ribosome complex were shown to undergo reversible exchange processes between kinetically and structurally distinct states. Using structural modeling and traditional mutagenic means, these configurations were shown to correspond to "Classical" positions, in which the 3′-CCA ends of A- and P-site tRNAs base pair with the A and P loop rRNA elements at the PTC. In addition, two distinguishable hybrid configurations arising from motions of each tRNA end were observed. Both hybrid configurations result from the motion of the deacylated P-site tRNA towards the E site, where it adopts a P/E configuration (Munro et al. 2007). Hybrid tRNA configurations, first demonstrated by Noller and coworkers using chemical footprinting methods (Moazed and Noller 1989), are key intermediates in the translocation process, lowering the activation barrier for translocation. Analytic treatment of the smFRET data revealed that A- and P-site tRNA motions can occur through independent or coupled processes. The rates of motions observed suggest that A- and P-site tRNA motions entailed large-scale conformational events in the ribosome.

3.3 Overall Dynamics of the Ribosome

In addition to the tRNA, the ribosome itself has been shown to have several dynamic regions. We refer to motion of a region of a subunit with respect to the same subunit as intrasubunit motion. We refer to the motion of the two subunits with respect to each other as intersubunit motion (Frank and Agrawal 2000). The most obvious motion performed by the ribosome is the so-called ratchet-like pivoting of the 30S subunit with respect to the 50S subunit (Fig. 3.6) (Frank and Agrawal 2000). Cryo-EM

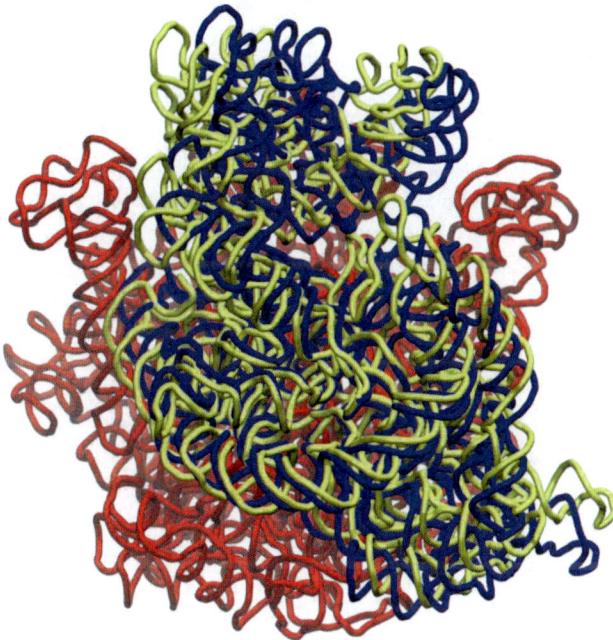

Fig. 3.6 Dynamic regions of the ribosome: pivoting of the 30S with respect to the 50S

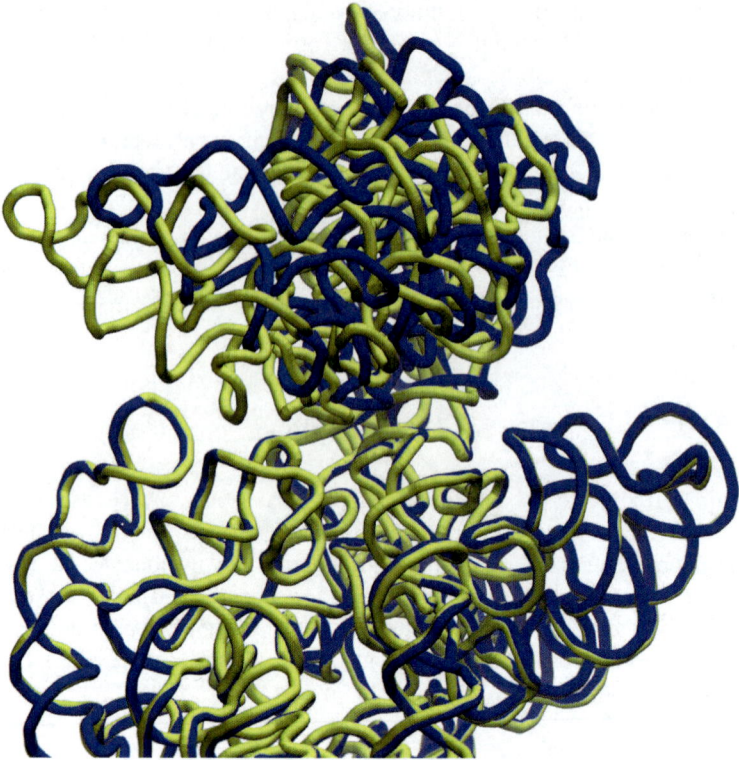

Fig. 3.7 Dynamics of the 30S head swivel in all-atom structure-based simulations

reconstructions of the 70S ribosome in the presence of EF-G showed a substantial rotation of the 30S relative to its initiator configuration upon EF-G binding (Frank and Agrawal 2000). It was previously thought that this motion causes translocation of the tRNA and mRNA through the ribosome. Recent FRET experiments have shown that ratchet-like fluctuations occur at equilibrium in absence of EF-G (Ermolenko et al. 2007). EF-G acts to trap the ribosome in the ratcheted position.

Several intrasubunit motions have been observed within the small subunit. The decoding center of the small subunit contains two universally conserved nucleotides: A1492 and A1493. These have been shown to be dynamic in NMR studies (Fourmy et al. 1998), fluorescence studies (Kaul and Pilch 2002) and are observed to be disordered in X-ray crystallography systems (Ogle et al. 2001; Berk et al. 2006).

The head domain of the small subunit has been shown to exist in multiple conformations (Schuwirth et al. 2005; Ratje et al. 2010) and also to be dynamic (Fig. 3.7) (Schuwirth et al. 2005; Ermolenko et al. 2007; Majumdar et al. 2005). Cate and coworkers first observed in alternative crystal forms that the head of the small subunit was rotated around a neck region with respect to its body. Spahn and coworkers have shown that this rotation is critical for translocation (Ratje et al. 2010). In particular, cryo-EM reconstructions of EF-G bound ribosomes have recently shown

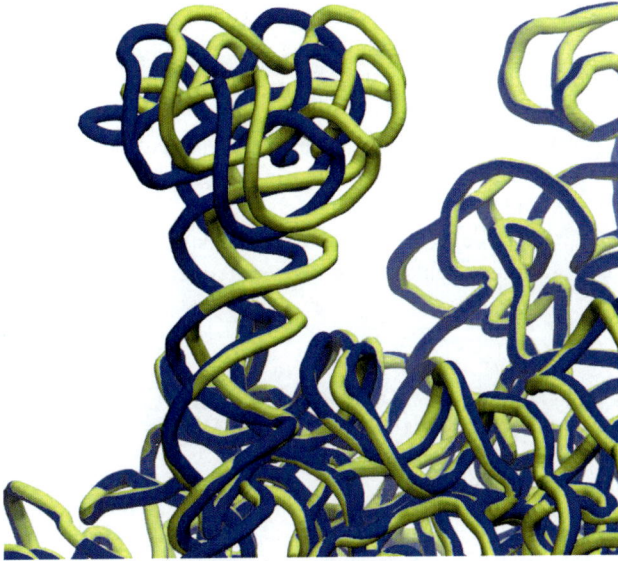

Fig. 3.8 Dynamics of the L1 stalk in all-atom structure-based simulations. *Blue* and *yellow* display snapshots of ribosome at different times during simulation

that two EF-G bound populations exist: one with the 30S body pivoted with respect to the 50S (ratcheted) and a second with the 30S body pivoted and the 30S head rotated (head swiveled). The study shows that while ratchet motion is necessary for translocation, the majority of codon–anticodon translocation movement occurs during 30S head swivel, an approximately orthogonal motion to the "ratchet-like" pivot of the 30S with respect to the 50S. FRET experiments containing donor-acceptor FRET pairs on small subunit proteins located on the head and body show that the head undergoes significant fluctuations in solution (Ermolenko et al. 2007; Majumdar et al. 2005).

The mRNA must also be dynamic, although this is difficult to measure. Clearly the mRNA scans the 30S for the correct initiation configuration. The mRNA also moves along the 30S during translocation. It is known that the mRNA is steadied by not only the tRNA anticodons, but also several universally conserved nucleotides on the 30S (e.g., G530, A1492, and A1493). However, it is not known what happens to this configuration during translocation and during initiation.

The 50S has been shown to undergo large intrasubunit conformational changes. The L1 stalk and L7/L12 stalk have been captured in two distinct configurations: factor bound and factor free (Frank and Agrawal 2000; Valle et al. 2003b). Upon ternary complex binding, the L7/L12 stalk moves towards the central protuberance. Upon EF-G binding and ratcheting, the L1 stalk moves towards the central protuberance. Single molecule FRET studies have shown the L1 stalk to undergo enormous and rapid fluctuations that are independent of tRNA fluctuations (Fig. 3.8) (Munro et al. 2010a, b; Cornish et al. 2009). However, upon EF-G binding, these fluctuations tend to be more synchronized. Because the L1 stalk interacts with the P/E position and E/E position tRNA, it is thought to be involved with E-site tRNA dissociation. Single molecule studies also show the L7/L12 stalk to fluctuate rapidly.

The L7/L12 protein complex itself is thought to be the most dynamic region of the ribosome. This consists of a long alpha helix (L10) bolstered by several L7/L12 NTD dimers (Diaconu et al. 2005; Bocharov et al. 2004). Each L7/L12 NTD domain is connected to its CTD by a long disordered flexible linker. These allow the CTD to undergo tremendous large fluctuations which may aid in capturing translation factors in a fly-casting mechanism (Shoemaker et al. 2000).

3.4 Molecular Dynamics Simulations and Other Computational Studies of the Ribosome

3.4.1 Background

Rapid kinetics, single molecule FRET, X-ray crystallography, and cryo-EM have provided an excellent glimpse into the dynamics of the ribosome. To obtain the free energy landscape of the ribosome and a more complete understanding of elongation, this large body of experiments must be integrated into a coherent picture. Molecular dynamics simulation is an excellent tool to provide a theoretical framework to ground interpretation of experimental results.

By labeling strategic regions of the ribosome, rapid kinetics experiments have been able to isolate the various substeps of elongation, giving us the framework for how translation proceeds (Wintermeyer et al. 2004). Single molecule FRET experiments use donor-acceptor label pairs to obtain time-dependent distance constraints on specific conformational changes (Blanchard 2009). X-ray crystallography structures produce high-resolution atomic models of the ribosome in its resting states (classical, EF-G bound, and EF-Tu bound) (Yusupova et al. 2006; Selmer et al. 2006; Schuwirth et al. 2000, 2005; Korostelev et al. 2006; Ban et al. 2000). Three-dimensional cryo-EM reconstructions call us to construct atomic models of the ribosome in a wide variety of functional states (Zhang et al. 2008; Fischer et al. 2010; Ratje et al. 2010). Molecular dynamics simulations integrate the above data into a single coherent mechanistic framework (Whitford et al. 2010b).

Modeling efforts (Malhotra et al. 1990, 1994) and advances in high-performance computing have produced an era in which molecular modeling and simulations can clarify the relationship between static ribosome structures (Ratje et al. 2010; Villa et al. 2009), structural fluctuations about these local energetic minima (Tama et al. 2003; Wang et al. 2004), transitions between minima, and ribosome function (Whitford et al. 2010a; Sanbonmatsu and Joseph 2003a; Sanbonmatsu et al. 2005). Molecular dynamics simulations and modeling studies have explored the decoding center (Sanbonmatsu and Joseph 2003b; Lim and Curran 2001; VanLoock et al.

1999, 2000), isolated tRNA (Auffinger et al. 1999) and the inner core of the 30S subunit (Li et al. 2003; Mears et al. 2002), kink turns (Razga et al. 2004, 2005), the ribosome tunnel (Ishida and Hayward 2008), and drug binding (Ge and Roux 2010; Vaiana and Sanbonmatsu 2009; Dlugosz and Trylska 2009). Adaptive-mesh refinement Poisson–Boltzmann studies based on phosphate/C-alpha structures have examined the electrostatic potential of the large and small subunits (Trylska et al. 2004). Coarse-grained normal mode studies have investigated the motions of the ribosome within the approximations of the Gaussian network model (Tama et al. 2003; Trylska et al. 2004; Chacon et al. 2003). Real-space refinement has been combined with cryo-EM structures to study the beginning and end states of a conformational change that occurs during translocation (Gao et al. 2003). Explicit solvent molecular dynamics simulations have previously helped elucidate functional mechanisms of biomolecular systems (Karplus and McCammon 2002; Berneche and Roux 2001; Young et al. 2001; Bockmann and Grubmuller 2002; McCammon et al. 1977) and have been validated with experiments (Sporlein et al. 2002). Large-scale supercomputers, together with the GROMACS program (Van Der Spoel et al. 2005), have enabled us to simulate the ribosome in explicit solvent (3.2×10^6 atoms) for an aggregate sampling of >2 μs (~2,100 ns), producing the largest simulation of a biomolecular complex to date (Whitford et al. 2010b), more than 20-fold larger than previous efforts, which simulated the ribosome for 85 ns (Trabuco et al. 2010a) and a small portion (1.25×10^5 atoms) for 2.1 μs (Trabuco et al. 2010b).

The petaFlop barrier was broken several years ago by the Los Alamos Roadrunner machine. Even with such fast supercomputers, a full simulation of the ribosome in explicit solvent for physiological time scales (~100 ms^{-1} s) is currently not possible. Using current molecular dynamics algorithms, we estimate this would require a ~10^3–10^4 fold speed-up or a 1–10 exaFlop machine, which may be available between 2,019 and 2,024 according to DOE SCIDAC projections. To examine short time scale stability of hydrogen bonds, we performed standard explicit solvent molecular dynamics simulations of localized regions of the ribosome (Sanbonmatsu and Joseph 2003c). To study localized conformational changes we have used exhaustive sampling techniques such as replica exchange explicit solvent molecular dynamics (Vaiana and Sanbonmatsu 2009; Garcia and Sanbonmatsu 2001; Sanbonmatsu 2006). To study the structural stability and to obtain a baseline for validation of reduced force field techniques, we performed multiple extensive explicit solvent simulations. Finally to study large-scale conformational changes we have used a variety of techniques. As a first step, we used targeted molecular dynamics simulations in explicit solvent to obtain a first look at steric interactions occurring during the large-scale conformational change. These results have been validated experimentally (Meskauskas and Dinman 2007; Baxter-Roshek et al. 2007). Next, we used a reduced-potential method, called structure-based simulation (Whitford et al. 2009a, b, 2010a; Noel et al. 2010). These unrestrained simulations allow us to observe spontaneous large-scale conformational changes with an estimated total sampling of ~200 ms. To validate, we compare against single molecule FRET, cryo-EM, and X-ray crystallography B-factors.

3.4.2   The Decoding Center

About a decade ago, we performed a limited explicit solvent study of the fidelity on the ribosome, simulating the tRNA–mRNA-ribosome decoding center complex for ultrafast time scales (~10 ns). This study was published several years later (Sanbonmatsu and Joseph 2003c). We were able to observe stable cognate (correct) codon–anticodon interactions and unstable near-cognate (incorrect) codon–anticodon interactions. We concluded that the decoding bases (G530, A1492, and A1493) measure the geometry of the codon–anticodon minihelix minor groove and test the stability of the codon–anticodon interaction. We emphasize that it is impossible to obtain accurate free energies from such short simulations; however this first functional study of the ribosome using molecular dynamics established the foundation for more extensive studies.

We have shown recently that at least 1 μs of replica sampling is required for exhaustive sampling of the decoding helix (Vaiana and Sanbonmatsu 2009). The 1 μs of replica sampling amounts to >>5 μs of standard molecular dynamics (Sanbonmatsu and Garcia 2002). Without exhaustive sampling, the entropic component of the free energy cannot be calculated accurately. In this study, the flipping of nucleotides A1492 and A1493 was studied. Our exhaustive sampling molecular dynamics simulations show that these nucleotides flip into and out of helix 44 of the small subunit when no ligands are bound to the small subunit. Several thousand flips occurred during our simulation. We estimate the barrier in free energy for base flipping of these nucleotides to be on the order of a few kcal/Mol, suggesting that the nucleotides continuously flip until ligands bind, trapping them in the flipped out configuration (Vaiana and Sanbonmatsu 2009; Sanbonmatsu 2006). We note that the study did not include the large subunit. The 50S may have a stabilizing role on A1492 and A1493.

The decoding center is the target of aminoglycoside antibiotics (Carter et al. 2000; Fourmy et al. 1998; Vicens and Westhof 2003). These antibiotics bind inside helix 44, preventing A1492 and A1493 from residing in their flipped in conformation. We have shown that rather than an induced-fit scenario, where the drug forces these nucleotides to flip out of their helix, a stochastic gating scenario is more likely. Here, the nucleotides continuously flip in and out. Drug binding traps the nucleotides in their flipped-out state. We note that upon correct tRNA binding, an intricate hydrogen bond network exists between G530, A1492, A1493, the tRNA anticodon, and the mRNA codon. Because G530, A1492, and A1493 sense the Watson–Crick geometry of the codon–anticodon interaction, these nucleotides act as a molecular reading head. Aminoglycosides effectively "jam" this reading head in the accepting configuration with A1492 and A1493 flipped out of their helix, allowing them to interact productively with the tRNA leading to acceptance of the tRNA into the ribosome ("accommodation"). Thus, the aminoglycoside interaction leads to the incorporation of incorrect amino acids into the nascent protein, producing widespread errors in proteins manufactured by the aminoglycoside-bound ribosome. This fidelity phenotype is known as misreading.

3.4.3 tRNA Accommodation

To investigate large-scale conformational changes, we first turned to accommodation, the large-scale conformational change involved in tRNA movement through the ribosome. During this conformational change, the aminoacyl-tRNA moves from the A/T state to the fully bound A/A state. This 70 Å movement requires the kinked anticodon arm to relax to its native, linear conformation. As a first step, in 2002–2003, we performed ultrafast explicit solvent targeted molecular dynamics simulations, with a total sampling of ~20 ns. These were published in 2005 (Sanbonmatsu et al. 2005). The simulations revealed several important features of accommodation that were not previously identified. Because the 3'-CCA end is single stranded, it is thought to be extremely flexible. Our study showed explicitly that flexibility of the 3'-CCA end is required for accommodation (Sanbonmatsu et al. 2005). In particular, the simulation revealed that many steric barriers are present in the accommodation pathway. The 3'-CCA end must be extremely flexible to move from the A/T state into the A/A state. In particular, the 3'-CCA end must navigate the intricate and universally conserved accommodation corridor consisting of large subunit rRNA helices H89, H90, and H92. The accommodation corridor was identified by our simulations (Sanbonmatsu et al. 2005). This could not be identified from X-ray crystallography or cryo-EM since these methods do not examine the transition between the A/T and the A/A positions.

In our most recent study, we have simulated spontaneous accommodation events. Here, a structure-based simulation approach was used (Whitford et al. 2009a, b, 2010a; Noel et al. 2010). In particular, a potential defined by the X-ray structure of the ribosome with the aminoacyl-tRNA occupying the A/A state was used. Much like an all-atom Go-model simulation, the tRNA was started in the A/T position. The tRNA and ribosome were then allowed to fluctuate freely under the A/A-based potential. Over 1,000 accommodation simulations were performed with an estimated total sampling on the order of hundreds of milliseconds. The tRNA was observed to be highly dynamic, undergoing reversible excursions from the A/T state towards the A/A state and back. These same reversible excursions were also observed in single molecule FRET studies of accommodation (Whitford et al. 2010a). We found that H89 must fluctuate to make way for the accommodating tRNA (Whitford et al. 2010a). The study also shows that the 3'-CCA end is so disordered that the 3'-CCA end itself presents an entropic barrier to accommodation.

To further validate our structure-based potential, extensive multiple explicit solvent simulations were performed with total sampling of ~2,100 ns (Whitford et al. 2010b). Simulations were performed starting the tRNA in the A/A state and also starting the tRNA in the A/T state. The configurational space sampled in the A/A state explicit solvent simulation was similar to the A/A state basin sampled by the structure-based simulation. Similar results were achieved for the A/T state simulations. We also note that our explicit solvent targeted molecular dynamics simulations sampled regions well within the configurational space defined by the structure-based accommodation study. Average fluctuations of the explicit solvent and structure-based simulations were similar. Both simulations showed agreement with X-ray B-factors (Korostelev et al. 2006), with the exception of the 50S stalks, which have been shown to undergo large amplitude fluctuations by single molecule FRET (Munro et al. 2010a, b). These fluctuations cannot be captured by X-ray crystallography.

3.4.4   The L1 Stalk

Finally, we used the structure-based method to investigate L1 stalk fluctuations. We found that enormous fluctuations both in displacement and twist, consistent with the large fluctuations observed by single molecule FRET.

3.4.5   Moving tRNAs Through the Ribosome

In the case of accommodation, our structure-based simulations display a wide variety of accommodation pathways. The shape of the pathway distribution is determined both by the intrinsic properties of the tRNA and by fluctuations of the ribosome. Furthermore, both the structure-based and explicit solvent simulations show large ensembles of configurations representing the A/A and A/T positions, consistent with single molecule FRET experiments. Our study reveals that large amplitude reversible excursions of the aminoacyl-tRNA combined with stochastic gating of helix 89 allow for accommodation. For accommodation to occur, the entropic barrier introduced by the 3'-CCA end must be overcome by (1) the enthalpic minima of the peptidyl transferase center, which includes a base pair between the 3'-CCA end and a ribosomal nucleotide, (2) the enthalpic minima of the anti-codon arm, which releases energy stored during ternary complex binding, and (3) an entropic "kick" produced by the presence of EF-Tu, which prevents the tRNA from escaping the ribosome. It should be emphasized that while extremely flexible components introduce entropic barriers, they also soften the system relative to a molecular machine consisting of more rigid components. This softening may act to fine-tune the free energy landscape. Overall, simulations have established several principles important for ribosome dynamics: multiple pathways, stochastic gating, large amplitude reversible excursions, and entropic barriers.

We suspect that similar principles operate during translocation. However, in this case, the energy landscape seen by the tRNA is itself dynamic: pivoting of the 30S body with respect to the 50S, swiveling of the 30S head, and the opening and closing of various gates will significantly alter the barriers in a dynamic fashion. Recent single molecule FRET experiments suggest that each component of the ribosome fluctuates independently and at different time scales. It is only during the rare moments that these fluctuations occur simultaneously in the proper directions that translocation may occur (Munro et al. 2010a, b). Such synchronization is enhanced by the presence of EF-G.

Beginning with energetics, the thermal bath in the environment surrounding the ribosome contains enough energy to move the tRNAs through the ribosome. In the case of factor-free translation, the only chemical reaction occurring is the peptidyl transferase reaction. Thermal fluctuations provide the energy to overcome enthalpic and entropic barriers involved in moving the aminoacyl-tRNA from solution to the accommodated A/A position. Once the peptidyl transferase reaction is complete, the system reaches the point of no return, where thermal fluctuations move the deacylated tRNA through the hybrid state to the exit site. The structure of the ribosome carefully controls this movement allowing the new peptidyl-tRNA to move by exactly one codon. Finally thermal fluctuations allow the deacylated tRNA to exit the ribosome to its entropically favorable state — in solution away from the ribosome. The introduction of factors serves to make this process more efficient.

References

Auffinger P, LouiseMay S, Westhof E (1999) Molecular dynamics simulations of solvated yeast tRNA(Asp). Biophys J 76:50–64

Ban N, Nissen P, Hansen J, Moore PB, Steitz TA (2000) The complete atomic structure of the large ribosomal subunit at 2.4 A resolution. Science 289:905–920

Baxter-Roshek JL, Petrov AN, Dinman JD (2007) Optimization of ribosome structure and function by rRNA base modification. PLoS One 2:e174

Berk V, Zhang W, Pai RD, Cate JH (2006) Structural basis for mRNA and tRNA positioning on the ribosome. Proc Natl Acad Sci USA 103:15830–15834

Berneche S, Roux B (2001) Energetics of ion conduction through the K+ channel. Nature 414:73–77

Blanchard SC (2009) Single-molecule observations of ribosome function. Curr Opin Struct Biol 19:103–109

Blanchard SC, Kim HD, Gonzalez RL Jr, Puglisi JD, Chu S (2004) tRNA dynamics on the ribosome during translation. Proc Natl Acad Sci USA 101:12893–12898

Blanchard SC, Cooperman BS, Wilson DN (2010) Probing translation with small-molecule inhibitors. Chem Biol 17:633–645

Bocharov EV, Sobol AG, Pavlov KV, Korzhnev DM, Jaravine VA, Gudkov AT, Arseniev AS (2004) From structure and dynamics of protein L7/L12 to molecular switching in ribosome. J Biol Chem 279:17697–17706

Bockmann RA, Grubmuller H (2002) Nanoseconds molecular dynamics simulation of primary mechanical energy transfer steps in F1-ATP synthase. Nat Struct Biol 9:198–202

Carter AP, Clemons WM, Brodersen DE, MorganWarren RJ, Wimberly BT, Ramakrishnan V (2000) Functional insights from the structure of the 30S ribosomal subunit and its interactions with antibiotics. Nature 407:340–348

Chacon P, Tama F, Wriggers W (2003) Mega-Dalton biomolecular motion captured from electron microscopy reconstructions. J Mol Biol 326:485–492

Connell SR, Topf M, Qin Y, Wilson DN, Mielke T, Fucini P, Nierhaus KH, Spahn CM (2008) A new tRNA intermediate revealed on the ribosome during EF4-mediated back-translocation. Nat Struct Mol Biol 15:910–915

Cornish PV, Ermolenko DN, Staple DW, Hoang L, Hickerson RP, Noller HF, Ha T (2009) Following movement of the L1 stalk between three functional states in single ribosomes. Proc Natl Acad Sci USA 106:2571–2576

Demeshkina N, Jenner L, Yusupova G, Yusupov M (2010) Interactions of the ribosome with mRNA and tRNA. Curr Opin Struct Biol 20:325–332

Diaconu M, Kothe U, Schlunzen F, Fischer N, Harms JM, Tonevitsky AG, Stark H, Rodnina MV, Wahl MC (2005) Structural basis for the function of the ribosomal L7/12 stalk in factor binding and GTPase activation. Cell 121:991–1004

Dlugosz M, Trylska J (2009) Aminoglycoside association pathways with the 30S ribosomal subunit. J Phys Chem B 113:7322–7330

Ermolenko DN, Majumdar ZK, Hickerson RP, Spiegel PC, Clegg RM, Noller HF (2007) Observation of intersubunit movement of the ribosome in solution using FRET. J Mol Biol 370:530–540

Feldman MB, Terry DS, Altman RB, Blanchard SC (2010) Aminoglycoside activity observed on single pre-translocation ribosome complexes. Nat Chem Biol 6:244

Fischer N, Konevega AL, Wintermeyer W, Rodnina MV, Stark H (2010) Ribosome dynamics and tRNA movement by time-resolved electron cryomicroscopy. Nature 466:329–333

Fourmy D, Yoshizawa S, Puglisi JD (1998) Paromomycin binding induces a local conformational change in the A-site of 16S rRNA. J Mol Biol 277:333–345

Frank J, Agrawal RK (2000) A ratchet-like inter-subunit reorganization of the ribosome during translocation. Nature 406:318–322

Frank J, Sengupta J, Gao H, Li W, Valle M, Zavialov A, Ehrenberg M (2005) The role of tRNA as a molecular spring in decoding, accommodation, and peptidyl transfer. FEBS Lett 579:959–962

Gao H, Sengupta J, Valle M, Korostelev A, Eswar N, Stagg SM, Van Roey P, Agrawal RK, Harvey SC, Sali A, Chapman MS, Frank J (2003) Study of the structural dynamics of the E. coli 70S ribosome using real-space refinement. Cell 113:789–801

Gao YG, Selmer M, Dunham CM, Weixlbaumer A, Kelley AC, Ramakrishnan V (2009) The structure of the ribosome with elongation factor G trapped in the posttranslocational state. Science 326:694–699

Garcia AE, Sanbonmatsu KY (2001) Exploring the energy landscape of a beta hairpin in explicit solvent. Proteins 42:345–354

Ge X, Roux B (2010) Absolute binding free energy calculations of sparsomycin analogs to the bacterial ribosome. J Phys Chem B 114:9525–9539

Geggier P, Dave R, Feldman MB, Terry DS, Altman RB, Munro JB, Blanchard SC (2010) Conformational sampling of aminoacyl-tRNA during selection on the bacterial ribosome. J Mol Biol 399:576–595

Ishida H, Hayward S (2008) Path of nascent polypeptide in exit tunnel revealed by molecular dynamics simulation of ribosome. Biophys J 95:5962–5973

Karplus M, McCammon JA (2002) Molecular dynamics simulations of biomolecules. Nat Struct Biol 9:646–652

Kaul M, Pilch DS (2002) Thermodynamics of aminoglycoside-rRNA Rrecognition: the binding of neomycin-class aminoglycosides of the A site of 16S rRNA. Biochemistry 41:7695–7706

Korostelev A, Trakhanov S, Laurberg M, Noller HF (2006) Crystal structure of a 70S ribosome-tRNA complex reveals functional interactions and rearrangements. Cell 126:1065–1077

Li W, Ma B, Shapiro B (2003) Binding interactions between the core central domain of 16S rRNA and the ribosomal protein S15 determined by molecular dynamics simulations. Nucleic Acids Res 31:629–638

Lim VI, Curran JF (2001) Analysis of codon: anticodon interactions within the ribosome provides new insights into codon reading and the genetic code structure. RNA 7:942–957

Majumdar ZK, Hickerson R, Noller HF, Clegg RM (2005) Measurements of internal distance changes of the 30S ribosome using FRET with multiple donor-acceptor pairs: quantitative spectroscopic methods. J Mol Biol 351:1123–1145

Malhotra A, Tan RK, Harvey SC (1990) Prediction of the three-dimensional structure of Escherichia coli 30S ribosomal subunit: a molecular mechanics approach. Proc Natl Acad Sci USA 87:1950–1954

Malhotra A, Tan RK, Harvey SC (1994) Modeling large RNAs and ribonucleoprotein particles using molecular mechanics techniques. Biophys J 66:1777–1795

Mankin A (2006) Antibiotic blocks mRNA path on the ribosome. Nat Struct Mol Biol 13:858–860

McCammon JA, Gelin BR, Karplus M (1977) Dynamics of folded proteins. Nature 267:585–590

Mears JA, Cannone JJ, Stagg SM, Gutell RR, Agrawal RK, Harvey SC (2002) Modeling a minimal ribosome based on comparative sequence analysis. J Mol Biol 321:215–234

Meskauskas A, Dinman JD (2007) Ribosomal protein L3: gatekeeper to the A site. Mol Cell 25:877–888

Moazed D, Noller HF (1989) Intermediate states in the movement of transfer RNA in the ribosome. Nature 342:142–148

Munro JB, Altman RB, O'Connor N, Blanchard SC (2007) Identification of two distinct hybrid state intermediates on the ribosome. Mol Cell 25:505–517

Munro JB, Altman RB, Tung CS, Cate JH, Sanbonmatsu KY, Blanchard SC (2010a) Spontaneous formation of the unlocked state of the ribosome is a multistep process. Proc Natl Acad Sci USA 107:709–714

Munro JB, Altman RB, Tung CS, Sanbonmatsu KY, Blanchard SC (2010b) A fast dynamic mode of the EF-G-bound ribosome. EMBO J 29:770–781

Noel JK, Whitford PC, Sanbonmatsu KY, Onuchic JN (2010) SMOG@ctbp: simplified deployment of structure-based models in GROMACS. Nucleic Acids Res 38(Suppl):W657–W661

Ogle JM, Brodersen DE, Clemons WM Jr, Tarry MJ, Carter AP, Ramakrishnan V (2001) Recognition of cognate transfer RNA by the 30S ribosomal subunit. Science 292:897–902

Pioletti M, Schlunzen F, Harms J, Zarivach R, Gluhmann M, Avila H, Bashan A, Bartels H, Auerbach T, Jacobi C, Hartsch T, Yonath A, Franceschi F (2001) Crystal structures of complexes of the small ribosomal subunit with tetracycline; edeine and IF3. EMBO J 20:1829–1839

Ratje R et al (2010) Head swivel on the ribosome facilitates translocation via intra-subunit tRNA hybrid sites. Nature 468:713–716

Razga F, Spackova N, Reblova K, Koca J, Leontis NB, Sponer J (2004) Ribosomal RNA kink-turn motif–a flexible molecular hinge. J Biomol Struct Dyn 22:183–194

Razga F, Koca J, Sponer J, Leontis NB (2005) Hinge-like motions in RNA kink-turns: the role of the second a-minor motif and nominally unpaired bases. Biophys J 88:3466–3485

Sanbonmatsu KY (2006) Energy landscape of the ribosomal decoding center. Biochimie 88:1053–1059

Sanbonmatsu KY, Garcia AE (2002) Structure of Met-enkephalin in explicit aqueous solution using replica exchange molecular dynamics. Proteins 46:225–234

Sanbonmatsu KY, Joseph S (2003) Understanding discrimination by the ribosome: stability testing and groove measurement of codon-anticodon pairs. J Mol Biol 328:33–47

Sanbonmatsu KY, Joseph S, Tung CS (2005) Simulating movement of tRNA into the ribosome during decoding. Proc Natl Acad Sci USA 102:15854–15859

Schluenzen F, Tocilj A, Zarivach R, Harms J, Gluehmann M, Janell D, Bashan A, Bartels H, Agmon I, Franceschi F, Yonath A (2000) Structure of functionally activated small ribosomal subunit at 3.3 angstrom resolution. Cell 102:615–623

Schlunzen F, Zarivach R, Harms R, Bashan A, Tocilj A, Albrecht R, Yonath A, Franceschi F (2001) Structural basis for the interaction of antibiotics with the peptidyl transferase centre in eubacteria. Nature 413:814–821

Schuwirth BS, Borovinskaya MA, Hau CW, Zhang W, Vila-Sanjurjo A, Holton JM, Cate JH (2005) Structures of the bacterial ribosome at 3.5 A resolution. Science 310:827–834

Selmer M, Dunham CM, Murphy FVt, Weixlbaumer A, Petry S, Kelley AC, Weir JR, Ramakrishnan V (2006) Structure of the 70S ribosome complexed with mRNA and tRNA. Science 313:1935–1942

Shoemaker BA, Portman JJ, Wolynes PG (2000) Speeding molecular recognition by using the folding funnel: the fly-casting mechanism. Proc Natl Acad Sci USA 97:8868–8873

Sporlein S, Carstens H, Satzger H, Renner C, Behrendt R, Moroder L, Tavan P, Zinth W, Wachtveitl J (2002) Ultrafast spectroscopy reveals subnanosecond peptide conformational dynamics and validates molecular dynamics simulation. Proc Natl Acad Sci USA 99:7998–8002

Tama F, Valle M, Frank J, Brooks CL 3rd (2003) Dynamic reorganization of the functionally active ribosome explored by normal mode analysis and cryo-electron microscopy. Proc Natl Acad Sci USA 100:9319–9323

Trabuco LG, Schreiner E, Eargle J, Cornish P, Ha T, Luthey-Schulten Z, Schulten K (2010a) The role of L1 stalk-tRNA interaction in the ribosome elongation cycle. J Mol Biol 402:741–760

Trabuco LG, Harrison CB, Schreiner E, Schulten K (2010b) Recognition of the regulatory nascent chain TnaC by the ribosome. Structure 18:627–637

Trylska J, Konecny R, Tama F, Brooks CL 3rd, McCammon JA (2004) Ribosome motions modulate electrostatic properties. Biopolymers 74:423–431

Vaiana AC, Sanbonmatsu KY (2009) Stochastic gating and drug-ribosome interactions. J Mol Biol 386:648–661

Valle M, Sengupta J, Swami NK, Grassucci RA, Burkhardt N, Nierhaus KH, Agrawal R, Frank J (2002) Cryo-EM reveals an active role for aminoacyl-tRNA in the accommodation process. EMBO J 21:3557–3567

Valle M, Zavialov A, Li W, Stagg SM, Sengupta J, Nielsen RC, Nissen P, Harvey SC, Ehrenberg M, Frank J (2003) Incorporation of aminoacyl-tRNA into the ribosome as seen by cryo-electron microscopy. Nat Struct Biol 10:899–906

Van Der Spoel D, Lindahl E, Hess B, Groenhof G, Mark AE, Berendsen HJ (2005) GROMACS: fast, flexible, and free. J Comput Chem 26:1701–1718

VanLoock MS, Easterwood TR, Harvey SC (1999) Major groove binding of the tRNA/mRNA complex to the 16S ribosomal RNA decoding site. J Mol Biol 285:2069–2078

VanLoock MS, Agrawal RK, Gabashvili IS, Qi L, Frank J, Harvey SC (2000) Movement of the decoding region of the 16S ribosomal RNA accompanies tRNA translocation. J Mol Biol 304:507–515

Vicens Q, Westhof E (2003) Crystal structure of geneticin bound to a bacterial 16S ribosomal RNA A site oligonucleotide. J Mol Biol 326:1175–1188

Villa E, Sengupta J, Trabuco LG, LeBarron J, Baxter WT, Shaikh TR, Grassucci RA, Nissen P, Ehrenberg M, Schulten K, Frank J (2009) Ribosome-induced changes in elongation factor Tu conformation control GTP hydrolysis. Proc Natl Acad Sci USA 106:1063–1068

Voorhees RM, Weixlbaumer A, Loakes D, Kelley AC, Ramakrishnan V (2009) Insights into substrate stabilization from snapshots of the peptidyl transferase center of the intact 70S ribosome. Nat Struct Mol Biol 16:528–533

Wang Y, Rader AJ, Bahar I, Jernigan RL (2004) Global ribosome motions revealed with elastic network model. J Struct Biol 147:302–314

Whitford PC, Noel JK, Gosavi S, Schug A, Sanbonmatsu KY, Onuchic JN (2009a) An all-atom structure-based potential for proteins: bridging minimal models with all-atom empirical forcefields. Proteins 75:430–441

Whitford PC, Schug A, Saunders J, Hennelly SP, Onuchic JN, Sanbonmatsu KY (2009b) Nonlocal helix formation is key to understanding S-adenosylmethionine-1 riboswitch function. Biophys J 96:L7–L9

Whitford PC, Geggier P, Altman RB, Blanchard SC, Onuchic JN, Sanbonmatsu KY (2010a) Accommodation of aminoacyl-tRNA into the ribosome involves reversible excursions along multiple pathways. RNA 16:1196–1204

Whitford PC, Onuchic JN, Sanbonmatsu KY (2010b) Connecting energy landscapes with experimental rates for aminoacyl-tRNA accommodation in the ribosome. J Am Chem Soc 132:13170–13171

Wintermeyer W, Peske F, Beringer M, Gromadski KB, Savelsbergh A, Rodnina MV (2004) Mechanisms of elongation on the ribosome: dynamics of a macromolecular machine. Biochem Soc Trans 32:733–737

Young MA, Gonfloni S, Superti-Furga G, Roux B, Kuriyan J (2001) Dynamic coupling between the SH2 and SH3 domains of c-Src and Hck underlies their inactivation by C-terminal tyrosine phosphorylation. Cell 105:115–126

Yusupova G, Jenner L, Rees B, Moras D, Yusupov M (2006) Structural basis for messenger RNA movement on the ribosome. Nature 444:391–394

Zhang W, Kimmel M, Spahn CM, Penczek PA (2008) Heterogeneity of large macromolecular complexes revealed by 3D cryo-EM variance analysis. Structure 16:1770–1776

Zhang W, Dunkle JA, Cate JH (2009) Structures of the ribosome in intermediate states of ratcheting. Science 325:1014–1017

Chapter 4
Structural Analyses of the Ribosome by Chemical Modification Methods

Jonathan A. Leshin, Arturas Meskauskas, and Jonathan D. Dinman

4.1 Introduction

4.1.1 What Is Chemical Modification and Why Is It Useful?

Imagine the following puzzle. Your "Sensei" tied a simple knot in a rope, dipped the knot in paint, allowed the paint to dry, and then untied it. Your challenge is to retie the knot based solely on the paint patterns. In essence this is a topological problem of dimensional reduction, in this case one in which the information about the three-dimensional structure of the knot is reduced to the two-dimensional linear rope that bears markings pertaining to its prior three-dimensional topology. The solution to this type of problem is based on the understanding that only those regions of the rope that lay on the surface were exposed to paint, while those that had been buried in its interior were not. In the case of a simple knot, a clever child should be able to solve this problem using the paint markings as guides to work through a set of possible solutions. While solving the structure of a ball of yarn that had been similarly processed would be much more challenging, the solution would lie in taking the same approach. It would merely require more time and computational power.

Nucleic acids are polymers of nucleic acid bases attached to ribose sugars that are in turn linked together by phosphodiester bonds. Each of these moieties, the bases, the sugars, and the phosphate groups, contains specific chemical groups that can be modified by specific chemicals. Referring back to the analogy above, the nucleic acid polymer is the rope, while the specific modifying chemicals constitute the palette of available paints. While double-stranded DNA tends to assume rigid linear topologies

J.A. Leshin • A. Meskauskas • J.D. Dinman (✉)
Department of Cell Biology and Molecular Genetics, Microbiology Building Room 2135,
University of Maryland, College Park, MD 20742, USA
e-mail: dinman@umd.edu

J.D. Dinman (ed.), *Biophysical Approaches to Translational Control of Gene Expression*,
Biophysics for the Life Sciences 1, DOI 10.1007/978-1-4614-3991-2_4,
© Springer Science+Business Media New York 2013

that do not easily condense into complex three-dimensional shapes, single-stranded RNA molecules do. Thus, if one were to expose a three-dimensionally complex RNA molecule to a selection of RNA modifying chemicals, only those portions of the RNA polymer exposed to the surface would react with the chemicals, while those buried in the interior would not. Given the tools to unravel (denature) the RNA and to detect and map the chemically modified RNA bases, it is possible to reconstruct the original three-dimensional structure using the basic strategy described above. Given the increasing importance of understanding the roles played by structurally complex RNA molecules on the biology of the cell, the utility of this approach becomes obvious.

4.1.2 Historical Background

The first major report detailing the use of chemical modification for mapping the structure of an RNA was published in 1980 (Peattie and Gilbert 1980). This work used a radioactively labeled RNA (yeast tRNA-Phe), modified with chemicals utilized in RNA sequencing methodologies employed at the time: dimethyl sulfate (DMS) and diethylpyrocarbonate (DEPC), followed by strand scission using sodium borohydride and aniline. The reactants were separated through large-format acrylamide gels, resolving the fragmented RNA species by molecular weight. Comparison of the banding patterns relative to native, semi-denatured and denatured tRNAs, enabled determination of whether or not a given base was likely involved in a base-pairing or base stacking interactions. The results were largely in agreement with crystal structures known at the time, but also enabled identification of additional tertiary interactions in this tRNA that were previously unknown. Furthermore, comments in the discussion section of the paper pointed to another innovation; by chemically modifying the RNA and then purifying and labeling it, an RNA molecule could be mapped in its native form without the need for renaturation. This would prove to be especially useful when dealing with large RNAs, e.g., ribosomal RNA (rRNA) or viral RNAs. Later groups (Stern et al. 1988) would expand the chemicals used to modify the RNA to those where strand scission would no longer be needed and which could interrogate a greater variety of bases.

The next major innovation in the field was introduced 3 years later (Hu and Dahlberg 1983). In addition to introducing a larger variety of modifying chemicals (discussed below), reverse transcription rather than end-labeling was employed to produce fragments for analysis. In this report, RNAs were chemically modified, denatured, and a DNA primer which had been radioactively labeled was annealed to the modified RNA species. Primer extension employed avian myeoloblastosis virus (AMV) reverse transcriptase, which produced strong stops 1 nucleotide 5′ of each modified base. The resulting cDNA products were separated through denaturing acrylamide gels alongside a reverse transcriptase Sanger sequencing reaction. By comparing the presence of bands in the chemically modified lanes against untreated samples, base-pairing, stacking, and tertiary interaction status could be determined for each nucleotide. Since primers can be designed for any sequence, extension reactions can be performed anywhere along an RNA, thus enabling the structure of any RNA to be examined, regardless of length. This represented a significant advancement, allowing for secondary and some tertiary structure analyses to be performed on RNAs which were too large for crystallographic studies. Additionally, examination of the chemical protection patterns of sequences conserved between species, differences in folding, and secondary structure were easily demonstrated.

4.2 Base-Specific Chemicals

4.2.1 Introduction

A variety of chemicals were employed to modify RNAs during the early period of chemically based structural analysis (Ehresmann et al. 1987; Weeks 2010). Each reagent was either base-specific or was intrinsically constrained by RNA structural elements, providing specificity. Most commonly, a series of chemicals were used to probe the same stretch of RNA, with each chemical providing different information. The resulting data were combined to give a series of potential structures or to further flesh out existing structural data. These chemicals and their reactions with RNAs are shown in Fig. 4.1.

4.2.2 Dimethyl Sulfate

DMS was one of the initial chemicals used to probe the paired and stacked nature of bases for structure determination (Peattie and Gilbert 1980). With regard to specificity, it can methylate the N-7 position of unpaired guanine, the N-3 position of unpaired cytosine, and the N-1 position of unpaired adenine. The methylated guanine is reactive to aniline cleavage, thus mainly of use in the end-labeled, strand scission method. An unreactive guanine indicates the presence of noncanonical base-pairing or Mg^{++} coordination. While the methylated cytosine requires further chemical treatment by hydrazine to be detected using strand scission, it can be detected directly using primer extension. The methylated adenines are only detectable using a primer extension-based method.

4.2.3 Diethylpyrocarbonate

DEPC was initially used in determining the secondary structure of yeast tRNA (Peattie and Gilbert 1980). It reacts with N-7 of adenine to create a site available for cleavage by aniline in the strand scission method. Unfortunately, all adenines, paired or unpaired, become unreactive in helices due to the higher sensitivity DEPC has for stacking interactions.

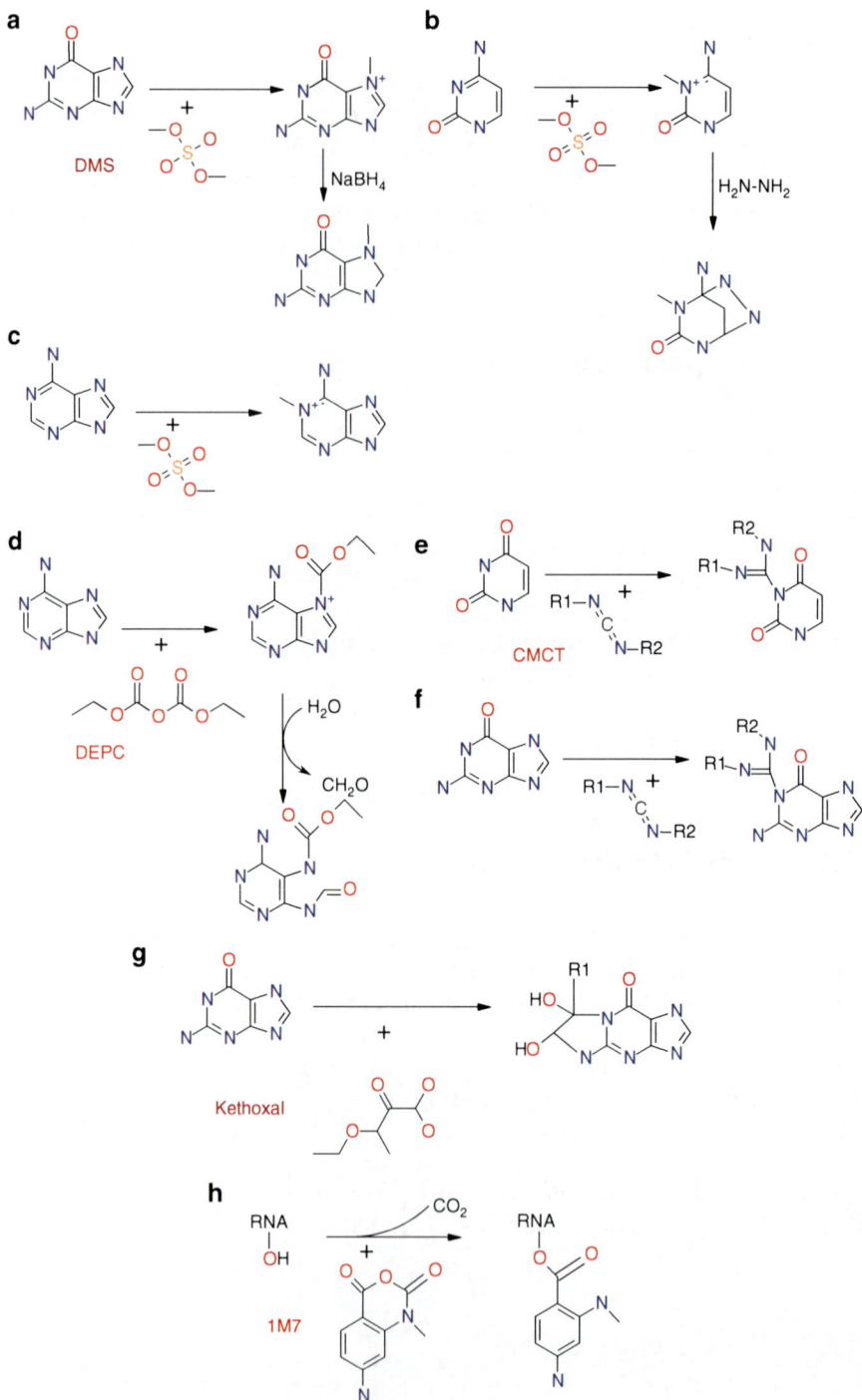

Fig. 4.1 Mechanism of chemical reactions. (**a**) Alkylation of N7-G by DMS followed by reduction by sodium borohydride; (**b**) alkylation of N3-C by DMS followed by hydrazine treatment; (**c**) alkylation of N1-A by DMS; (**d**) carbethoxylation of N7-A by DEPC; (**e**) modification of N3-U by CMCT; (**f**) modification of N1-G by CMCT; (**g**) modification of N1-G and N2-G by kethoxal; (**h**) acylation of 2'-OH of the ribose backbone by 1M7. Note: N-methyl isatoic anhydride and BzCN work via similar mechanisms to modify the 2'-OH of the ribose backbone

4.2.4 1-Cyclohexyl-3-(2-Morpholinoethyl) Carbodiimide Metho-p-Toluenesulfonate

Carbodiimide Metho-*p*-toluenesulfonate (CMCT) reacts with both uracil at N-3 and guanine at N-1, although the reaction is slightly preferential for N3-U. This chemical may be used to map unpaired guanines and uracils. However, this reaction only occurs at slightly basic pH 8.0 and is not amenable for use in neutral or acidic solutions.

4.2.5 β-Etoxy-α-Ketobutyraldehyde (Kethoxal)

Kethoxal reacts with unpaired guanines, forming a kethoxy-guanine adduct. In contrast to CMCT, kethoxal adducts are primarily stable under slightly acidic conditions but decompose in basic conditions. This problem is partially solved by addition of borate ions.

Analyses of RNA structure using these chemical reagents are usually performed in one of the two fashions: either by comparing native RNA to denatured RNA or by comparing any chemical modification to a nonreactive chemical such as DMSO. On an autoradiogram, the presence of a band at a given sequence position indicates that the base is available for modification and, as such, is not paired or otherwise in a tertiary structure interaction.

4.3 Problems Associated with Traditional Chemical Probing Reagents and Methods

4.3.1 Introduction

While each of the chemicals and methods described above represented significant advances towards bettering our understanding of RNA structure, each has its own disadvantage as follows:

4.3.2 End-Labeling

For end-labeling, analyses are limited to 150–200 bases from the 3′ end of an RNA. This problem was largely solved by using radioactively labeled primers, which can hybridize anywhere along the sequence of the RNA species to be probed, enabling primer walking along the length of the RNA. Thus, end-labeling is not routinely used, except to probe very short RNAs such as 5S rRNA and tRNAs, where primer extension methods do not resolve important structural elements close to the 3′ ends.

4.3.3   Radioactive Primer Labeling

While radioactive primer labeling is much more versatile than end-labeling, it has its own inherent limitations. The use of radioactivity is the first disadvantage. The mostly commonly used radioisotope is ^{32}P. While ^{32}P provides excellent signal strength, its relatively short half-life limits the ability to store primers for long periods of time, thus requiring continuous relabeling of primers in order to replicate experiments. This can become labor intensive and costly. Other radioisotopes offer longer half-lives (such as ^{14}C or ^{35}S), although primers labeled with these isotopes have significantly weaker signals. Additionally, autoradiolytic decay products can produce artifactual signals and increase background noise. An additional disadvantage lies in identification of appropriate primers. For example, hybridization of primers to A-U rich regions of RNA requires lower annealing temperatures, which can result in nonspecific hybridization, again increasing background noise. Lastly, at the time that this technique was first developed, primers were derived from restriction digestion products, thus making it difficult to guarantee that the primer had the correct sequence. This problem was solved with the advent of cost-effective DNA oligonucleotide synthesis technologies.

4.3.4   Urea Polyacrylamide Gel Electrophoresis

The method of separation of the fragments is another area for improvement. For urea polyacrylamide gel electrophoresis to be useful, large-format gels must be used (as large as 45 cm). Unfortunately, these large gels are extremely fragile and prone to problems with pouring and breakage. Length of run represents another potential problem with this method. Even with extremely large gel formats, only 150–200 base pairs can be resolved with any level of accuracy. Therefore, for large RNAs such as the ribosomal RNA or viral genomes, many gels are required to obtain full coverage. Lastly, quantification of the autoradiographic data generated from these gels is imprecise. Each band for which quantifiable data is desired must be assessed by computer software with high level of human manipulation for accurate results. The labor intensiveness of this kind of analysis led more often to simple qualitative interpretation of results based on perceived differences in band intensities.

4.3.5   Chemical Modifications

The chemicals used to modify the RNA represent another area for improvement. While each chemical has its positives and negatives, multiple chemicals are required to fully interrogate a region of RNA. This results in an intense time requirement to fully interrogate large RNAs. Compounding this is the fact that many of these chemicals are dangerous to work with and require special precautions. Additionally, DEPC cannot be used with buffers containing Tris or HEPES, as the presence of large numbers of amino groups in these reagents serves as a sink for the chemical probe.

These problems aside, radioactively labeled primer extension was an extremely useful tool for many years. However, as discussed next, a number of technological advancements have enabled researchers to overcome many of these limitations, adding significant power and cost savings to chemical approaches to probing RNA structure.

4.4   Modern Techniques

4.4.1   Introduction

Since 1998 a host of new technologies to circumvent the existing limitations of chemical probing have been devised. The first of these technologies, in-line cleavage or in-line probing, involves backbone-based breakage of the RNA. This technique uses either spontaneous cleavage (Soukup and Breaker 1999) or uranyl-based photocleavage (Wittberger et al. 2000) to locate breakages in the RNA. These are followed by the same reverse primer extension as before. While useful for locating certain points of weakness in the RNA structure, they tend to be most useful on small molecules.

4.4.2   SHAPE (Selective 2'-Hydroxyl Acylation Analyzed by Primer Extension)

In 2000 a new variant of chemical probing was developed (Chamberlin and Weeks 2000). This group created RNAs with a special 2' amine substitution and reacted them with activated esters. Bases that had the 2' amine substitution in single-stranded regions reacted faster than those in duplex or other base-pairing situations. They found that this reactivity was not due to electrostatics or solvent accessibility, but rather to local nucleotide flexibility. While an interesting finding, it was constrained by having to create synthetic RNAs with this substitution. However, in 2005, a further breakthrough was developed (Merino et al. 2005). Rather than having to create synthetic RNAs, a chemical (N-methyl isatoic anhydride [NMIA]) is used to modify flexible 2' hydroxyl groups in the sugar backbone of RNAs. Since this is a conserved feature in every base, it allows the simultaneous interrogation of all bases in a given molecule, identifying those bases with the greatest flexibility, i.e., those not involved in base-pairing or other intermolecular interactions. The chemically modified bases are subsequently identified by their ability to promote strong stops in primer extension reactions. This represented a huge step forward for chemical modification of RNAs for structural analysis.

The upside of SHAPE is simple; all bases can be examined in a single reaction, as opposed to the multiple chemical reactions previously required by DMS, kethoxal, and CMCT. The downsides are similar however. The use of large-format gel electrophoresis was still required, making examining large RNAs difficult. A new computer program called SAFA (semi-automated footprinting analysis) was developed (Das et al. 2005) which somewhat lowered the amount of work required to perform quantitative analysis, but this still remained a time-consuming task for large-scale work. Radioactivity was also still employed, with its attendant problems, though help was on the way. In 2007, a technique was developed whereby radioactivity was swapped for IR-dyes (Ying et al. 2007). In addition to being safer than radioactive labeling they had immense advantage of being more sensitive. This greater sensitivity allows for finer distinctions to be drawn from any given reaction. IR dyes also have long half-lives, allowing primers to be stored for long periods of time, reducing a recurring cost.

4.4.3 hSHAPE

In 2008, the Weeks lab published a study in which they were able to measure the nucleotide flexibility of the first 10% of the HIV genome in variety of potential states (Wilkinson et al. 2008). The key innovation in this research was an updated form SHAPE called hSHAPE (high-throughput SHAPE) which used a new program called ShapeFinder (Vasa et al. 2008). By coupling SHAPE chemical interrogation to dye labeled primer extension to automated capillary sequencing, they were able to perform a quantitative measurement across a large portion of viral genome. This new technique incorporated a number of different methodologies to lead to a far faster technique for performing structural analysis than previously available.

4.4.4 New Chemicals

One of the improved components is a series of new chemicals. These chemicals all act on the 2′ hydroxyl of the sugar backbone of RNA to create a $2'$-O-methyl group, which causes elongating reverse transcriptase to arrest transcription. As every base of RNA has a 2′ hydroxyl group, these chemicals will interrogate each base simultaneously. An additional benefit of these chemicals is that they are rapidly self-quenching process in aqueous solutions, obviating the need for a separate quenching step. The first of these chemicals was NMIA. While it works effectively, it has a relatively long reaction time (about 20 min), which can allow degradation of the target RNA by RNases present in any of the components present in the reaction. A further refinement of NMIA is 1M7 (1-methyl 7-nitroisatoic anhydride) (Mortimer and Weeks 2007). 1M7 has a half-life of 14 s and reactions are complete in approximately 70 s. This allows sufficient time for already folded molecules to react, while minimizing the time that contaminating nucleases have to react with substrate RNAs. An even faster reagent exists in BzCn (benzoyl cyanide) which has a half-life of 0.2 s in aqueous solution (Weeks 2010). BzCn allows researchers to take "snapshots" of evolving folding reactions (Mortimer and Weeks 2007). Studies in 2009 showed that the chemicals are equally reactive across all base types and that any changes in reactivity between purines and pyrimidines are smaller than those between flexible and inflexible (Wilkinson et al. 2009).

4.4.5 Capillary vs. Gel Electrophoresis

Capillary electrophoresis had for a long time been used in sequencing and fragment analysis applications. Since SHAPE is a specialized type of fragment analysis, albeit one with fragments at every base, it seemed logical to conclude that capillary electrophoresis could be used to interrogate these types of data as well. Data generated from capillary electrophoresis is also better formatted for quantitation, with ShapeFinder replacing SAFA as the analysis program of choice. In comparison to gel electrophoresis's relatively short read length of ~150 bases, capillary electrophoresis could read extension products as long as ~600 bases, thus making analyses of large molecules such as rRNAs or viral RNAs a much more reasonable task. Use of capillary electrophoresis also makes using fluorescent primers a more practical option.

4.4.6 ShapeFinder

ShapeFinder is a program developed in the Morgan Giddings lab in 2008 (Vasa et al. 2008). It allows users to convert capillary electrophoresis trace data into numerical data. Unlike SAFA, where users must manually interact with each point to be studied, ShapeFinder uses a number of automated steps to decrease processing time.

1. Fitted baseline adjust—Since different dyes have different background levels, this tool equilibrates all traces to a common background. It zeroes each channel over a range of detector reading to between 10 and 20 "peak" windows.
2. Matrixing—Each dye employed emits light over a spectral range. This step helps reduce spectral overlap and isolates the signals from each dye. Applied Biosystems sequencers perform this automatically as the traces are recorded and this step is omitted when using data from these machines.
3. Mobility shift—Capillary electrophoresis separates fragments of DNA based on size. However, the fluorescent dyes used for detection also have magnetic charges, which result in dye-specific mobility shifts. The mobility shift step in ShapeFinder allows the user to correct for the different charge strengths and line up each sequence so that peaks match. This is done by performing an extension where each dye receives the same dideoxynucleotide. The peaks are then lined up and the shift files created from this line up are used in subsequent experiments.

Every so often new mobility shifts have to be created as capillary decay will cause minor changes in the shift speed.

4. Signal decay correction—hSHAPE works by detecting reverse transcriptase stops caused by adducts formed by 1M7. Since adducts may form anywhere downstream of a primer, this tends to favor the creation of shorter cDNA fragments. Additionally, reverse transcriptase is imperfect and will not perfectly extend infinitely. Thus, as the length of a trace increases, the signal height decreases. Signal decay correction corrects for this by applying a statistical model to correct peaks heights. After correction, intense peaks at the beginning, middle, and end of a given selection should be approximately the same.

5. Scaling—Any of the previous steps can affect the scale of each of the primers relative to one another. The Scaling option allows the user to correct trace heights so that the lowest 5–10% of the experimental and control peaks are approximately the same. Once this correction is applied, peaks heights are quantitatively equalized relative to each other.

6. Align and Integrate—This is the core of the ShapeFinder program. By this point, all peaks are appropriately relative, peaks are aligned with equivalent peaks in each channel, and baselines have been set to zero. After the region of interest is chosen, the range of bases to which they correspond are entered and an appropriate sequence file is selected. The program attempts to find all peaks, match the equivalent peak in each channel, and align them to the chosen sequence. Next, the user is able to manually remove or add peaks that were missed during the automated selection and find corresponding peaks in the sequencing channels. Once all peaks are matched and aligned to the appropriate sequence, the program defines the area under each peak. A file is produced containing the area under each curve, both experimental and control, their standard deviations, and the absolute SHAPE reactivity of each position, defined by subtracting the control area of a peak from the equivalent experimental area. This file also determines the sequence identity of each peak. From this point further downstream analyses may be performed by the user.

4.4.7 Completed hSHAPE Projects

Since its development, the hSHAPE protocol has been applied to numerous structural challenges. In 2009, the entire HIV-1 genome was analyzed in vitro (Watts et al. 2009). This examination helped show how secondary and tertiary structure elements of the HIV-1 genome are involved in numerous regulatory functions, including the gag-pol frameshift signal. The reactivity and flexibility of the entire salt-washed yeast ribosome was completed in 2011 (Leshin et al. 2011). A complete salt-washed structure map is the first step in better understanding the structure of eukaryotic translation Fig. 4.2.

Fig. 4.2 Example of ribosome colored to show SHAPE reactivity. Bases were colored based on their relative reactivity to 1M7 and mapped onto a recent X-ray crystal structure (Yusupov Science 2010). *Grey*, unmeasured base; *black*, unreactive (0); *yellow*, weak (1); *light orange*, moderate (2); *dark orange*, strong (3); *red*, highly reactive (4)

4.5 Future Directions

hSHAPE provides the technical platform upon which a new generation of RNA structural modeling is being built. hSHAPE is being used to examine RNA structure at extremely fast time scales (Mortimer and Weeks 2007). By using BzCN as the adduct-forming reagent it is possible to examine how RNA molecules fold, interconvert between forms, and how RNA interacts with proteins on a time scale of seconds. Most recently, this has been used to examine how a single nucleotide acts as the rate-limiting step for folding of an RNase P ribozyme (Mortimer and Weeks 2009).

The data provided by hSHAPE combined with computationally based molecular modeling will enable RNA structures to solve de novo. More importantly, the quantitative nature of this technique provides empirical data which can be used to impose constraints on RNA bases for molecular modeling. The initial steps were taken in 2010, where computer modeling was used to predict the structure of RNAs such as the HCV IRES and *E. coli* 16S rRNA (Low and Weeks 2010). In the absence of hSHAPE-based constraints, the models' prediction of structure was ~50% correct. The addition of constraints based on hSHAPE flexibility data increased the accurate prediction of structure to over 95%. Future work along these lines will allow hSHAPE to be used in conjunction with X-ray crystallography, NMR, and cryo-EM to help model interactions and structural changes of RNA induced by various transacting factors.

Similarly, application of hSHAPE to intact ribosomes will provide a critical dataset by providing quantitative data for the flexibility of every base in the ribosome. When integrated into molecular dynamics simulations, these data will provide the basis for filling the gaps left by X-ray crystallographic and cryo-EM methods, enabling the entire dynamic range of the ribosome to be modeled. The availability of these models will further the understanding of how translation occurs and what the intermediate states of translation look like. This data will open up important paths in drug development and understanding of ribosome-based disease states.

Acknowledgments We would like to acknowledge members of the Dinman lab past and present for their contributions to our understanding of how ribosome structure influences its function. This work was supported by grants to JDD from the National Institutes of Health (R01 GM058859 and 3R01GM058859-10A1S1).

References

Chamberlin SL, Weeks KM (2000) Mapping local nucleotide flexibility by selective acylation of 2'-amine substituted RNA. J Am Chem Soc 122:216–224

Das R, Laederach A, Pearlman SM, Herschlag D, Altman RB (2005) SAFA: semi-automated footprinting analysis software for high-throughput quantification of nucleic acid footprinting experiments. RNA 11:344–354

Ehresmann C, Baudin F, Mougel M, Romby P, Ebel JP, Ehresmann B (1987) Probing the structure of RNAs in solution. Nucleic Acids Res 15:9109–9128

Hu JC, Dahlberg JE (1983) Structural features required for the binding of tRNATrp to avian myeloblastosis virus reverse transcriptase. Nucleic Acids Res 11:4823–4833

Leshin JA, Heselpoth R, Belew AT, Dinman JD (2011) High throughput structural analysis of yeast ribosomes using hSHAPE. RNA Biol 8:478–487

Low JT, Weeks KM (2010) SHAPE-directed RNA secondary structure prediction. Methods 52:150–158

Merino EJ, Wilkinson KA, Coughlan JL, Weeks KM (2005) RNA structure analysis at single nucleotide resolution by selective 2'-hydroxyl acylation and primer extension (SHAPE). J Am Chem Soc 127:4223–4231

Mortimer SA, Weeks KM (2007) A fast-acting reagent for accurate analysis of RNA secondary and tertiary structure by SHAPE chemistry. J Am Chem Soc 129:4144–4145

Mortimer SA, Weeks KM (2009) C2′-endo nucleotides as molecular timers suggested by the folding of an RNA domain. Proc Natl Acad Sci USA 106:15622–15627

Peattie DA, Gilbert W (1980) Chemical probes for higher-order structure in RNA. Proc Natl Acad Sci USA 77:4679–4682

Soukup GA, Breaker RR (1999) Relationship between internucleotide linkage geometry and the stability of RNA. RNA 5:1308–1325

Stern S, Moazed D, Noller HF (1988) Structural analysis of RNA using chemical and enzymatic probing monitored by primer extension. Methods Enzymol 164:481–489

Vasa SM, Guex N, Wilkinson KA, Weeks KM, Giddings MC (2008) ShapeFinder: a software system for high-throughput quantitative analysis of nucleic acid reactivity information resolved by capillary electrophoresis. RNA 14:1979–1990

Watts JM, Dang KK, Gorelick RJ, Leonard CW, Bess JW Jr, Swanstrom R, Burch CL, Weeks KM (2009) Architecture and secondary structure of an entire HIV-1 RNA genome. Nature 460:711–716

Weeks KM (2010) Advances in RNA structure analysis by chemical probing. Curr Opin Struct Biol 20:295–304

Wilkinson KA, Gorelick RJ, Vasa SM, Guex N, Rein A, Mathews DH, Giddings MC, Weeks KM (2008) High-throughput SHAPE analysis reveals structures in HIV-1 genomic RNA strongly conserved across distinct biological states. PLoS Biol 6:e96

Wilkinson KA, Vasa SM, Deigan KE, Mortimer SA, Giddings MC, Weeks KM (2009) Influence of nucleotide identity on ribose 2′-hydroxyl reactivity in RNA. RNA 15:1314–1321

Wittberger D, Berens C, Hammann C, Westhof E, Schroeder R (2000) Evaluation of uranyl photo-cleavage as a probe to monitor ion binding and flexibility in RNAs. J Mol Biol 300:339–352

Ying BW, Fourmy D, Yoshizawa S (2007) Substitution of the use of radioactivity by fluorescence for biochemical studies of RNA. RNA 13:2042–2050

Chapter 5
Methods for Studying the Interactions of Translation Factors with the Ribosome

Assen Marintchev

Abbreviations

4-thioU	4-Thiouracil
5'-UTR	5'-Untranslated region
6-thioG	6-Thioguanine
Cryo-EM	Cryo-electron microscopy
CTD	C-terminal domain
CTT	C-terminal tail
dsRNA	Double-stranded RNA
eIF	Eukaryotic translation initiation factor
E-site	Exit site
Fe(II)-BABE	Fe(II) 1-(p-bromoacetamidobenzyl)-EDTA
FRET	Fluorescence resonance energy transfer
IRES	Internal ribosome entry site
NTD	N-terminal domain
NTT	N-terminal tail
OB-fold	Oligonucleotide/oligosaccharide binding fold
P-site	Peptidyl-tRNA site
Ribosomal A-site	Ribosomal aminoacyl-tRNA site
rRNA	Ribosomal RNA
RT	Reverse transcriptase
ssRNA	Single-stranded RNA
tRNA	Transfer RNA

A. Marintchev (✉)
Department of Physiology and Biophysics, Boston University School of Medicine,
East Concord Street, 02118-2526 Boston, MA, USA
e-mail: amarint@bu.edu

J.D. Dinman (ed.), *Biophysical Approaches to Translational Control of Gene Expression*,   83
Biophysics for the Life Sciences 1, DOI 10.1007/978-1-4614-3991-2_5,
© Springer Science+Business Media New York 2013

5.1   Introduction

Translation is a multistep process involving a number of RNA and protein molecules. At every stage, individual proteins bind to the ribosome, change their position, or are released from the complex. Therefore, to understand the mechanisms of translation, one would need to know the structure of the translation complexes at every step of the process. Recent breakthroughs have yielded a number of high-resolution crystal structures of ribosomes and ribosomal subunits (reviewed in Schmeing and Ramakrishnan 2009), including the first crystal structures of eukaryotic ribosomes (Ben-Shem et al. 2010; Rabl et al. 2011). However, only a handful of structures are available for translation factor—ribosome complexes. To a great extent, this is due to the fact that most of these complexes are dynamic and/or unstable, making it difficult to obtain crystals or even cryo-electron microscopy (Cryo-EM) reconstructions; although the increasing number of crystal structures of ribosomal complexes reported in recent years is a source of optimism (reviewed in Schmeing and Ramakrishnan 2009).

The limited information about the structure of the ribosomal complexes has always been an obstacle in our understanding of the mechanisms of translation. Therefore, a number of alternative biochemical and biophysical approaches have been used successfully to study the interactions of translation factors with the ribosome and determine their approximate binding sites. These include cross-linking, footprinting, directed hydroxyl radical probing, and more recently fluorescence resonance energy transfer (FRET). The goal of this chapter is to provide a description of these methods, the information obtained with them, their limitations, and how they complement the results from Cryo-EM and crystallography. Other binding assays, such as ultracentrifugation or fluorescence anisotropy, which do not provide information about the ribosomal location of the proteins, are not discussed.

5.2   Cross-Linking

5.2.1   Method Description

Chemical and UV cross-linking has been used for several decades to determine the positions of ribosomal proteins, translation factors, and RNAs on the ribosome (reviewed in Fraser and Doudna 2007; Green and Noller 1997). Individual components of molecular complexes directly interact with one another, or at least close in space, are linked with covalent bonds using a compound that reacts with both molecules simultaneously (chemical cross-linking) or through direct covalent bond formation induced by UV irradiation (UV cross-linking). Depending on the specific reagent used, protein–protein, protein–RNA, or RNA–RNA cross-links can be observed. Multiple factors are taken into consideration in selection of an appropriate cross-linking reagent. For example, if an interaction involves a surface rich in Lys and Arg and a surface rich in Asp and Glu, then a bifunctional cross-linker that can bridge an amino and a carboxyl groups would be appropriate. Cross-linking reagents have variable spacer lengths. UV cross-linking involves no spacer and is thus zero-length cross-linking. A cross-linking agent with a long spacer (long distance between its two reactive sites) can bridge the distance between reactive groups located on two different molecules, yielding higher efficiency of cross-linking. Increased cross-linker length can also allow cross-linking of molecules that are not in direct contact with each other, leading to lower specificity. It should be noted however that specificity may not be a major issue with long-length cross-linkers because comparisons of cross-linking data with the crystal structure of the ribosome did not find significant correlation between the length of the cross-linker and the distance between the cross-linked species within the ribosome (Whirl-Carrillo et al. 2002).

5.3   Limitations and Challenges

Any two noninteracting proteins can be nonspecifically cross-linked to each other if mixed at high enough concentration and under appropriate conditions, just from random collisions. Therefore, proper controls are required to avoid false positive results. Conversely, absence of cross-linking is not in itself proof that two molecules do not interact with each other.

With the ribosomal complexes containing large ribosomal RNAs (rRNAs) and tens of proteins, identifying the cross-linked molecules and the specific positions can be challenging. Cross-links in proteins can be identified using 2D gels and Mass Spectrometry. The exact position of cross-links in RNA are easier to determine using reverse transcriptase (RT) since the cross-links block the enzyme, causing termination in front of the cross-link, in the same way as footprinting (see below). Several approaches are used to simplify the task of identifying cross-links in complex mixtures:

– Adding an affinity tag on a specific molecule in order to be able to isolate cross-linked species that contain it and/or labeling the molecule in order to be able to detect the complexes.
– Using reversible cross-linkers or cleavable cross-linkers. The latter type can be used to transfer a label (e.g. biotin) to the molecules that were cross-linked, in order to aid in their further identification.
– Using an experimental setup where a specific molecule is activated for cross-linking. This can be achieved by pretreatment of a protein with a bifunctional cross-linker before assembling the ribosomal complexes, ensuring that the only cross-linkers available in the reaction mix are already covalently attached to the molecule of interest. For RNAs, specific cross-linking can be achieved by incorporating modified nucleotides, such as 4-thiouracil (4-thioU) or 6-thioguanine (6-thioG) at selected positions. Cross-linking with halogenated nucleotide analogs can be performed using longer-wavelength UV light (360 nm) that does not affect the protein or RNA backbone, both ensuring specific cross-linking and minimizing damage to the interacting molecules (Favre et al. 1998).

5.3.1 Applications

Cross-linking with halogenated nucleotide analogs was used successfully to study the interactions of mRNA with the mammalian ribosome and components of the translation initiation complex (Pisarev et al. 2006, 2008). In eukaryotes, to be efficiently recognized by the translation initiation machinery, the start codon needs to be in an optimum sequence context: GCC(A/G)CCAUGG (Kozak 1986). The nucleotides at positions −3 and +4 (where the A in AUG is +1) are the most important determinants. Cross-linking using mRNA containing 4-thioU or 6-thioG at position −3 or +4 allowed identifying nucleotides in 18S rRNA and ribosomal proteins interacting with the mRNA at these positions, with some of the interactions being dependent on, or modulated by the nature of the nucleotide at these mRNA positions. In addition to ribosomal components, eukaryotic translation initiation factor (eIF) 2α was also efficiently cross-linked to mRNA at position −3. The authors showed that eIF2α has a role in recognition of the proper sequence context, thus obtaining information both about the ribosomal position of eIF2α and its function (Pisarev et al. 2006). The same approach was also used to determine the path of mRNA in the eukaryotic 48S complex (Pisarev et al. 2008). Cross-linking in 48S ribosomal complexes with a series of model mRNAs with 4-thioU incorporated at specific positions showed that the mRNA path in eukaryotic 48S initiation complexes is very similar to that in elongating bacterial 70S ribosomes (Selmer et al. 2006; Yusupova et al. 2001). Two eIF3 subunits, eIF3a and eIF3d, were found to cross-link to mRNA upstream from the start codon and extend the mRNA Exit channel on the 40S ribosomal subunit (Pisarev et al. 2008). Footprinting and hydroxyl radical probing (see below) showed that eIF3 also interacts with 18S rRNA helix 16 (h16) near the mRNA Entry channel of the small ribosomal subunit (Pisarev et al. 2008), which together with yeast two-hybrid assays, and in vitro binding data (Chiu et al. 2010; Elantak et al. 2010; Valasek et al. 2003) identified several points of contact between eIF3 subunits and ribosomal components spanning the solvent surface of the 40S subunit between the mRNA Entry and Exit channels. Analysis of these results in the context of the Cryo-EM reconstruction of eIF3 (Siridechadilok et al. 2005) provided valuable information about the overall orientation of eIF3 on the 40S ribosomal subunit.

5.3.2 Outlook

In summary, cross-linking can be used to identify pairs of molecules that are in close proximity on the ribosome or directly interacting with each other. Observing cross-links between a translation factor and a ribosomal component or another translation factor allows mapping that protein's approximate binding site in ribosomal complexes. Where the exact positions of the cross-links can be determined, the approach also provides some information about the orientation of the bound translation factor.

While cross-linking was one of the first approaches in studying the architecture of ribosomes and ribosomal complexes, its potential utility is even greater now that high-resolution structures of both bacterial and eukaryotic ribosomes have been solved. Work from the 1970s and 1980s reported cross-linking of eIF2 (Westermann et al. 1979), eIF3 (Westermann and Nygard 1983), and Met-tRNAi (Westermann et al. 1981) to several ribosomal proteins. While these reports cannot be used directly, due to possible difficulties in identification of individual ribosomal proteins and changes in the nomenclature over the last three decades, they demonstrate the valuable information to be gained by identifying contacts of individual eIF2 and eIF3 subunits with ribosomal proteins and nucleotides in rRNA in the context of the high-resolution crystal structure of the eukaryotic 40S ribosomal subunit (Rabl et al. 2011) and the Cryo-EM reconstruction of eIF3 (Siridechadilok et al. 2005). Specifically, it should be possible to tentatively assign eIF3 subunits to segments of electron density and model their orientation on the ribosome.

5.4 Footprinting

5.4.1 Method Description

Binding of a protein to RNA can protect the RNA segment covered by the protein from nuclease cleavage, chemical cleavage, or modification. The binding site can then be identified using RT primer extension. The RNA breaks or modified nucleotides stop the reverse transcriptase, and the resulting products are visualized on a sequencing gel. Protein binding to the RNA results in disappearance or weakening of bands corresponding to the protein binding site (footprint). Thus, the protein binding site can be identified by comparing primer extension results in the presence and absence of the protein of interest. In a complex RNA with a well-defined tertiary structure, protein binding often indirectly affects RNA segments away from the binding site and, therefore, not all footprints are necessarily at the binding site. Of course, since the protein must bind directly in order to cause allosteric effects, at least a large portion of the footprints are at the protein binding site.

Protein binding can also cause enhanced cleavage at certain positions through changes in the conformation of RNA or its accessibility. Cleavage by endonucleases and many chemicals is dependent on the conformation of RNA, particularly whether it is single-stranded (ssRNA) or double-stranded (dsRNA). Thus, changes in cleavage patterns can be used to detect conformational changes in RNA. For the purposes of identifying a protein binding site, however, it is preferable to use footprinting reagents that cleave both ssRNA and dsRNA, such as hydroxyl radicals. It should be noted that the sensitivity of individual nucleotides in rRNA to hydroxyl radical cleavage varies significantly (see the discussion of directed hydroxyl radical probing below). Hydroxyl radicals are generated typically by mixing Fe(II)-EDTA, H_2O_2, and ascorbate, where the interaction of Fe(II) with H_2O_2 releases the hydroxyl radicals while converting Fe(II) into Fe(III). The role of ascorbate is to regenerate Fe(II)-EDTA by reducing Fe(III) back to Fe(II), and allow continuous hydroxyl radical production (reviewed in Ehresmann et al. 1987; Kolupaeva et al. 2007).

5.4.2 Applications

Before high-resolution structures became available, footprinting studies were instrumental in mapping contact interfaces between the two ribosomal subunits and binding sites for tRNAs, translation factors, and antibiotics (reviewed in Fraser and Doudna 2007). When the crystal structures of the ribosome were reported, comparative analysis of the available footprinting data with the actual positions of ribosomal proteins and RNAs confirmed that footprinting can provide reliable information about the overall position of proteins on the ribosome (Whirl-Carrillo et al. 2002). More recently, Pestova and co-workers showed that the helicase DHX29 is required for scanning through stable stems in the 5'-untranslated region (5'-UTR) of mRNA. Chemical and enzymatic footprinting indicated that DHX29 likely binds to 18S rRNA h16 near the mRNA Entry channel of the small ribosomal subunit, in line with its function (Pisareva et al. 2008).

Another application of footprinting that has become popular in recent years is tracking conformational rearrangements in ribosomal complexes at different stages of translation or upon binding of antibiotics using changes in footprinting patterns. An important feature of the approach is that when the structure of the ribosome at specific stages is known from X-ray crystallography or Cryo-EM, the observed footprinting patterns can be readily correlated to specific ribosomal conformations. For example, Noller and co-workers used chemical footprinting, in combination with other assays, to identify the effects of EF-G in complex with GTP, GDPNP (a non-hydrolyzable GTP analog), and GDP on the positions of mRNA and tRNAs in translocation. They found that GTP is required for complete translocation to occur and for the release of Exit site (E-site) tRNA, whereas EF-G in complex with GDPNP or GDP·fusidic acid only causes movement of the Aminoacyl-tRNA site (A-site) and Peptidyl-tRNA site (P-site) tRNAs to hybrid A/P and P/E states, respectively (Spiegel et al. 2007). Green and co-workers reported that stop codon recognition by release factors induced specific conformational changes in the decoding center of the small ribosomal subunit, which may help in promoting peptide release (Youngman et al. 2007). A combination of footprinting and FRET was used to elucidate the mechanism of action of the antibiotic viomycin, which appears to lock the ribosome in a translocation intermediate (Ermolenko et al. 2007).

Time-resolved hydroxyl radical footprinting is a variant of the method that allows studying the dynamics of ribosomal complexes. Synchrotron X-ray irradiation is typically used to generate hydroxyl radicals, since the approach requires large amounts of hydroxyl radicals to be produced over a short time period (reviewed in Fraser and Doudna 2007).

Footprinting has also been successfully used to study the mechanisms of translation initiation at internal ribosome entry sites (IRES) in mRNA. IRESs are typically large RNA segments with complex tertiary structures that recruit the translation initiation complex directly to the start codon, without scanning from the 5'-end of mRNA. Accordingly, similar to rRNA, chemical and enzymatic footprinting has been used extensively to study the binding of translation initiation factors, accessory proteins, and ribosomal complexes to IRESs, as well as the conformational changes they induce in the IRES structure (Sizova et al. 1998; Kolupaeva et al. 2000a, b, 2003, 2007; Pilipenko et al. 2000; Vallejos et al. 2011).

5.4.3 Outlook

Since high-resolution structures of ribosome:translation factor complexes are more difficult to obtain than structures of free ribosomes and ribosomal subunits, such studies traditionally lag behind. So far, there is only one crystal structure of a ribosome-bound eukaryotic translation factor: the 40S:eIF1 complex (Rabl et al. 2011) and a handful of Cryo-EM reconstructions and models (Passmore et al. 2007; Siridechadilok et al. 2005; Taylor et al. 2007). Therefore footprinting, which poses much smaller technical challenges, has an important role to play in mapping the ribosomal locations of translation factors and accessory proteins. Conversely, where the structures of individual ribosomal complexes are known, footprinting and time-resolved footprinting are promising tools for tracking conformational rearrangements along the translation pathway.

5.5 Directed Hydroxyl Radical Probing

5.5.1 Method Description

Unlike cross-linking and footprinting, which only identify the binding site, directed hydroxyl radical probing can also yield a low-resolution docking model for the position of a protein on the ribosome. The method is based on the release of hydroxyl radicals from a specific position on a protein surface resulting in cleavage of nearby RNA. Typically, the cysteine-reactive reagent Fe(II) 1-(p-bromoacetamidobenzyl)-EDTA (Fe(II)-BABE) is covalently attached to solvent-exposed cysteine side chains on the protein. First, a cysteine-less mutant of the protein of interest is generated. Then, a series of single-cysteine mutants are made allowing attachment of a single Fe(II)-BABE at desired positions on the protein surface. Next, the individual Fe(II)-BABE-modified mutant proteins are used to assemble ribosomal complexes. H_2O_2 and ascorbate are then added to preformed complexes to initiate production of hydroxyl radicals. The hydroxyl radicals diffusing from the point of origin cleave nearby exposed bonds. Since the radicals are short-lived and diffusion into wider space quickly lowers their concentration, the cleavage intensity depends on the distance from the source. Cleavage is strongest within ~15 Å from the cysteine side chain, with the Fe(II)-BABE moiety itself being ~10 Å long. Weak cleavages can be observed up to ~40 Å (Joseph et al. 1997). The positions in rRNA cleaved from individual single-cysteine mutants are identified by RT primer extension, which also serves to amplify the signal, since the same rRNA molecule can serve as a template multiple times. The cleavages are converted into distance restraints between the modified amino acid in the protein and the cleaved positions in the rRNA, which allows docking the protein onto the ribosome structure. The distance restraints from directed hydroxyl radical probing, sometimes combined with any other available structural information, can be used as input for interactive or automated docking with a number of software programs for molecular modeling and structure visualization (Pisarev et al. 2007). The binding interfaces predicted from the docking model can be examined further, e.g., by site-directed mutagenesis.

5.5.2   Factors Affecting Cleavage Efficiency

Cleavage is nonuniform in that only a fraction of all rRNA positions in the vicinity of the Fe(II)-BABE moiety are cleaved. Cleavage efficiency depends on a number of factors:

- Whether a given bond in rRNA is exposed, partially exposed or buried.
- Whether there is an uninterrupted open space between the source of hydroxyl radicals and the bond in the rRNA. For example, if a protein fragment lies directly between the Fe(II)-BABE and a fully exposed RNA segment, very little or no cleavage would be observed, because diffusion of free radicals along the shortest path from the source to the RNA target would be blocked.
- If the mobility of the Fe(II)-BABE is restricted, e.g., by being trapped between two surfaces, the hydroxyl radicals are released from a discreet position ~10 Å away from the cysteine side chain, instead of from a broad area around it.

The last two factors can result in distinct, often nonoverlapping, cleavage patterns from residues close to each other on the protein surface, providing valuable information for docking the protein onto the ribosome (Fig. 5.1).

- The intrinsic susceptibility of the bond to cleavage. While hydroxyl radicals are in general nonspecific, it appears that certain nucleotides in rRNA are cleaved much more efficiently, often corresponding to some of the functionally important nucleotides in the ribosome. For example, contamination with trace amounts of free Fe(II)-BABE yields detectable cleavages in specific positions in rRNA, underscoring the need to ensure complete removal of all unreacted Fe(II)-BABE from the protein sample before using it for complex assembly (T. V. Pestova, personal communication).

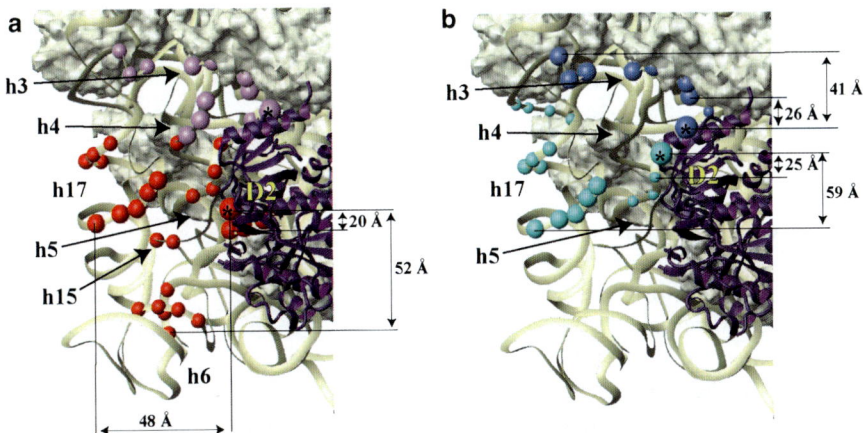

Fig. 5.1 Directed hydroxyl radical cleavages of 18S rRNA from specific positions of human eIF5B. The rRNA is shown as beige ribbon and ribosomal proteins are shown in surface representation. The docked model of eIF5B is shown in *purple* ribbon. The positions of cysteines on the surface of domain 2 of eIF5B are shown as colored spheres with *asterisks*. The cleavage positions in 18S rRNA are shown as spheres with the same color as that of the cysteine from which they are cleaved, and the radii of the spheres correspond to the cleavage intensities: weak, medium, and strong. (a) Cleavages from C884 (*red*) and C894 (*violet*). (b) Cleavages from C952 (*cyan*) and C961 (*blue*). The distances between several cleavage positions and corresponding eIF5B residues are shown. Some of the helices in rRNA are labeled, e.g., h3 is helix 3. eIF5B domain 2, where the above cysteines are located is labeled with "D2." Note that the pairs of mutants shown in panels (a) and (b) produce nonoverlapping sets of cleavages in rRNA, including strong cleavages of nucleotides expected to lie as far as ~50 Å away. Since the distances between C884 and C894 and between C952 and C961 are only ~27 Å and ~10 Å, respectively, and the length of the Fe(II)-BABE linker itself is ~10 Å, these cleavage patterns indicate that: (1) the Fe(II)-BABE linker is not moving freely in the ribosomal complexes; and (2) the cleaved rRNA segments that appear distant in the docking model are likely brought closer to eIF5B in the complex through conformational changes. Reproduced from Unbehaun et al. (2007) with permission

5.5.3 Cysteine-Less vs. Partially Cysteine-Less Mutants

It is often impossible to mutate all existing cysteines in the protein of interest, because the resulting cysteine-less mutant protein is not active and/or unstable. In most cases, that is not a problem, as long as derivatization with Fe(II)-BABE itself does not inactivate the protein. A partially cysteine-less protein can be used, instead by mutating only those cysteines that are not essential for structure or function. Cysteines that are important for structure and stability are usually buried and not accessible for derivatization. Even if some of the endogenous cysteines in a partially cysteine-less mutant do get modified, that would not invalidate the results, because any cleavages from such cysteines would be present in the negative control (an Fe(II)-BABE-treated partially cysteine-less mutant) as well as in all complexes with individual cysteine mutants.

5.5.4   Impact of Conformational Changes in the Ribosomal Complexes

The observed cleavage patterns often may indicate conformational changes in the protein and/or the ribosome, e.g., if segments cleaved efficiently from the same position or from multiple different positions on the protein are too far apart from each other, making it impossible to generate a docking model that satisfies all distance restraints simultaneously (Fig. 5.1). However, due to the limited number of distance restraints, it is impossible to model the actual conformational changes in the absence of additional information, such as Cryo-EM data. Therefore, docking a protein using distance restraints from directed hydroxyl radical probing is done by treating both the protein and the ribosome as rigid bodies. Similarly, if a protein fragment is flexible in the free protein, its conformation cannot be inferred from the directed radical cleavage data and its position on the ribosome cannot be mapped with the same degree of certainty as that of a globular domain with known structure. It is, however, possible to qualitatively determine whether the fragment remains flexible on the ribosome: a flexible fragment samples multiple conformations and positions and yields a diffuse cleavage pattern, whereas if it becomes immobilized, more discreet sets of cleavages are observed (Yu et al. 2009). The docking model of eIF1A on the small ribosomal subunit, based on directed hydroxyl radical probing (Yu et al. 2009), shown in Fig. 5.2, exemplifies both the power and the limitations of the method. The model includes docking onto the ribosome of the folded domain of eIF1A, the unfolded flexible N-terminal tail (NTT) that becomes immobilized upon ribosome binding, and the unfolded flexible C-terminal tail (CTT) that remains mobile upon binding. Whereas the eIF1A folded domain could be reliably docked, only the approximate position of the NTT could be determined, while the flexible CTT (that has no set position or conformation on the ribosome) was modeled within the open space it could occupy (Fig. 5.2). The mutual orientation of the head and body of the small ribosomal subunit changes between different states of the ribosome and could be influenced by eIF1A binding (see, e.g., Passmore et al. 2007). Therefore, the head orientation in the model in Fig. 5.2 need not be the same as that in the actual 40S:eIF1A complex, which offers an example of uncertainties due to possible conformational changes in the ribosome itself. As explained above, the cleavage patterns can only qualitatively indicate conformational changes in the ribosome but those changes would be impossible to model based on directed hydroxyl radical cleavage data alone. Thus when looking at a model, it is critical to pay attention to the "fine print." A molecule in the model may be known to occupy that specific position or may be modeled in a hypothetical position it could be occupying, but which remains to be further tested and refined.

5.5.5   Applications

Directed hydroxyl radical probing was first used to map the ribosomal positions of tRNAs and ribosomal proteins by the group of Harry Noller (Heilek et al. 1995;

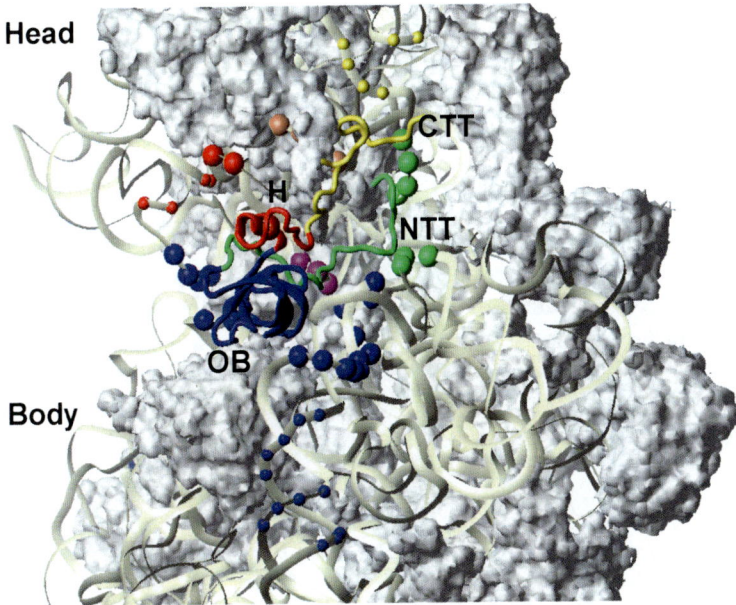

Fig. 5.2 Model for the position of human eIF1A on the small ribosomal subunit based on directed hydroxyl radical probing, and its limitations. The NMR solution structure of human eIF1A (Battiste et al. 2000) was docked onto the crystal structure of the *Thermus thermophilus* 30S ribosomal subunit (Carter et al. 2001). rRNA is shown as *light yellow* ribbon. Ribosomal proteins, in surface representation, are colored *grey* (only a portion of the small ribosomal subunit is shown). eIF1A is shown as ribbon. The oligonucleotide/oligosaccharide (OB)-fold subdomain is *blue*; the helical subdomain (H) is *red*; the N-terminal tail (NTT) is *green*, and the C-terminal tail (CTT) is *yellow*. Positions of cleavage in rRNA are shown as colored spheres: *blue* (from the OB-fold), *red* (from the helical subdomain), *yellow* (from the CTT only), *magenta* (from both the NTT and the OB-fold), *green* (from both the NTT and the CTT), and *orange* (from the NTT, the CTT and the helical subdomain). The radii of the spheres are proportional to the strongest cleavage observed: weak or medium/strong. The folded domain of eIF1A (OB-fold and helical subdomains) was docked as a rigid body. The natively unfolded flexible NTT and CTT tails (Battiste et al. 2000) were modeled based on directed hydroxyl radical probing. The position of the eIF1A folded domain on the body of the small ribosomal subunit is supported by multiple distance restraints. The head of the small subunit can rotate with respect to the body (see, e.g., Passmore et al. 2007) and therefore, its orientation in the 40S:eIF1A complex may be different from the model. The observed cleavage patterns can only qualitatively indicate possible head rotation. The eIF1A-NTT appears to become at least partially immobilized upon binding to the ribosome. However, its conformation is unknown and the directed cleavage data are not sufficient to pinpoint its exact orientation. Therefore in the model, the eIF1A-NTT position is only approximate and its backbone conformation is essentially random. eIF1A-CTT appears to remain mobile in binary 40S:eIF1A complexes and its mobility is only partially restricted in 43S preinitiation complexes, presumably by the initiator tRNA that occupies large portion of the ribosomal P-site. Therefore, eIF1A-CTT does not have a set position or conformation and is modeled within the space it appears to occupy in a random conformation. Reproduced from Yu et al. (2009) with permission

Heilek and Noller 1996a, b; Joseph et al. 1997). The ribosomal positions of several bacterial translation factors have been determined using this approach, sometimes combined with footprinting data: EF-G (Wilson and Noller 1998), RF1 (Wilson et al. 2000), RRF (Lancaster et al. 2002), IF3 (Dallas and Noller 2001), IF2 (Marzi et al. 2003). More recently, the positions of several eukaryotic translation factors were also determined by directed hydroxyl radical probing: eIF1 (Lomakin et al. 2003), eIF5B (Unbehaun et al. 2007), eIF3j (Fraser et al. 2007), and eIF1A (Yu et al. 2009). The overall location of the middle domain of human eIF4G (eIF4Gm) on the 40S subunit was also mapped (Yu et al. 2011).

Directed hydroxyl radical probing was used for mapping the binding sites of translation factors or accessory proteins on IRES RNAs, which have well-defined tertiary structures. However, in the absence of a three-dimensional structure of the respective IRESs, a docking model could not be achieved (Yu et al. 2011; Kafasla et al. 2009, 2010; Kolupaeva et al. 2003).

5.5.6   Comparisons with X-Ray Crystallography and Cryo-EM Data

In cases where crystal structures or Cryo-EM reconstructions have become available, the ribosomal positions of translation factors determined by directed hydroxyl radical probing have been found to be correct overall. In certain cases, different conformations and/or orientations of individual domains have been observed, as was for example the case with RRF (Agrawal et al. 2004; Lancaster et al. 2002), which is not surprising, considering the fact that hydroxyl radical probing does not provide direct information about conformational changes. Sometimes, directed hydroxyl radical cleavage data can provide information complementary to Cryo-EM reconstructions. Cleavages from domains 1 and 2 of human eIF5B on the 80S ribosome (Unbehaun et al. 2007) and domains 1 and 2 of EF-G on the bacterial 70S ribosome (Wilson and Noller 1998), for instance, are very similar with one another but are not fully compatible with the Cryo-EM reconstructions of EF-G (Agrawal et al. 1998) and the bacterial eIF5B homolog IF2 (Allen et al. 2005; Myasnikov et al. 2005). These observations indicate possible transient conformational changes in the ribosome induced by both eIF5B and EF-G (Unbehaun et al. 2007).

Directed hydroxyl radical probing (Dallas and Noller 2001) allowed resolution of the discrepancy observed between the structures of the IF3:30S ribosomal subunit complexes obtained by Cryo-EM (McCutcheon et al. 1999) and X-ray crystallography (Pioletti et al. 2001). They showed that the C-terminal domain of IF3 (IF3-CTD) binds to the interface surface of the 30S subunit, near the ribosomal P-site, the position identified by Cryo-EM, but which was inaccessible in the 30S ribosomal subunit crystals into which IF3 had been soaked (Pioletti et al. 2001), due to crystal packing. The authors also reassigned the position of the IF3 N-terminal domain (IF3-NTD), compared to that proposed based on the low-resolution Cryo-EM data. Human eIF1 binds to the same site on the human 40S ribosomal subunit (Lomakin et al. 2003) as that of IF3 on the bacterial 30S subunit (Dallas and Noller 2001). Remarkably, Pestova and co-authors used directed hydroxyl radical cleavage (Lomakin et al. 2006) to find that bacterial IF3 and human eIF1 can bind to each other's sites on mammalian and bacterial ribosomes, respectively. eIF1 bound not only to the principal IF3 binding site on the platform, but also to the binding site on the solvent face of the bacterial 30S ribosomal subunit observed by crystallography (Pioletti et al. 2001), indicating that the IF3 binding site on the solvent face of the 30S subunit is likely a bona fide secondary binding site for IF3.

The docking models obtained from directed radical probing may only be correct to within a few Å, especially if the protein and/or the ribosome undergo conformational changes upon binding or if the actual ribosomal structure differs from the structure used for the docking. For example, in the recently published crystal structure of the eukaryotic 40S:eIF1 complex (Rabl et al. 2011), the position of eIF1 differs by several Å from the model obtained earlier with directed hydroxyl radical probing (Lomakin et al. 2003) and clashes with the position of the P-site tRNA inferred from the structure of the elongating ribosome. Thus, the 40S:eIF1 structure confirms earlier speculations that the position of the initiator tRNA in the P-site of initiation complexes differs from that of P-site tRNA in elongating ribosomes (Lomakin et al. 2003; Marintchev and Wagner 2004). More importantly, the eIF1 position found in the crystal structure indicates that eIF1 contacts eIF1A on the ribosome (Rabl et al. 2011). This finding is consistent with the observation that eIF1 and eIF1A bind cooperatively to the 40S subunit (Maag and Lorsch 2003; Majumdar et al. 2003), whereas the inability to detect binding between free eIF1 and eIF1A could be explained with an affinity too low to be observed in solution. The 40S:eIF1 crystal structure raises another intriguing possibility. Since IF3-CTD and eIF1 bind to the same sites on both the bacterial and eukaryotic small ribosomal subunits (Lomakin et al. 2006), and since eIF1A and its bacterial homolog IF1 bind to the same site in the ribosomal A-site (Carter et al. 2001; Yu et al. 2009), it is likely that if eIF1 and eIF1A contact each other on the eukaryotic 40S subunit, IF3-CTD, and IF1 may contact each other on the bacterial 30S subunit. Dallas and Noller used the position of the P-site tRNA as a constraint in docking (Dallas and Noller 2001) and thus IF3-CTD could not have been docked to the position occupied by eIF1, even if the cleavage data were consistent with it. In support of this possibility, IF1 and IF3 bind cooperatively to the ribosome (Zucker and Hershey 1986), similar to eIF1A and eIF1. However, cross-linking studies have failed to detect cross-links between IF1 and IF3 on the ribosome (Boileau et al. 1983).

5.5.7 Outlook

Although directed hydroxyl radical probing is much more challenging and labor-intensive than cross-linking and footprinting, it can provide structural models for the ribosomal positions of proteins, comparable to low-resolution Cryo-EM reconstructions. It is therefore the method of choice for systems where Cryo-EM and

X-ray-structures are not available. As described above, directed hydroxyl radical probing can provide information about possible conformational changes in the ribosomal complexes. Thus, cleavage patterns could in principle be used to aid in the analysis of Cryo-EM data, e.g., by providing indications for the presence of certain complexes/conformations. However, this possibility has not been explored.

Hydroxyl radicals cleave proteins just as efficiently as RNA, and directed hydroxyl radical probing has been used to study protein–protein interactions (Datwyler and Meares 2000). In theory, the same approach could be used to study ribosomal complexes. However, in a multicomponent system such as the ribosome, identifying cleavages in proteins is not as simple as in rRNA, where reverse transcription also serves to amplify the signal. The task of identifying cleavages in proteins can be simplified if the target protein has an affinity tag, which can be added using commercially available kits. Thus, directed hydroxyl radical probing could be used to determine the mutual orientations and distances between translation factors. In systems that are easily genetically manipulated, such as bacteria and yeast, this strategy could also be used to map interactions between translation factors and ribosomal proteins. In principle, Mass Spectrometry could be used to identify cleavages in untagged native ribosomal proteins, but this remains a technically challenging and nontrivial task.

5.6 Fluorescence Resonance Energy Transfer

5.6.1 *Method Description*

FRET is used to measure distances between two fluorophores within a molecule or a complex. It relies on energy transfer between the two fluorophores. FRET occurs when the emission spectrum of one fluorophore (donor) overlaps with the excitation spectrum of the other (acceptor) and the two fluorophores are close enough in space. The FRET efficiency is 50 % when the distance between the two fluorophores is equal to the Förster distance R_0, which is in the range of 20–90 Å, depending on the pair of fluorophores. R_0 itself is a function of the efficiency of energy transfer, which in turn depends on the degree of overlap between the emission spectrum of the donor and the excitation spectrum of the acceptor. The FRET efficiency is inversely proportional to the sixth power of the distance between the two fluorophores. When the distance between the two fluorophores is comparable to R_0, it can often be measured by FRET with <20 % error with a proper set of controls (Lakowicz 1999). For distances much shorter or longer than R_0, the FRET efficiency is too close to 100 % or 0 %, respectively, for quantitative measurements and only upper or lower distance restraints can be obtained. FRET is also used to study binding between two fluorescently labeled molecules if binding brings the fluorophores within FRET distance. Titration of one molecule with increasing amounts of the other leads to gradual increase in FRET, until saturation is reached when one of the molecules is completely bound (Lakowicz 1999).

5.6.2 Applications

FRET and single-molecule FRET have been used extensively in kinetic and mechanistic studies of translation (see Chaps. 3 and 7). Using FRET, Lorsch and co-workers were able to show that the eIF1A C-terminal tail (eIF1A-CTT) moves away from eIF1 in ribosomal complexes upon start codon recognition (Maag et al. 2005).

5.6.3 Outlook

FRET can in principle provide complementary information to that obtained with directed hydroxyl radical probing. The same sets of single-cysteine mutants can be used with a thiol-reactive reagent to attach a fluorophore to specific positions on the protein surface. FRET is not limited to RNA and can be used to measure distances between specific positions on two proteins. Furthermore, depending on the fluorophore pairs used, greater distances can be measured. A major limitation of the approach, compared to directed hydroxyl radical probing, is that only one distance restraint is obtained per complex assembled with a pair of fluorescently labeled proteins. However, it is sometimes possible to measure the distance to within $\pm 20\%$ (Lakowicz 1999), instead of only obtaining upper distance limits as with directed hydroxyl radical probing. Thus, with sufficient number of fluorescently labeled pairs, FRET could in theory yield a low-resolution docking model for a ribosome-bound translation factor. Efforts to use FRET to obtain structural information about ribosomal complexes are underway in the McCarthy group (Stevenson et al. 2010).

5.7 Summary

Over the last several decades, cross-linking, footprinting, and directed hydroxyl radical probing have provided information about the structure of ribosomal complexes and helped identify the positions of multiple translation factors. In addition to yielding structural information about complexes whose structures are not available, these methods can also help validate low-resolution Cryo-EM reconstructions and/or help assign proteins to the correct regions of electron density. A number of new applications have been developed in recent years or are yet to be explored. Fluorescence techniques, and in particular FRET has the potential to complement the more "traditional" methods by providing structural information about the ribosomal complexes, although it remains to be seen whether they can yield useful experimental results.

Acknowledgements This work was supported by a Howard Temin award K01 CA119107 from the NCI.

References

Agrawal RK, Penczek P, Grassucci RA, Frank J (1998) Visualization of elongation factor G on the Escherichia coli 70S ribosome: the mechanism of translocation. Proc Natl Acad Sci U S A 95(11):6134–6138

Agrawal RK, Sharma MR, Kiel MC, Hirokawa G, Booth TM, Spahn CM, Grassucci RA, Kaji A, Frank J (2004) Visualization of ribosome-recycling factor on the Escherichia coli 70S ribosome: functional implications. Proc Natl Acad Sci U S A 101(24):8900–8905. doi:10.1073/pnas.0401904101 0401904101 [pii]

Allen GS, Zavialov A, Gursky R, Ehrenberg M, Frank J (2005) The cryo-EM structure of a translation initiation complex from Escherichia coli. Cell 121(5):703–712. doi:S0092-8674(05)00295-3 [pii] 10.1016/j.cell.2005.03.023

Battiste JL, Pestova TV, Hellen CU, Wagner G (2000) The eIF1A solution structure reveals a large RNA-binding surface important for scanning function. Mol Cell 5(1):109–119. doi: S1097-2765(00)80407-4 [pii]

Ben-Shem A, Jenner L, Yusupova G, Yusupov M (2010) Crystal structure of the eukaryotic ribosome. Science 330(6008):1203–1209. doi:330/6008/1203 [pii]10.1126/science.1194294

Boileau G, Butler P, Hershey JW, Traut RR (1983) Direct cross-links between initiation factors 1, 2, and 3 and ribosomal proteins promoted by 2-iminothiolane. Biochemistry 22(13):3162–3170

Carter AP, Clemons WM Jr, Brodersen DE, Morgan-Warren RJ, Hartsch T, Wimberly BT, Ramakrishnan V (2001) Crystal structure of an initiation factor bound to the 30S ribosomal subunit. Science 291(5503):498–501

Chiu WL, Wagner S, Herrmannova A, Burela L, Zhang F, Saini AK, Valasek L, Hinnebusch AG (2010) The C-terminal region of eukaryotic translation initiation factor 3a (eIF3a) promotes mRNA recruitment, scanning, and, together with eIF3j and the eIF3b RNA recognition motif, selection of AUG start codons. Mol Cell Biol 30(18):4415–4434. doi:MCB.00280-10 [pii] 10.1128/MCB.00280-10

Dallas A, Noller HF (2001) Interaction of translation initiation factor 3 with the 30S ribosomal subunit. Mol Cell 8(4):855–864. doi:S1097-2765(01)00356-2 [pii]

Datwyler SA, Meares CF (2000) Protein-protein interactions mapped by artificial proteases: where sigma factors bind to RNA polymerase. Trends Biochem Sci 25(9):408–414. doi:S0968-0004(00)01652-2 [pii]

Ehresmann C, Baudin F, Mougel M, Romby P, Ebel JP, Ehresmann B (1987) Probing the structure of RNAs in solution. Nucleic Acids Res 15(22):9109–9128

Elantak L, Wagner S, Herrmannova A, Karaskova M, Rutkai E, Lukavsky PJ, Valasek L (2010) The indispensable N-terminal half of eIF3j/HCR1 cooperates with its structurally conserved binding partner eIF3b/PRT1-RRM and with eIF1A in stringent AUG selection. J Mol Biol 396(4):1097–1116. doi:S0022-2836(09)01554-X [pii] 10.1016/j.jmb.2009.12.047

Ermolenko DN, Spiegel PC, Majumdar ZK, Hickerson RP, Clegg RM, Noller HF (2007) The antibiotic viomycin traps the ribosome in an intermediate state of translocation. Nat Struct Mol Biol 14(6):493–497. doi:nsmb1243 [pii] 10.1038/nsmb1243

Favre A, Saintome C, Fourrey JL, Clivio P, Laugaa P (1998) Thionucleobases as intrinsic photoaffinity probes of nucleic acid structure and nucleic acid-protein interactions. J Photochem Photobiol B 42(2):109–124. doi:S1011134497001164 [pii]

Fraser CS, Doudna JA (2007) Quantitative studies of ribosome conformational dynamics. Q Rev Biophys 40(2):163–189. doi:S0033583507004647 [pii] 10.1017/S0033583507004647

Fraser CS, Berry KE, Hershey JW, Doudna JA (2007) eIF3j is located in the decoding center of the human 40S ribosomal subunit. Mol Cell 26(6):811–819. doi:S1097-2765(07)00320-6 [pii] 10.1016/j.molcel.2007.05.019

Green R, Noller HF (1997) Ribosomes and translation. Annu Rev Biochem 66:679–716. doi:10.1146/annurev.biochem.66.1.679

Heilek GM, Noller HF (1996a) Directed hydroxyl radical probing of the rRNA neighborhood of ribosomal protein S13 using tethered Fe(II). RNA 2(6):597–602

Heilek GM, Noller HF (1996b) Site-directed hydroxyl radical probing of the rRNA neighborhood of ribosomal protein S5. Science 272(5268):1659–1662

Heilek GM, Marusak R, Meares CF, Noller HF (1995) Directed hydroxyl radical probing of 16S rRNA using Fe(II) tethered to ribosomal protein S4. Proc Natl Acad Sci U S A 92(4):1113–1116

Joseph S, Weiser B, Noller HF (1997) Mapping the inside of the ribosome with an RNA helical ruler. Science 278(5340):1093–1098

Kafasla P, Morgner N, Poyry TA, Curry S, Robinson CV, Jackson RJ (2009) Polypyrimidine tract binding protein stabilizes the encephalomyocarditis virus IRES structure via binding multiple sites in a unique orientation. Mol Cell 34(5):556–568. doi:S1097-2765(09)00268-8 [pii] 10.1016/j.molcel.2009.04.015

Kafasla P, Morgner N, Robinson CV, Jackson RJ (2010) Polypyrimidine tract-binding protein stimulates the poliovirus IRES by modulating eIF4G binding. EMBO J 29(21):3710–3722. doi:emboj2010231 [pii] 10.1038/emboj.2010.231

Kolupaeva VG, Pestova TV, Hellen CU (2000a) An enzymatic footprinting analysis of the interaction of 40S ribosomal subunits with the internal ribosomal entry site of hepatitis C virus. J Virol 74(14):6242–6250

Kolupaeva VG, Pestova TV, Hellen CU (2000b) Ribosomal binding to the internal ribosomal entry site of classical swine fever virus. RNA 6(12):1791–1807

Kolupaeva VG, Lomakin IB, Pestova TV, Hellen CU (2003) Eukaryotic initiation factors 4G and 4A mediate conformational changes downstream of the initiation codon of the encephalomyocarditis virus internal ribosomal entry site. Mol Cell Biol 23(2):687–698

Kolupaeva VG, de Breyne S, Pestova TV, Hellen CU (2007) In vitro reconstitution and biochemical characterization of translation initiation by internal ribosomal entry. Methods Enzymol 430:409–439. doi:S0076-6879(07)30016-5 [pii] 10.1016/S0076-6879(07)30016-5

Kozak M (1986) Point mutations define a sequence flanking the AUG initiator codon that modulates translation by eukaryotic ribosomes. Cell 44(2):283–292. doi:0092-8674(86)90762-2 [pii]

Lakowicz JR (1999) Energy transfer. In: Lakowicz JR (ed) principles of fluorescence spectroscopy, 2nd edn. Kluwer Academic, New York, NY, pp 368–394

Lancaster L, Kiel MC, Kaji A, Noller HF (2002) Orientation of ribosome recycling factor in the ribosome from directed hydroxyl radical probing. Cell 111(1):129–140. doi:S0092867402009388 [pii]

Lomakin IB, Kolupaeva VG, Marintchev A, Wagner G, Pestova TV (2003) Position of eukaryotic initiation factor eIF1 on the 40S ribosomal subunit determined by directed hydroxyl radical probing. Genes Dev 17(22):2786–2797. doi:10.1101/gad.1141803 1141803 [pii]

Lomakin IB, Shirokikh NE, Yusupov MM, Hellen CU, Pestova TV (2006) The fidelity of translation initiation: reciprocal activities of eIF1, IF3 and YciH. EMBO J 25(1):196–210. doi:7600904 [pii] 10.1038/sj.emboj.7600904

Maag D, Lorsch JR (2003) Communication between eukaryotic translation initiation factors 1 and 1A on the yeast small ribosomal subunit. J Mol Biol 330(5):917–924. doi:S002228360300665X [pii]

Maag D, Fekete CA, Gryczynski Z, Lorsch JR (2005) A conformational change in the eukaryotic translation preinitiation complex and release of eIF1 signal recognition of the start codon. Mol Cell 17(2):265–275. doi:S1097276504007737 [pii] 10.1016/j.molcel.2004.11.051

Majumdar R, Bandyopadhyay A, Maitra U (2003) Mammalian translation initiation factor eIF1 functions with eIF1A and eIF3 in the formation of a stable 40S preinitiation complex. J Biol Chem 278(8):6580–6587. doi:10.1074/jbc.M210357200 M210357200 [pii]

Marintchev A, Wagner G (2004) Translation initiation: structures, mechanisms and evolution. Q Rev Biophys 37(3–4):197–284. doi:S0033583505004026 [pii] 10.1017/S0033583505004026

Marzi S, Knight W, Brandi L, Caserta E, Soboleva N, Hill WE, Gualerzi CO, Lodmell JS (2003) Ribosomal localization of translation initiation factor IF2. RNA 9(8):958–969

McCutcheon JP, Agrawal RK, Philips SM, Grassucci RA, Gerchman SE, Clemons WM Jr, Ramakrishnan V, Frank J (1999) Location of translational initiation factor IF3 on the small ribosomal subunit. Proc Natl Acad Sci U S A 96(8):4301–4306

Myasnikov AG, Marzi S, Simonetti A, Giuliodori AM, Gualerzi CO, Yusupova G, Yusupov M, Klaholz BP (2005) Conformational transition of initiation factor 2 from the GTP- to GDP-bound state visualized on the ribosome. Nat Struct Mol Biol 12(12):1145–1149. doi:nsmb1012 [pii] 10.1038/nsmb1012

Passmore LA, Schmeing TM, Maag D, Applefield DJ, Acker MG, Algire MA, Lorsch JR, Ramakrishnan V (2007) The eukaryotic translation initiation factors eIF1 and eIF1A induce an open conformation of the 40S ribosome. Mol Cell 26(1):41–50. doi:S1097-2765(07)00188-8 [pii] 10.1016/j.molcel.2007.03.018

Pilipenko EV, Pestova TV, Kolupaeva VG, Khitrina EV, Poperechnaya AN, Agol VI, Hellen CU (2000) A cell cycle-dependent protein serves as a template-specific translation initiation factor. Genes Dev 14(16):2028–2045

Pioletti M, Schlunzen F, Harms J, Zarivach R, Gluhmann M, Avila H, Bashan A, Bartels H, Auerbach T, Jacobi C, Hartsch T, Yonath A, Franceschi F (2001) Crystal structures of complexes of the small ribosomal subunit with tetracycline, edeine and IF3. EMBO J 20(8): 1829–1839. doi:10.1093/emboj/20.8.1829

Pisarev AV, Kolupaeva VG, Pisarev VP, Merrick WC, Hellen CU, Pestova TV (2006) Specific functional interactions of nucleotides at key −3 and +4 positions flanking the initiation codon with components of the mammalian 48S translation initiation complex. Genes Dev 20(5): 624–636. doi:20/5/624 [pii] 10.1101/gad.1397906

Pisarev AV, Unbehaun A, Hellen CU, Pestova TV (2007) Assembly and analysis of eukaryotic translation initiation complexes. Methods Enzymol 430:147–177. doi:S0076-6879(07)30007-4 [pii] 10.1016/S0076-6879(07)30007-4

Pisarev AV, Kolupaeva VG, Yusupov MM, Hellen CU, Pestova TV (2008) Ribosomal position and contacts of mRNA in eukaryotic translation initiation complexes. EMBO J 27(11):1609–1621. doi:emboj200890 [pii] 10.1038/emboj.2008.90

Pisareva VP, Pisarev AV, Komar AA, Hellen CU, Pestova TV (2008) Translation initiation on mammalian mRNAs with structured 5′UTRs requires DExH-box protein DHX29. Cell 135(7):1237–1250. doi:S0092-8674(08)01374-3 [pii] 10.1016/j.cell.2008.10.037

Rabl J, Leibundgut M, Ataide SF, Haag A, Ban N (2011) Crystal structure of the eukaryotic 40S ribosomal subunit in complex with initiation factor 1. Science 331(6018):730–736. doi:science.1198308 [pii] 10.1126/science.1198308

Schmeing TM, Ramakrishnan V (2009) What recent ribosome structures have revealed about the mechanism of translation. Nature 461(7268):1234–1242. doi:nature08403 [pii] 10.1038/nature08403

Selmer M, Dunham CM, Murphy FVt, Weixlbaumer A, Petry S, Kelley AC, Weir JR, Ramakrishnan V (2006) Structure of the 70S ribosome complexed with mRNA and tRNA. Science 313(5795):1935–1942. doi:1131127 [pii] 10.1126/science.1131127

Siridechadilok B, Fraser CS, Hall RJ, Doudna JA, Nogales E (2005) Structural roles for human translation factor eIF3 in initiation of protein synthesis. Science 310(5753):1513–1515. doi:310/5753/1513 [pii] 10.1126/science.1118977

Sizova DV, Kolupaeva VG, Pestova TV, Shatsky IN, Hellen CU (1998) Specific interaction of eukaryotic translation initiation factor 3 with the 5′ nontranslated regions of hepatitis C virus and classical swine fever virus RNAs. J Virol 72(6):4775–4782

Spiegel PC, Ermolenko DN, Noller HF (2007) Elongation factor G stabilizes the hybrid-state conformation of the 70S ribosome. RNA 13(9):1473–1482. doi:rna.601507 [pii] 10.1261/rna.601507

Stevenson AL, Juanes PP, McCarthy JE (2010) Elucidating mechanistic principles underpinning eukaryotic translation initiation using quantitative fluorescence methods. Biochem Soc Trans 38(6):1587–1592. doi:BST0381587 [pii] 10.1042/BST0381587

Taylor DJ, Nilsson J, Merrill AR, Andersen GR, Nissen P, Frank J (2007) Structures of modified eEF2 80S ribosome complexes reveal the role of GTP hydrolysis in translocation. EMBO J 26(9):2421–2431. doi:7601677 [pii] 10.1038/sj.emboj.7601677

Unbehaun A, Marintchev A, Lomakin IB, Didenko T, Wagner G, Hellen CU, Pestova TV (2007) Position of eukaryotic initiation factor eIF5B on the 80S ribosome mapped by directed hydroxyl radical probing. EMBO J 26(13):3109–3123. doi:7601751 [pii] 10.1038/sj.emboj.7601751

Valasek L, Mathew AA, Shin BS, Nielsen KH, Szamecz B, Hinnebusch AG (2003) The yeast eIF3 subunits TIF32/a, NIP1/c, and eIF5 make critical connections with the 40S ribosome in vivo. Genes Dev 17(6):786–799. doi:10.1101/gad.1065403

Vallejos M, Deforges J, Plank TD, Letelier A, Ramdohr P, Abraham CG, Valiente-Echeverria F, Kieft JS, Sargueil B, Lopez-Lastra M (2011) Activity of the human immunodeficiency virus type 1 cell cycle-dependent internal ribosomal entry site is modulated by IRES trans-acting factors. Nucleic Acids Res 39(14):6186–6200. doi:gkr189 [pii]10.1093/nar/gkr189

Westermann P, Nygard O (1983) The spatial arrangement of the complex between eukaryotic initiation factor eIF-3 and 40S ribosomal subunit. Cross-linking between factor and ribosomal proteins. Biochim Biophys Acta 741(1):103–108

Westermann P, Heumann W, Bommer UA, Bielka H, Nygard O, Hultin T (1979) Crosslinking of initiation factor eIF-2 to proteins of the small subunit of rat liver ribosomes. FEBS Lett 97(1):101–104. doi:0014-5793(79)80061-7 [pii]

Westermann P, Nygard O, Bielka H (1981) Cross-linking of Met-tRNAf to eIF-2 beta and to the ribosomal proteins S3a and S6 within the eukaryotic inhibition complex, eIF-2.GMPPCP.Met-tRNAf.small ribosomal subunit. Nucleic Acids Res 9(10):2387–2396

Whirl-Carrillo M, Gabashvili IS, Bada M, Banatao DR, Altman RB (2002) Mining biochemical information: lessons taught by the ribosome. RNA 8(3):279–289

Wilson KS, Noller HF (1998) Mapping the position of translational elongation factor EF-G in the ribosome by directed hydroxyl radical probing. Cell 92(1):131–139. doi:S0092-8674(00)80905-8 [pii]

Wilson KS, Ito K, Noller HF, Nakamura Y (2000) Functional sites of interaction between release factor RF1 and the ribosome. Nat Struct Biol 7(10):866–870. doi:10.1038/82818

Youngman EM, He SL, Nikstad LJ, Green R (2007) Stop codon recognition by release factors induces structural rearrangement of the ribosomal decoding center that is productive for peptide release. Mol Cell 28(4):533–543. doi:S1097-2765(07)00627-2 [pii] 10.1016/j.molcel.2007.09.015

Yu Y, Marintchev A, Kolupaeva VG, Unbehaun A, Veryasova T, Lai SC, Hong P, Wagner G, Hellen CU, Pestova TV (2009) Position of eukaryotic translation initiation factor eIF1A on the 40S ribosomal subunit mapped by directed hydroxyl radical probing. Nucleic Acids Res 37(15):5167–5182. doi:gkp519 [pii] 10.1093/nar/gkp519

Yu Y, Abaeva IS, Marintchev A, Pestova TV, Hellen CU (2011) Common conformational changes induced in type 2 picornavirus IRESs by cognate trans-acting factors. Nucleic Acids Res 39(11):4851–4865. doi:gkr045 [pii] 10.1093/nar/gkr045

Yusupova GZ, Yusupov MM, Cate JH, Noller HF (2001) The path of messenger RNA through the ribosome. Cell 106(2):233–241. doi:S0092-8674(01)00435-4 [pii]

Zucker FH, Hershey JW (1986) Binding of Escherichia coli protein synthesis initiation factor IF1 to 30S ribosomal subunits measured by fluorescence polarization. Biochemistry 25(12):3682–3690

Chapter 6
Riboproteomic Approaches to Understanding IRES Elements

Encarnacion Martinez-Salas, David Piñeiro, and Noemi Fernandez

Abbreviations

CBV3	Coxsackie B3 virus
DHX9	DEAH-box polypeptide 9
DRBP76	Double-stranded RNA-binding protein 76
Ebp1	erbB-3-binding protein 1
eIFs	Translation initiation factors
EMCV	Encephalomyocarditis virus
FBP2	Binding protein 2
FMDV	Foot-and-mouth disease virus
FUSE	Far upstream element
G3BP	Ras-GTPase-activating protein
HCV	Hepatitis C virus
HIV-1	Human immunodeficiency virus
hnRNPs	Heterogeneous ribonucleoproteins
HRV	Human rhinovirus
IGF2BP1	Insulin-like growth factor II mRNA-binding protein 1
IRES	Internal ribosome entry site
ITAFs	IRES transacting factors
LEF-1	Lymphoid enhancer factor
NF45	Nuclear factor of activated T cells
PABP1	Poly(A) binding protein

E. Martinez-Salas (✉) • D. Piñeiro • N. Fernandez
Encarnacion Martinez-Salas, Centro de Biologia Molecular Severo Ochoa, CSIC-UAM,
Nicolas Cabrera 1, Cantoblanco, 28049 Madrid, Spain
e-mail: emartinez@cbm.uam.es

J.D. Dinman (ed.), *Biophysical Approaches to Translational Control of Gene Expression*,   103
Biophysics for the Life Sciences 1, DOI 10.1007/978-1-4614-3991-2_6,
© Springer Science+Business Media New York 2013

PCBP1-2   Poly(rC) binding protein
PTB   Polypyrimidine tract-binding protein
PV   Poliovirus
RRM   RNA recognition motif
SILAC   Stable isotopic labeling with amino acid in cell culture
Unr   Upstream of N-ras

6.1   Introduction: IRES Elements and Their *Trans*-Acting Factors

The process of RNA translation consists of sequential steps, initiation, elongation, termination, and ribosome recycling. In eukaryotes, translational control is mainly exerted at the initiation step, assisted by proteins termed translation initiation factors (eIFs). Most eukaryotic mRNAs initiate translation using a cap-dependent mechanism, by which the m^7GpppN (cap) at the 5' end of the mRNA is recognized by the eIF4F complex (consisting of the cap-binding factor eIF4E, the RNA helicase eIF4A, and the scaffold protein eIF4G that in turn interacts with eIF4E, eIF4A, eIF3, and the poly(A) binding protein (PABP1)). The pre-initiation complex scans in 5' to 3' direction until an appropriate initiation codon is encountered, leading to the 48S complex formation assisted by eIF1, eIF2, and eIF5. Finally, eIF5B mediates joining of the 60S and 40S ribosomal subunits, generating the 80S complex competent for protein synthesis.

In contrast, initiation of protein synthesis in mRNAs translated under stress conditions is often driven by internal ribosome entry site (IRES) elements. IRES are *cis*-acting sequences usually located in the 5' untranslated region (UTR) of mRNAs, that recruit the translation machinery using a 5' end-independent mechanism with the help of eIFs and RNA-binding proteins termed IRES transacting factors (ITAFs). The process of internal initiation operates in RNA viruses as well as in cellular mRNAs translated under stress conditions that compromise cap-dependent translation (Martinez-Salas et al. 2008; Spriggs et al. 2010). However, there is great diversity regarding their primary sequences, secondary structures, and ITAFs required for activity. The large degree of diversity within viral IRES is typically illustrated by two distantly related IRES elements, the one present in the intergenic region of the dicistroviruses genome, whose activity depends exclusively on its three-dimensional RNA structure, and those present in the genome of picornaviruses that have significantly longer nucleotide sequences and depend on both RNA structure and ITAFs (Filbin and Kieft 2009; Martinez-Salas 2008). It is thought that RNA-binding proteins assist picornavirus IRES activity through their involvement in the acquisition of a proper RNA structure.

The lack of apparent conserved features has led to the view that different IRES elements could recruit the ribosomal subunits assisted by unique sets of ITAFs. Thus, understanding the role played by ITAFs is crucial to unravel the strategies employed by mRNAs to recruit the translation machinery under conditions of cap-dependent translation repression. Identification of ITAFs, initially carried out using biochemical approaches (Fitzgerald and Semler 2009; Pestova et al. 2001), has received great attention from recent advances in RNA affinity purification followed by mass spectrometry and further validation by functional assays (Cobbold et al. 2008; Pacheco and Martinez-Salas 2010; Pacheco et al. 2008; Tsai et al. 2011; Weinlich et al. 2009; Yu et al. 2005). Here we discuss the impact of proteomic approaches to characterize the function of ITAFs mediating the expression of proteins in eukaryotic cells.

6.2   ITAFs Interacting with Picornavirus IRES

Picornavirus IRES, which were the first internal initiation elements described (Jang et al. 1988; Pelletier and Sonenberg 1988), span about 450 nucleotides upstream of the functional start codon. Despite performing the same function, less than 50% of primary sequence is conserved between picornavirus IRES. Nevertheless, conservation of RNA secondary structure and ITAFs requirement allows them to be grouped into four types (Belsham 2009; Martinez-Salas 2008). IRES types I and II require eIF4G, eIF4A, eIF2, and eIF3 (de Breyne et al. 2009; Kolupaeva et al. 1998; Lopez de Quinto et al. 2001; Lopez de Quinto and Martinez-Salas 2000), and addition of auxiliary factors stimulates complex formation in reconstitution assays (Andreev et al. 2007; Pilipenko et al. 2000). These data together with the fact that a partial IRES sequence having the capacity to interact with eIF4G, eIF3, eIF4B, and polypyrimidine tract-binding protein (PTB) is not sufficient to promote IRES activity (Fernandez-Miragall et al. 2009) demonstrates that additional factors are necessary in this process.

Early after the discovery of picornavirus IRES, UV-cross-linking procedures allowed the identification of the PTB as an ITAF (Jang and Wimmer 1990). PTB is a multifunctional RNA-binding protein with four RNA recognition motifs (RRMs) that recognize U/C-rich sequences (Conte et al. 2000). Picornavirus IRES elements usually have two polypyrimidine tracts located at each end of the IRES region (Martinez-Salas et al. 2008). It has been recently shown that a single PTB molecule binds to the encephalomyocarditis virus (EMCV) IRES, with RRM1-2 contacting the 3' end, and RRM3 contacting the 5' end of the IRES, thereby constraining the IRES structure in a unique orientation (Kafasla et al. 2009).

Riboproteomic approaches based on mass spectrometry analysis of affinity-purified ribonucleoprotein complexes assembled on tagged RNAs with cytoplasmic cell extracts have also facilitated the identification of RNA-binding proteins interacting with several picornavirus IRES elements (Table 6.1). In these approaches, proteins that interact in an unspecific manner with any RNA sequence can be recognized using appropriate RNA controls and total cytoplasmic RNA as nonspecific competitor (Pacheco et al. 2008). The large majority of ITAFs identified using proteomic procedures are RNA-binding proteins, indicating that there is no gross contamination with other factors present in the cell extracts. Furthermore, and validating

Table 6.1 RNA-binding proteins interacting with viral IRES

ITAF	IRES activity	Targets	Protein function
PTB	Stimulation	FMDV, EMCV, TMEV, PV, HRV, HAV, EV71, HCV	Splicing, RNA stability, RNA localization
PCBP2	Stimulation	PV, HRV, HAV, CVB3, EV71, FMDV[a], EMCV[a], HCV	RNA stability, translation
SRp20	Stimulation	PV	Splicing
Nucleolin	Stimulation	PV, HRV, FMDV, HCV, HIV-1	rRNA maturation, transport
FBP2	Repressor	EV71	RNA stability
FBP1	Stimulation	EV71	Translation control
Gemin5	Repressor	FMDV, HCV	snRNAs biogenesis, translation
hnRNP K	–	FMDV, EV71	RNA stability, translation
hnRNPU	–	FMDV, HIV-1	RNA stability
DHX9	–	HRV, FMDV, HCV	RNA helicase
DDX1	–	FMDV	RNA helicase
DDX3	Stimulation	HCV, HIV-1	RNA helicase
Unr	Stimulation	PV, HRV, EV71, HCV	Translation control
Ebp1/PA2G4/ITAF$_{45}$	Stimulation	FMDV, EMCV[a], EV71	Transcription regulator
DRBP76:NF45	Repressor	HRV, HCV	Transcription, RNA stability
G3BP	–	FMDV	Stress granules
RACK1	–	HCV	Ribosomal subunits joining
IGF2BP1/IMP1	Stimulation	EV71[a], HCV	RNA stability, translation
La	Stimulation	PV, EMCV, HAV[b], HCV	Transcription, translation control
hnRNPA1/A2	Stimulation	EV71, HCV, HIV-1	RNA processing, translation
NSAP1/hnRNP Q	Stimulation	HCV	RNA processing, translation
hnRNP L, D	Stimulation	HCV, HIV-1	RNA stability, translation
GAPDH	Repressor	HAV	RNA transport, translation control
YB-BP1	–	HCV	Transcription, RNA stability
Tubulin, tropomyosin	–	HCV, HIV-1	Cytoskeleton
HuR	Stimulation	HCV, HIV-1[b]	RNA stability
hRIP	Stimulation	HIV-1	RNA localization, export
Sam68	Repressor	HCV, HIV-1	Splicing regulator
Activ transc coact. p15	–	HIV-1	Transcription regulator
DNA topoisomerase	–	HIV-1	DNA replication
Mt-ssb	–	HIV-1	Single-stranded DNA-binding
DNA-binding protein A	–	HIV-1	DNA replication

[a]No effect
[b]Supression

this type of approaches, proteins reported to interact with IRES elements by biochemical methods (for example, eIF4B and eIF3) were identified exclusively bound to specific domains of the foot-and-mouth disease virus (FMDV) or hepatitis C virus (HCV) IRES, respectively, by mass spectrometry following RNA affinity purification (Pacheco et al. 2008; Yu et al. 2005). These approaches have been helpful to identify ribonucleoprotein complexes assembled on individual domains of one particular IRES, on two different IRES elements, or in RNAs carrying two cis-acting regions, for example the IRES and the 3′ UTR (Weinlich et al. 2009). The possibility of applying the riboproteomic methodology to long RNAs with different cis-acting regions is highly relevant for isolating factors that mediate the formation of functional bridges between distant mRNA regions.

A complex network of interactions among gene expression pathways is suggested from the observation that ITAFs are often proteins previously identified as transcription regulators, splicing factors, RNA transport, RNA stability, or translation control proteins (Table 6.1). Examples of this group of proteins are the polyr(C) binding protein (PCBP2), the SR splicing factor (SRp20), nucleolin, the far upstream element-binding protein 2 (FBP2) or Gemin5, among others. PCBP2 stimulates the activity of poliovirus (PV), human rhinovirus (HRV), and coxsackie B3 virus (CBV3) IRES (Choi et al. 2004; Gamarnik et al. 2000; Sean et al. 2009); SRp20 up-regulates PV IRES-mediated translation via its interaction with PCBP2 (Bedard et al. 2007); nucleolin is a protein involved in rDNA transcription, rRNA maturation, ribosome assembly, and nucleo-cytoplasmic transport (Waggoner and Sarnow 1998) that interacts with HRV, FMDV, and PV IRES; the nuclear protein FBP2 is a KH protein that shuttles to the cytoplasm in infected cells negatively regulating enterovirus (EV71) IRES activity (Lin et al. 2009a). Gemin5 is the RNA-binding factor of the survival of motor neurons (SMN) complex that assembles the seven member (Sm) proteins on snRNAs, which are essential components of the splicing machinery (Battle et al. 2006). Gemin5 binds directly to FMDV and HCV IRES down-regulating translation efficiency (Pacheco et al. 2009); it is predominantly located in the cell cytoplasm (Hao le et al. 2007) and has the capacity to bind m^{7}GTP (Bradrick and Gromeier 2009), explaining its down-regulatory role of cap-dependent translation (Pacheco et al. 2009).

Other proteins with important roles in RNA transport and stabilization, including heterogeneous ribonucleoproteins (hnRNPs) and RNA helicases, have been identified in riboproteomic approaches bound to various picornavirus IRES elements (Table 6.1). The modular organization of RNA-binding proteins, which confer the capacity to recognize large numbers of targets, raises the possibility that binding of a particular RNA-binding protein to a given RNA could facilitate different sorts of regulation depending on the other partners on the complex. Conversely, secondary protein–protein bridges could also be the source of factors identified in these approaches. This could be the case of hnRNPs, helicases, and unr (upstream of N-ras). The later is a cold-shock RNA-binding protein that interacts with PABP1 and stimulates picornavirus IRES-dependent translation (Boussadia et al. 2003; Hunt et al. 1999).

HnRNPs, often identified in proteomic approaches associated to different sorts of ribonucleoprotein complexes, are a family of nuclear proteins (named from hnRNP A1 to hnRNP U) with RNA-binding and protein–protein binding motifs (Lunde et al. 2007) that shuttle with the RNA to the cytoplasm (Kim et al. 2000a). Various members of the hnRNP family associate with viral IRES elements (Table 6.1). HnRNP K, PCBP1 (hnRNP E1), and PCBP2 (hnRNP E2) share the KH RNA-binding domain (Makeyev and Liebhaber 2002). HnRNP K is the most abundant protein that recognizes poly(rC) regions and regulates transcription, RNA turnover, and translation (Lee et al. 2007). Both hnRNP K and PCBP1-2 have been identified by mass spectrometry associated to the FMDV and EV71 IRES (Pacheco et al. 2008; Shih et al. 2011).

Due to their modular organization, ITAFs act in large complexes with various other factors. Hence, proteins interacting with different targets may result in distinct effects depending on the target RNA and the other partners of the complex. The IRES elements of EMCV and FMDV, with similar secondary RNA structure, have different requirements in terms of functional RNA-protein association. One example is Ebp1 protein (erbB-3-binding protein 1, also known as proliferation-associated factor (PA2G4) and $ITAF_{45}$) identified with domain 3 of the FMDV IRES in proteomic analysis (Pacheco et al. 2008). Ebp1 cooperates with PTB to stimulate FMDV IRES activity in reconstitution studies (Pilipenko et al. 2000; Andreev et al. 2007), but its depletion does not affect EMCV IRES activity (Monie et al. 2007).

The expression pattern and abundance of ITAFs have been proposed to mediate cell-type specificity, as in the case of the neural form of PTB (Pilipenko et al. 2001). On the other hand, the double-stranded RNA-binding protein 76 (DRBP76, also termed NF90/NFAR-1) that forms a heterodimer with NF45 (nuclear factor of activated T cells) (DRBP76:NF45) and differs in subcellular distribution in neuronal and non-neuronal malignant cells, represses HRV translation in neuronal cells (Merrill et al. 2006). Finally, a group of factors interacting with picornavirus IRES include stress granules proteins, such as the Ras-GTPase-activating protein (G3BP) and PCBP2 (Pacheco et al. 2008).

6.3   ITAFs Interacting with HCV and HIV IRES

The IRES elements of HCV and the human immunodeficiency virus (HIV-1) differ profoundly from picornavirus IRES in RNA structure and factor requirement (Balvay et al. 2009; Lukavsky 2009). The HCV IRES is organized in three structural domains whose integrity is required for protein synthesis. Although the basal portion of domain III is sufficient to form a high-affinity interaction with the 40S ribosomal subunit in the absence of eIFs (Kieft et al. 2001), eIF3 enhances formation of the 48S initiation complex interacting with domain III (Buratti et al. 1998; Pestova et al. 1998). Additionally, under conditions of inactivation of eIF2 by phosphorylation, the HCV IRES forms a pre-initiation complex with eIF3 and eIF5B alone (Terenin et al. 2008). Consistent with these reconstitution studies, the interaction of eIF3 with the

HCV IRES has also been identified by mass spectrometry of IRES-bound protein complexes and cryo-electron microscopy (Siridechadilok et al. 2005; Pacheco et al. 2008; Yu et al. 2005); the ribosomal proteins that participate in IRES-40S interaction have been identified by cross-linking and mass spectrometry (Fukushi et al. 2001; Ji et al. 2004).

Besides ribosomal proteins and eIF3 subunits, the receptor for activated protein kinase C (RACK1) and nucleolin were identified in IRES-40S ribosomal complexes (Yu et al. 2005). RACK1 regulates translation initiation by recruiting protein kinase C to the 40S subunit (Sengupta et al. 2004). A mass spectrometry study identified 39 different proteins in ribonucleoprotein complexes assembled with tobramycin (Tob)-aptamer tagged RNAs that contained both the IRES and the 3′ UTR of HCV. One of the proteins was the insulin-like growth factor II mRNA-binding protein 1 (IGF2BP1) (Weinlich et al. 2009). Further studies showed that IGF2BP1 coimmunoprecipitates with eIF3 and the 40S subunit, suggesting that it enhances HCV IRES-dependent translation by recruiting the ribosomal subunits to a pseudo-circularized RNA.

ITAFs shared between HCV and picornavirus are PTB, PCBP2, nucleolin, Gemin5, unr, hnRNPA1/A2, La autoantigen (La), NS1-associated protein (NSAP1, also known as hnRNP D) (Gosert et al. 2000; Liu et al. 2009; Meerovitch et al. 1993; Mondal et al. 2008; Pacheco et al. 2009; Song et al. 2006; Florez et al. 2005; Hahm et al. 1998; Lin et al. 2009b; Park et al. 2011). Protein–protein association during mRNA transport, such as hnRNP U or hnRNP A/B (Kukalev et al. 2005; Percipalle et al. 2002), can explain the identification of cytoskeleton proteins by mass spectrometry analysis of factors associated to FMDV, HCV, and HIV-1 IRES, YB-1/ PTB with HCV IRES (Lu et al. 2004; Pacheco et al. 2008; Vallejos et al. 2011; Weinlich et al. 2009) or glyceraldehyde 3-phosphate dehydrogenase (GAPDH) with hepatitis A virus IRES (Schultz et al. 1996).

Proteins identified by mass spectrometry using a discrete domain of the HCV IRES as target have also been identified interacting with the entire HCV IRES, providing information about the binding site of the protein. This is the case of RNA helicases DEAH-box polypeptide 9 (DHX9, also known as RNA helicase A) or DEAD-box polypeptide 1 (DDX1), which is a dual interactor between hnRNP K and poly(A)-mRNA (Chen et al. 2002). The DDX/DHX family of proteins plays important roles in pre-mRNA processing, ribosome biogenesis, RNA turnover, RNA export, translation, and association/dissociation of large RNP complexes. RNA helicase A also recognizes a complex structure at the 5′-UTR of retrovirus mRNA precursors, facilitating its association to polyribosomes (Hartman et al. 2006).

Proteomic studies of factors interacting with the HIV-1 IRES were focused on extracts prepared from G2/M synchronized cells (Vallejos et al. 2011), owing to the observation that this IRES element is specifically activated in this phase of the cell cycle. The function of the recently identified HIV-1 ITAFs, which includes several hnRNPs, splicing factors, and RNA helicases (Table 6.1), remains to be determined. A second group of proteins are transcriptional regulators and DNA-binding proteins required for viral replication, presumably reflecting the nuclear phase of this particular RNA virus. However, it would not be surprising if opposing activities are found, as already shown for the Hu antigen R (HuR) protein that represses the activity of the HIV-1 IRES while it stimulates that of HCV (Rivas-Aravena et al. 2009). HIV-1 IRES-mediated translation is also stimulated by hnRNP A1/2 or the RNA helicase DDX3, hRIP, and Sam68 (Liu et al. 2011).

6.4   ITAFs Interacting with Cellular IRES

Cellular IRES are typically present in mRNAs encoding stress response proteins; hence, they have evolved mechanism to evade global repression of translation (Holcik and Korneluk 2000; Spriggs et al. 2005). A characteristic of cellular IRES elements is to depend on its interaction with ITAFs to facilitate the binding of the mRNA to 40S subunits. In this way, changes in the abundance, posttranslational modifications, or subcellular location of ITAFs contribute to control IRES-mediated translation during stress conditions, including nutrient stress, temperature shock, hypoxia, cell cycle arrest, or apoptosis (Conte et al. 2008; Komar and Hatzoglou 2011; Lewis and Holcik 2008; Spriggs et al. 2010).

Conventional biochemical approaches carried out to identify ITAFs interacting with cellular IRES, in conjunction with functional characterization, have provided a long list of factors (Table 6.2), basically consisting in abundant RNA-binding proteins that shuttle between the nucleus and the cytoplasm. Identification of ITAFs relied basically on the characterization of ribonucleoprotein complexes assembled with the IRES of interest (Fox and Stover 2009; Graber et al. 2010; Lewis et al. 2008; Majumder et al. 2009; Spriggs et al. 2009).

The identification of factors controlling the expression of apoptotic proteins as well as the myc family of transcription factors has been cumbersome in gathering information of proteins mediating IRES-dependent translation (Bushell et al. 2006; Cobbold et al. 2008, 2010; Henis-Korenblit et al. 2002; Mitchell et al. 2001). Many of these factors have been also identified associated to other cellular IRES (Table 6.2). This is the case of PTB (hnRNP I) that stimulates translation driven by IRES of mRNAs encoding proteins required for cell survival under apoptosis, hypoxia, nutrient deprivation, or cell growth dysregulation (the apoptotic protease activating factor 1 (apaf-1), BCL2-associated athanogene (BAG)-1, p53, hypoxia-inducible factor (HIF1a), among others) (Dobbyn et al. 2008; Grover et al. 2008; Schepens et al. 2005), although it represses translation initiation driven by the glucose-regulated protein GRP78 immunoglobulin heavy chain-binding protein (BiP) IRES (Kim et al. 2000b).

Recently, a proteomic approach devoted to isolate RNA-protein complexes from living cells has been applied to obtain quantitative profiles of proteins interacting with the lymphoid enhancer factor (LEF-1) IRES (Tsai et al. 2011). This approach is based on the use of biotin-tagged transfected RNAs expressing the bacteriophage protein MS2, which allows affinity purification of UV-cross-linked ribonucleoprotein complexes assembled in living cells, combined with stable isotope labeling with amino acids in cell culture (SILAC)-based quantitative mass spectrometry.

Table 6.2 RNA-binding proteins interacting with cellular IRES

ITAF	activity	IRES targets	Protein function
PTB	Stimulation	Apaf-1, BAG-1, c-myc, cat-1, p53, HIF-1α, INR, Rev-erbα LEF-1, BiP[a]	Splicing, RNA stability, localization
SFPQ/PSF, nonO/p54nrb	Stimulation	c-myc, LEF-1	Splicing, nuclear retention
PCBP2	Stimulation	LEF-1[a], c-myc	RNA processing, translation
PCBP1	Stimulation	BAG-1	RNA processing, translation
hnRNP K	–	LEF-1, c-myc	RNA stability, translation
HuR	Stimulation	XIAP, BCL-2, LEF-1, CAT-1, HIF-1	RNA stability
	Repressor	IGF1 receptor, TM	
YB-BP1	Stimulation	c-myc, LEF-1[b], Apaf-1[b], BAG-1[b]	Transcription, RNA stability, translation control
DEK	–	LEF-1	Exon-junction complexes
hnRNP Q	–	LEF-1, Rev-erbα	RNA stability, translation
hnRNP L	–	c-myc, VEGF-A, cat-1	RNA stability, translation
hnRNP A1	Stimulation	LEF-1[b], FGF-2, VEGF, XIAP, Bcl-x$_L$, Apaf-1, c-myc, unr, cyclin D1	RNA processing, export, translation
hnRNP A2	–	HIF-1α	RNA processing
hnRNP C1/C2	Stimulation	LEF-1[b], XIAP, c-myc, unr, PDGF2/c-Sis, IGF1 receptor	RNA processing, transport
NF45	Stimulation	cIAP1	Transcription control
Unr	Stimulation	Apaf-1, PITSLREp58, ODC	Translation control
DAP5	Stimulation	Apaf-1, c-IAP1, c-myc, DAP5	Translation control, apoptosis
GRSF	Stimulation	c-myc, Apaf-1[b], BAG-1[b]	Splicing, transport
Nucleolin	Stimulation	HIF-1α	rRNA maturation, transport
DKC1	Stimulation	XIAP, Bcl-XL	rRNA modification
FMRP	Stimulation	SOD1	Neuron translation control
JKTBP1/hnRNP D-like	–	NRF	RNA stability
ZNF9	–	ODC	Transcription, translation

[a]Suppression
[b]No effect

Comparison of the profiles obtained in tagged-IRES to tagged-cap RNAs revealed the enrichment of 36 factors in the IRES-tagged RNAs (Table 6.2). Some of the validated IRES-interacting proteins by western blot analysis of UV-cross-linked complexes under stringent conditions, the splicing factor-related protein proline and glutamine-rich SFPQ/PSF, the non-POU domain-containing octamer binding nuclear RNA-binding protein (nonO/p54nrb), PCBP2, and HuR are multifunctional nuclear proteins found in complexes associated with various RNA processes.

Interestingly, some of these factors are also known to control the activity of viral IRES elements (Table 6.1) and various cellular IRES elements. Factors with important activities in translation, eIF4A, eIF2A, eIF3g, and ribosomal proteins RPS7, RPS19, RPL26, were also enriched with the IRES-tagged RNA. A functional link between validated proteins and IRES activity, measured by siRNA depletion, verified the involvement of factor SFPQ in translation control (Tsai et al. 2011). This approach also led to the identification of novel IRES-binding factors, illustrated by the DNA-binding oncogene DEK (Table 6.2), suggesting that the identified proteins might work within large RNA-protein complexes. As already mentioned, protein–protein interactions, such as Y-box-binding protein (YB-1)/PTB, may contribute to the upregulation of c-myc expression (Cobbol et al. 2010).

Different cellular IRES elements reveal distinct responses to stress conditions inhibitory of global translation. For example, the unr, c-myc, and serine/threonine-protein kinase PITSLREp58 IRES elements are activated during mitosis (Schepens et al. 2007; Tinton et al. 2005), while during apoptosis the Apaf IRES is activated and the XIAP is inhibited (Ungureanu et al. 2006). These differential responses have been attributed to changes in regulatory ITAFs although the precise mechanism underlying ITAF function remains elusive. Similarly, stress-dependent modifications or relocalization of hnRNP A1 has been suggested to mediate internal initiation of c-myc, unr, cyclin D1, vascular endothelial growth factor (VEGF), fibroblast growth factor (FGF-2), Apaf-1, and XIAP mRNAs (Bonnal et al. 2005; Cammas et al. 2007; Jo et al. 2008; Lewis et al. 2007; Shi et al. 2011). On the other hand, proteins such as HuR appear to exert a stimulatory role on apoptotic genes (Durie et al. 2011), while IGF1 receptor or the thrombomodulin (TM) are repressed (Blume et al. 2010; Yeh et al. 2008) by unknown mechanisms.

The frequent identification of multifunctional proteins, such as PTB or PCBP2, interacting with different IRES elements suggests a mechanism for the coordinated regulation of translation initiation of a subset of mRNAs bound by shuttling proteins such as hnRNPs and splicing factors. Conversely, IRES activity of c-myc, Apaf-1, unr, and PITSLRE mRNAs is differentially regulated depending on the unr-partners, hnRNP K/polyr(C) binding protein PCBP1-2, or hnRNP C1-2, respectively (Evans et al. 2003; Mitchell et al. 2003; Schepens et al. 2007). This sort of differential regulation suggested that specialized subsets of RNA-binding proteins might exert their function in translation control by binding to the IRES region of specific cellular mRNAs during splicing complex assembly before nuclear export (Sawicka et al. 2008).

In contrast to the general ITAFs mentioned above, death-associated protein (DAP5), NF45, G-rich RNA sequence binding factor (GRSF-1), fragile-X mental retardation protein (FMRP), dyskeratosis congenita (DKC1), heterogeneous nuclear ribonucleoprotein D-like protein (JKTBP1), or zinc-finger protein (ZNF9) seems to be more specific (Gerbasi and Link 2007; Graber et al. 2010; Lewis et al. 2008; Omnus et al. 2011; Sammons et al. 2010). Examples are DAP5 and DKC1 that appear to mediate apoptotic IRES activity, whereas FMRP seems to be specific for translation control in neurons (Bechara et al. 2009). In summary, individual mRNAs seem to use different mechanisms to evade the repression of protein synthesis.

6.5 Concluding Remarks

ITAFs are RNA-binding proteins that shuttle between the nucleus and the cytoplasm, activating or repressing the expression of proteins critical in the cellular response to growth, nutritional, environmental, and proliferation signals. With a few exceptions of ITAFs interacting with viral IRES, the precise role played by ITAFs is not well understood. It has been proposed that ITAFs help to remodel the RNA structure in a way that enhances its affinity for components of the translation machinery, or that they substitute some canonical factors providing functional bridges between the mRNA and the ribosomal subunits. Elucidating the function of ITAFs demands a deep understanding of their RNA targets and their protein partners with potential modifications. In this regard, implementation of proteomic approaches will continue to help understanding the integrated action of ITAFs on mRNA targets.

Acknowledgments This work was supported by grant BFU2011-25437, CSD2009-00080, and by an Institutional grant from Fundación Ramón Areces.

References

Andreev DE, Fernandez-Miragall O, Ramajo J, Dmitriev SE, Terenin IM, Martinez-Salas E, Shatsky IN (2007) Differential factor requirement to assemble translation initiation complexes at the alternative start codons of foot-and-mouth disease virus RNA. RNA 13:1366–1374

Balvay L, Soto Rifo R, Ricci EP, Decimo D, Ohlmann T (2009) Structural and functional diversity of viral IRESes. Biochim Biophys Acta 1789:542–557

Battle DJ, Lau CK, Wan L, Deng H, Lotti F, Dreyfuss G (2006) The Gemin5 protein of the SMN complex identifies snRNAs. Mol Cell 23:273–279

Bechara EG, Didiot MC, Melko M, Davidovic L, Bensaid M, Martin P, Castets M, Pognonec P, Khandjian EW, Moine H, Bardoni B (2009) A novel function for fragile X mental retardation protein in translational activation. PLoS Biol 7:e16

Bedard KM, Daijogo S, Semler BL (2007) A nucleo-cytoplasmic SR protein functions in viral IRES-mediated translation initiation. EMBO J 26:459–467

Belsham GJ (2009) Divergent picornavirus IRES elements. Virus Res 139:183–192

Blume SW, Jackson NL, Frost AR, Grizzle WE, Shcherbakov OD, Choi H, Meng Z (2010) Northwestern profiling of potential translation-regulatory proteins in human breast epithelial cells and malignant breast tissues: evidence for pathological activation of the IGF1R IRES. Exp Mol Pathol 88:341–352

Bonnal S, Pileur F, Orsini C, Parker F, Pujol F, Prats AC, Vagner S (2005) Heterogeneous nuclear ribonucleoprotein A1 is a novel internal ribosome entry site trans-acting factor that modulates alternative initiation of translation of the fibroblast growth factor 2 mRNA. J Biol Chem 280:4144–4153

Boussadia O, Niepmann M, Creancier L, Prats AC, Dautry F, Jacquemin-Sablon H (2003) Unr is required in vivo for efficient initiation of translation from the internal ribosome entry sites of both rhinovirus and poliovirus. J Virol 77:3353–3359

Bradrick SS, Gromeier M (2009) Identification of gemin5 as a novel 7-methylguanosine cap-binding protein. PLoS One 4:e7030

Buratti E, Tisminetzky S, Zotti M, Baralle FE (1998) Functional analysis of the interaction between HCV 5′UTR and putative subunits of eukaryotic translation initiation factor eIF3. Nucleic Acids Res 26:3179–3187

Bushell M, Stoneley M, Kong YW, Hamilton TL, Spriggs KA, Dobbyn HC, Qin X, Sarnow P, Willis AE (2006) Polypyrimidine tract binding protein regulates IRES-mediated gene expression during apoptosis. Mol Cell 23:401–412

Cammas A, Pileur F, Bonnal S, Lewis SM, Leveque N, Holcik M, Vagner S (2007) Cytoplasmic relocalization of heterogeneous nuclear ribonucleoprotein A1 controls translation initiation of specific mRNAs. Mol Biol Cell 18:5048–5059

Chen HC, Lin WC, Tsay YG, Lee SC, Chang CJ (2002) An RNA helicase, DDX1, interacting with poly(A) RNA and heterogeneous nuclear ribonucleoprotein K. J Biol Chem 277: 40403–40409

Choi K, Kim JH, Li X, Paek KY, Ha SH, Ryu SH, Wimmer E, Jang SK (2004) Identification of cellular proteins enhancing activities of internal ribosomal entry sites by competition with oligodeoxynucleotides. Nucleic Acids Res 32:1308–1317

Cobbol LC, Wilson LA, Sawicka K, King HA, Kondrashov AV, Spriggs KA, Bushell M, Willis AE (2010) Upregulated c-myc expression in multiple myeloma by internal ribosome entry results from increased interactions with and expression of PTB-1 and YB-1. Oncogene 29: 2884–2891

Cobbold LC, Spriggs KA, Haines SJ, Dobbyn HC, Hayes C, de Moor CH, Lilley KS, Bushell M, Willis AE (2008) Identification of internal ribosome entry segment (IRES)-trans-acting factors for the Myc family of IRESs. Mol Cell Biol 28:40–49

Conte MR, Grune T, Ghuman J, Kelly G, Ladas A, Matthews S, Curry S (2000) Structure of tandem RNA recognition motifs from polypyrimidine tract binding protein reveals novel features of the RRM fold. EMBO J 19:3132–3141

Conte C, Riant E, Toutain C, Pujol F, Arnal JF, Lenfant F, Prats AC (2008) FGF2 translationally induced by hypoxia is involved in negative and positive feedback loops with HIF-1alpha. PLoS One 3:e3078

de Breyne S, Yu Y, Unbehaun A, Pestova TV, Hellen CU (2009) Direct functional interaction of initiation factor eIF4G with type 1 internal ribosomal entry sites. Proc Natl Acad Sci U S A 106:9197–9202

Dobbyn HC, Hill K, Hamilton TL, Spriggs KA, Pickering BM, Coldwell MJ, de Moor CH, Bushell M, Willis AE (2008) Regulation of BAG-1 IRES-mediated translation following chemotoxic stress. Oncogene 27:1167–1174

Durie D, Lewis SM, Liwak U, Kisilewicz M, Gorospe M, Holcik M (2011) RNA-binding protein HuR mediates cytoprotection through stimulation of XIAP translation. Oncogene 30:1460–1469

Evans JR, Mitchell SA, Spriggs KA, Ostrowski J, Bomsztyk K, Ostarek D, Willis AE (2003) Members of the poly (rC) binding protein family stimulate the activity of the c-myc internal ribosome entry segment in vitro and in vivo. Oncogene 22:8012–8020

Fernandez-Miragall O, Lopez de Quinto S, Martinez-Salas E (2009) Relevance of RNA structure for the activity of picornavirus IRES elements. Virus Res 139:172–182

Filbin ME, Kieft JS (2009) Toward a structural understanding of IRES RNA function. Curr Opin Struct Biol 1:267–276

Fitzgerald KD, Semler BL (2009) Bridging IRES elements in mRNAs to the eukaryotic translation apparatus. Biochim Biophys Acta 1789:518–528

Florez PM, Sessions OM, Wagner EJ, Gromeier M, Garcia-Blanco MA (2005) The polypyrimidine tract binding protein is required for efficient picornavirus gene expression and propagation. J Virol 79:6172–6179

Fox JT, Stover PJ (2009) Mechanism of the internal ribosome entry site-mediated translation of serine hydroxymethyltransferase 1. J Biol Chem 284:31085–31096

Fukushi S, Okada M, Stahl J, Kageyama T, Hoshino FB, Katayama K (2001) Ribosomal protein S5 interacts with the internal ribosomal entry site of hepatitis C virus. J Biol Chem 276: 20824–20826

Gamarnik AV, Boddeker N, Andino R (2000) Translation and replication of human rhinovirus type 14 and mengovirus in Xenopus oocytes. J Virol 74:11983–11987

Gerbasi VR, Link AJ (2007) The myotonic dystrophy type 2 protein ZNF9 is part of an ITAF complex that promotes cap-independent translation. Mol Cell Proteomics 6:1049–1058

Gosert R, Chang KH, Rijnbrand R, Yi M, Sangar DV, Lemon SM (2000) Transient expression of cellular polypyrimidine-tract binding protein stimulates cap-independent translation directed by both picornaviral and flaviviral internal ribosome entry sites in vivo. Mol Cell Biol 20:1583–1595

Graber TE, Baird SD, Kao PN, Mathews MB, Holcik M (2010) NF45 functions as an IRES trans-acting factor that is required for translation of cIAP1 during the unfolded protein response. Cell Death Differ 17:719–729

Grover R, Ray PS, Das S (2008) Polypyrimidine tract binding protein regulates IRES-mediated translation of p53 isoforms. Cell Cycle 7:2189–2198

Hahm B, Kim YK, Kim JH, Kim TY, Jang SK (1998) Heterogeneous nuclear ribonucleoprotein L interacts with the 3' border of the internal ribosomal entry site of hepatitis C virus. J Virol 72:8782–8788

Hao le T, Fuller HR, le Lam T, Le TT, Burghes AH, Morris GE (2007) Absence of gemin5 from SMN complexes in nuclear Cajal bodies. BMC Cell Biol 8:28

Hartman TR, Qian S, Bolinger C, Fernandez S, Schoenberg DR, Boris-Lawrie K (2006) RNA helicase A is necessary for translation of selected messenger RNAs. Nat Struct Mol Biol 13:509–516

Henis-Korenblit S, Shani G, Sines T, Marash L, Shohat G, Kimchi A (2002) The caspase-cleaved DAP5 protein supports internal ribosome entry site-mediated translation of death proteins. Proc Natl Acad Sci U S A 99:5400–5405

Holcik M, Korneluk RG (2000) Functional characterization of the X-linked inhibitor of apoptosis (XIAP) internal ribosome entry site element: role of La autoantigen in XIAP translation. Mol Cell Biol 20:4648–4657

Hunt SL, Hsuan JJ, Totty N, Jackson RJ (1999) unr, a cellular cytoplasmic RNA-binding protein with five cold-shock domains, is required for internal initiation of translation of human rhinovirus RNA. Genes Dev 13:437–448

Jang SK, Wimmer E (1990) Cap-independent translation of encephalomyocarditis virus RNA: structural elements of the internal ribosomal entry site and involvement of a cellular 57-kD RNA-binding protein. Genes Dev 4:1560–1572

Jang SK, Krausslich HG, Nicklin MJ, Duke GM, Palmenberg AC, Wimmer E (1988) A segment of the 5' nontranslated region of encephalomyocarditis virus RNA directs internal entry of ribosomes during in vitro translation. J Virol 62:2636–2643

Ji H, Fraser CS, Yu Y, Leary J, Doudna JA (2004) Coordinated assembly of human translation initiation complexes by the hepatitis C virus internal ribosome entry site RNA. Proc Natl Acad Sci U S A 101:16990–16995

Jo OD, Martin J, Bernath A, Masri J, Lichtenstein A, Gera J (2008) Heterogeneous nuclear ribonucleoprotein A1 regulates cyclin D1 and c-myc internal ribosome entry site function through Akt signaling. J Biol Chem 283:23274–23287

Kafasla P, Morgner N, Poyry TA, Curry S, Robinson CV, Jackson RJ (2009) Polypyrimidine tract binding protein stabilizes the encephalomyocarditis virus IRES structure via binding multiple sites in a unique orientation. Mol Cell 34:556–568

Kieft JS, Zhou K, Jubin R, Doudna JA (2001) Mechanism of ribosome recruitment by hepatitis C IRES RNA. RNA 7:194–206

Kim JH, Hahm B, Kim YK, Choi M, Jang SK (2000a) Protein-protein interaction among hnRNPs shuttling between nucleus and cytoplasm. J Mol Biol 298:395–405

Kim YK, Hahm B, Jang SK (2000b) Polypyrimidine tract-binding protein inhibits translation of bip mRNA. J Mol Biol 304:119–133

Kolupaeva VG, Pestova TV, Hellen CU, Shatsky IN (1998) Translation eukaryotic initiation factor 4G recognizes a specific structural element within the internal ribosome entry site of encephalomyocarditis virus RNA. J Biol Chem 273:18599–18604

Komar AA, Hatzoglou M (2011) Cellular IRES-mediated translation: the war of ITAFs in pathophysiological states. Cell Cycle 10:229–240

Kukalev A, Nord Y, Palmberg C, Bergman T, Percipalle P (2005) Actin and hnRNP U cooperate for productive transcription by RNA polymerase II. Nat Struct Mol Biol 12:238–244

Lee PT, Liao PC, Chang WC, Tseng JT (2007) Epidermal growth factor increases the interaction between nucleolin and heterogeneous nuclear ribonucleoprotein K/poly(C) binding protein 1 complex to regulate the gastrin mRNA turnover. Mol Biol Cell 18:5004–5013

Lewis SM, Holcik M (2008) For IRES trans-acting factors, it is all about location. Oncogene 27:1033–1035

Lewis SM, Veyrier A, Hosszu Ungureanu N, Bonnal S, Vagner S, Holcik M (2007) Subcellular relocalization of a trans-acting factor regulates XIAP IRES-dependent translation. Mol Biol Cell 18:1302–1311

Lewis SM, Cerquozzi S, Graber TE, Ungureanu NH, Andrews M, Holcik M (2008) The eIF4G homolog DAP5/p97 supports the translation of select mRNAs during endoplasmic reticulum stress. Nucleic Acids Res 36:168–178

Lin JY, Li ML, Shih SR (2009a) Far upstream element binding protein 2 interacts with enterovirus 71 internal ribosomal entry site and negatively regulates viral translation. Nucleic Acids Res 37:47–59

Lin JY, Shih SR, Pan M, Li C, Lue CF, Stollar V, Li ML (2009b) hnRNP A1 interacts with the 5′ untranslated regions of enterovirus 71 and Sindbis virus RNA and is required for viral replication. J Virol 83:6106–6114

Liu HM, Aizaki H, Choi KS, Machida K, Ou JJ, Lai MM (2009) SYNCRIP (synaptotagmin-binding, cytoplasmic RNA-interacting protein) is a host factor involved in hepatitis C virus RNA replication. Virology 386:249–256

Liu J, Henao-Mejia J, Liu H, Zhao Y, He JJ (2011) Translational regulation of HIV-1 replication by HIV-1 Rev cellular cofactors Sam68, eIF5A, hRIP, and DDX3. J Neuroimmune Pharmacol 6:308–321

Lopez de Quinto S, Martinez-Salas E (2000) Interaction of the eIF4G initiation factor with the aphthovirus IRES is essential for internal translation initiation in vivo. RNA 6:1380–1392

Lopez de Quinto S, Lafuente E, Martinez-Salas E (2001) IRES interaction with translation initiation factors: functional characterization of novel RNA contacts with eIF3, eIF4B, and eIF4GII. RNA 7:1213–1226

Lu H, Li W, Noble WS, Payan D, Anderson DC (2004) Riboproteomics of the hepatitis C virus internal ribosomal entry site. J Proteome Res 3:949–957

Lukavsky PJ (2009) Structure and function of HCV IRES domains. Virus Res 139:166–171

Lunde BM, Moore C, Varani G (2007) RNA-binding proteins: modular design for efficient function. Nat Rev Mol Cell Biol 8:479–490

Majumder M, Yaman I, Gaccioli F, Zeenko VV, Wang C, Caprara MG, Venema RC, Komar AA, Snider MD, Hatzoglou M (2009) The hnRNA-binding proteins hnRNP L and PTB are required for efficient translation of the Cat-1 arginine/lysine transporter mRNA during amino acid starvation. Mol Cell Biol 29:2899–2912

Makeyev AV, Liebhaber SA (2002) The poly(C)-binding proteins: a multiplicity of functions and a search for mechanisms. RNA 8:265–278

Martinez-Salas E (2008) The impact of RNA structure on picornavirus IRES activity. Trends Microbiol 16:230–237

Martinez-Salas E, Pacheco A, Serrano P, Fernandez N (2008) New insights into internal ribosome entry site elements relevant for viral gene expression. J Gen Virol 89:611–626

Meerovitch K, Svitkin YV, Lee HS, Lejbkowicz F, Kenan DJ, Chan EK, Agol VI, Keene JD, Sonenberg N (1993) La autoantigen enhances and corrects aberrant translation of poliovirus RNA in reticulocyte lysate. J Virol 67:3798–3807

Merrill MK, Dobrikova EY, Gromeier M (2006) Cell-type-specific repression of internal ribosome entry site activity by double-stranded RNA-binding protein 76. J Virol 80:3147–3156

Mitchell SA, Brown EC, Coldwell MJ, Jackson RJ, Willis AE (2001) Protein factor requirements of the Apaf-1 internal ribosome entry segment: roles of polypyrimidine tract binding protein and upstream of N-ras. Mol Cell Biol 21:3364–3374

Mitchell SA, Spriggs KA, Coldwell MJ, Jackson RJ, Willis AE (2003) The Apaf-1 internal ribosome entry segment attains the correct structural conformation for function via interactions with PTB and unr. Mol Cell 11:757–771

Mondal T, Ray U, Manna AK, Gupta R, Roy S, Das S (2008) Structural determinant of human La protein critical for internal initiation of translation of hepatitis C virus RNA. J Virol 82:11927–11938

Monie TP, Perrin AJ, Birtley JR, Sweeney TR, Karakasiliotis I, Chaudhry Y, Roberts LO, Matthews S, Goodfellow IG, Curry S (2007) Structural insights into the transcriptional and translational roles of Ebp1. EMBO J 26:3936–3944

Omnus DJ, Mehrtens S, Ritter B, Resch K, Yamada M, Frank R, Nourbakhsh M, Reboll MR (2011) JKTBP1 is involved in stabilization and IRES-dependent translation of NRF mRNAs by binding to 5′ and 3′ untranslated regions. J Mol Biol 407:492–504

Pacheco A, Martinez-Salas E (2010) Insights into the biology of IRES elements through riboproteomic approaches. J Biomed Biotechnol 2010:458927

Pacheco A, Reigadas S, Martinez-Salas E (2008) Riboproteomic analysis of polypeptides interacting with the internal ribosome-entry site element of foot-and-mouth disease viral RNA. Proteomics 8:4782–4790

Pacheco A, Lopez de Quinto S, Ramajo J, Fernandez N, Martinez-Salas E (2009) A novel role for Gemin5 in mRNA translation. Nucleic Acids Res 37:582–590

Park SM, Paek KY, Hong KY, Jang CJ, Cho S, Park JH, Kim JH, Jan E, Jang SK (2011) Translation-competent 48S complex formation on HCV IRES requires the RNA-binding protein NSAP1. Nucleic Acids Res 39(17):7791–7802

Pelletier J, Sonenberg N (1988) Internal initiation of translation of eukaryotic mRNA directed by a sequence derived from poliovirus RNA. Nature 334:320–325

Percipalle P, Jonsson A, Nashchekin D, Karlsson C, Bergman T, Guialis A, Daneholt B (2002) Nuclear actin is associated with a specific subset of hnRNP A/B-type proteins. Nucleic Acids Res 30:1725–1734

Pestova TV, Shatsky IN, Fletcher SP, Jackson RJ, Hellen CU (1998) A prokaryotic-like mode of cytoplasmic eukaryotic ribosome binding to the initiation codon during internal translation initiation of hepatitis C and classical swine fever virus RNAs. Genes Dev 12:67–83

Pestova TV, Kolupaeva VG, Lomakin IB, Pilipenko EV, Shatsky IN, Agol VI, Hellen CU (2001) Molecular mechanisms of translation initiation in eukaryotes. Proc Natl Acad Sci U S A 98:7029–7036

Pilipenko EV, Pestova TV, Kolupaeva VG, Khitrina EV, Poperechnaya AN, Agol VI, Hellen CU (2000) A cell cycle-dependent protein serves as a template-specific translation initiation factor. Genes Dev 14:2028–2045

Pilipenko EV, Viktorova EG, Guest ST, Agol VI, Roos RP (2001) Cell-specific proteins regulate viral RNA translation and virus-induced disease. EMBO J 20:6899–6908

Rivas-Aravena A, Ramdohr P, Vallejos M, Valiente-Echeverria F, Dormoy-Raclet V, Rodriguez F, Pino K, Holzmann C, Huidobro-Toro JP, Gallouzi IE, Lopez-Lastra M (2009) The Elav-like protein HuR exerts translational control of viral internal ribosome entry sites. Virology 392:178–185

Sammons MA, Antons AK, Bendjennat M, Udd B, Krahe R, Link AJ (2010) ZNF9 activation of IRES-mediated translation of the human ODC mRNA is decreased in myotonic dystrophy type 2. PLoS One 5:e9301

Sawicka K, Bushell M, Spriggs KA, Willis AE (2008) Polypyrimidine-tract-binding protein: a multifunctional RNA-binding protein. Biochem Soc Trans 36:641–647

Schepens B, Tinton SA, Bruynooghe Y, Beyaert R, Cornelis S (2005) The polypyrimidine tract-binding protein stimulates HIF-1alpha IRES-mediated translation during hypoxia. Nucleic Acids Res 33:6884–6894

Schepens B, Tinton SA, Bruynooghe Y, Parthoens E, Haegman M, Beyaert R, Cornelis S (2007) A role for hnRNP C1/C2 and Unr in internal initiation of translation during mitosis. EMBO J 26:158–169

Schultz DE, Hardin CC, Lemon SM (1996) Specific interaction of glyceraldehyde 3-phosphate dehydrogenase with the 5′-nontranslated RNA of hepatitis A virus. J Biol Chem 271:14134–14142

Sean P, Nguyen JH, Semler BL (2009) Altered interactions between stem-loop IV within the 5′ noncoding region of coxsackievirus RNA and poly(rC) binding protein 2: effects on IRES-mediated translation and viral infectivity. Virology 389:45–58

Sengupta J, Nilsson J, Gursky R, Spahn CM, Nissen P, Frank J (2004) Identification of the versatile scaffold protein RACK1 on the eukaryotic ribosome by cryo-EM. Nat Struct Mol Biol 11:957–962

Shi Y, Frost P, Hoang B, Benavides A, Gera J, Lichtenstein A (2011) IL-6-induced enhancement of c-Myc translation in multiple myeloma cells: critical role of cytoplasmic localization of the RNA-binding protein hnRNP A1. J Biol Chem 286:67–78

Shih SR, Stollar V, Li ML (2011) Host factors in EV71 replication. J Virol 85:9658–9666 June 29 (ahead of print)

Siridechadilok B, Fraser CS, Hall RJ, Doudna JA, Nogales E (2005) Structural roles for human translation factor eIF3 in initiation of protein synthesis. Science 310:1513–1515

Song Y, Friebe P, Tzima E, Junemann C, Bartenschlager R, Niepmann M (2006) The hepatitis C virus RNA 3′-untranslated region strongly enhances translation directed by the internal ribosome entry site. J Virol 80:11579–11588

Spriggs KA, Bushell M, Mitchell SA, Willis AE (2005) Internal ribosome entry segment-mediated translation during apoptosis: the role of IRES-trans-acting factors. Cell Death Differ 12: 585–591

Spriggs KA, Cobbold LC, Jopling CL, Cooper RE, Wilson LA, Stoneley M, Coldwell M, Poncet D, Shen YC, Morley SJ, Bushell M, Willis AE (2009) Canonical initiation factor requirements of the Myc family of internal ribosome entry site segments. Mol Cell Biol 29:1565–1574

Spriggs KA, Bushell M, Willis AE (2010) Translational regulation of gene expression during conditions of cell stress. Mol Cell 40:228–237

Terenin IM, Dmitriev SE, Andreev DE, Shatsky IN (2008) Eukaryotic translation initiation machinery can operate in a bacterial-like mode without eIF2. Nat Struct Mol Biol 15:836–841

Tinton SA, Schepens B, Bruynooghe Y, Beyaert R, Cornelis S (2005) Regulation of the cell-cycle-dependent internal ribosome entry site of the PITSLRE protein kinase: roles of Unr (upstream of N-ras) protein and phosphorylated translation initiation factor eIF-2alpha. Biochem J 385:155–163

Tsai BP, Wang X, Huang L, Waterman ML (2011) Quantitative profiling of in vivo-assembled RNA-protein complexes using a novel integrated proteomic approach. Mol Cell Proteomics 10(M110):007385

Ungureanu NH, Cloutier M, Lewis SM, de Silva N, Blais JD, Bell JC, Holcik M (2006) Internal ribosome entry site-mediated translation of Apaf-1, but not XIAP, is regulated during UV-induced cell death. J Biol Chem 281:15155–15163

Vallejos M, Deforges J, Plank TD, Letelier A, Ramdohr P, Abraham CG, Valiente-Echeverria F, Kieft JS, Sargueil B, Lopez-Lastra M (2011) Activity of the human immunodeficiency virus type 1 cell cycle-dependent internal ribosomal entry site is modulated by IRES trans-acting factors. Nucleic Acids Res 108(5):1839–1844

Waggoner S, Sarnow P (1998) Viral ribonucleoprotein complex formation and nucleolar-cytoplasmic relocalization of nucleolin in poliovirus-infected cells. J Virol 72:6699–6709

Weinlich S, Huttelmaier S, Schierhorn A, Behrens SE, Ostareck-Lederer A, Ostareck DH (2009) IGF2BP1 enhances HCV IRES-mediated translation initiation via the 3′UTR. RNA 15: 1528–1542

Yeh CH, Hung LY, Hsu C, Le SY, Lee PT, Liao WL, Lin YT, Chang WC, Tseng JT (2008) RNA-binding protein HuR interacts with thrombomodulin 5′ untranslated region and represses internal ribosome entry site-mediated translation under IL-1 beta treatment. Mol Biol Cell 19:3812–3822

Yu Y, Ji H, Doudna JA, Leary JA (2005) Mass spectrometric analysis of the human 40S ribosomal subunit: native and HCV IRES-bound complexes. Protein Sci 14:1438–1446

Chapter 7
Rapid Kinetic Analysis of Protein Synthesis

Marina V. Rodnina and Wolfgang Wintermeyer

7.1 Introduction

Protein synthesis comprises a sequence of consecutive reactions that entail recognition of substrates by the ribosome, synthesis of chemical bonds, and release of the reaction products. Each of the translation phases—initiation, elongation, termination, and ribosome recycling—is itself a multistep process. To achieve the overall high rate of protein synthesis, with ribosomes initiating roughly every 3 s on an mRNA and about 0.1 s required for each cycle of nascent peptide elongation, each of the individual reactions on the ribosome must be rapid. This implies that the reaction intermediates of protein synthesis are short-lived, with lifetimes in the few milliseconds range. The overall process is strongly biased towards forward reactions due to the combination of dynamic fluctuations with irreversible steps of GTP hydrolysis and peptide bond formation. To study such reactions, transient kinetic techniques are particularly suitable, as they can reveal sequences of interactions and allow monitoring the formation and consumption of the reaction intermediates of protein synthesis in real time.

Rapid kinetic analysis has been utilized to study essentially every step of proteins synthesis. In the first part of this review, we consider the basics of transient kinetic techniques. The second part gives an overview of two cases where the results of rapid kinetic studies provided essential insights into the mechanisms of partial reactions of protein synthesis. We will describe how rapid kinetic analyses contributed to understanding the mechanisms of chemical reactions catalyzed by the active site on the 50S ribosomal subunit, i.e., peptide bond formation and peptidyl-tRNA hydrolysis, during the elongation and termination phases of protein synthesis,

M.V. Rodnina (✉) • W. Wintermeyer
Max Planck Institute for Biophysical Chemistry, 37077 Goettingen, Germany
e-mail: rodnina@mpibpc.mpg.de; wolfgang.wintermeyer@mpibpc.mpg.de

J.D. Dinman (ed.), *Biophysical Approaches to Translational Control of Gene Expression*, 119
Biophysics for the Life Sciences 1, DOI 10.1007/978-1-4614-3991-2_7,
© Springer Science+Business Media New York 2013

respectively. To illustrate the power of the stopped-flow technique, we describe fluorescence experiments that lead to understanding the distinct roles of the elongation factor G (EF-G) in the catalysis of tRNA-mRNA translocation during elongation and subunit splitting in the ribosome recycling phase.

7.2   Rapid Kinetic Techniques

Rapid kinetic measurements allow for the detection of transient reaction intermediates. Following the time courses of the formation of intermediates and their consumption in subsequent reactions provides the sequence of events as well as reaction rate constants (for a comprehensive overview, see Johnson 2003). Among the rapid kinetic methods that have been used to study protein synthesis, stopped-flow and quench-flow techniques are particularly important. In a stopped-flow apparatus, two reactants are rapidly mixed and a change in an optical parameter is monitored, such as fluorescence, absorption, or light scattering. Commercially available machines provide a reliable signal 1 ms after mixing (the dead-time of the apparatus) and allows to follow reactions with rates of up to 500 s^{-1}. In quench-flow experiments, which in studies of protein synthesis are usually used to determine the rates of chemical reactions, reactions are initiated and stopped (quenched) rapidly, and the products are quantified offline by a variety of analytical methods. The minimum reaction time of commercial quench-flow machines is close to 3 ms; higher time resolution can be reached by using the continuous-flow mode, albeit at the expense of the consumption of large amounts of substrate. Recently, a number of additional rapid kinetic techniques have been established, such as rapid filtration, time-resolved cross-linking, or footprinting (Fabbretti et al. 2007). Flash photolysis, which has been successfully utilized in other fields, is another yet underexploited method to study translation intermediates. In the following, we focus on the standard stopped-flow and quench-flow setups which are broadly used in ribosome research.

Dissecting a reaction mechanism by means of transient kinetics entails several steps. The first step is the choice of the technique, stopped-flow or quench flow, and the reporter groups appropriate to study the particular reaction, for stopped-flow usually fluorescent groups. The positions of labels should be chosen using structural information (wherever available) and sequence conservation alignments. Because the quality of the signal depends critically on the position and the spectroscopic properties of the fluorescence label, it is advantageous to try different fluorophores and various positions in proteins/RNA for labeling. The functional properties of the labeled molecules have to be characterized by biochemical methods. The second step is to carefully choose the reaction conditions for the stopped-flow or quench-flow experiments. This step is very important because the reaction conditions, in particular the concentrations of reactants, determine the complexity of the kinetic data obtained. Finally, the evaluation of the data should result in a kinetic model that can be tested by further experiments.

7.3   Fluorescence Labeling

For proteins, the most common approach is to introduce single cysteine residues, preferably at non-conserved surface positions, and use thiol-reactive derivatives of fluorophores to attach fluorescent groups at the desired positions. Intrinsic cysteines in the protein to be labeled should be removed, unless they are buried and inaccessible to the labeling reagents. Another approach, which avoids mutagenesis, is to introduce fluorescence labels in a random way by labeling lysine or cysteine residues at the surface of the protein. Usually, there are many potentially reactive residues; to avoid detrimental effects of extensive modification on the activity of the protein, labeling should be limited to a few residues per molecule on the average. One potential disadvantage of the approach is that the labeling is likely to result in a heterogeneous pool of molecules with dyes attached at different positions which may report different rearrangements, resulting in multiple, often poorly defined kinetic phases. Alternatively, fluorescence tags can be introduced at the N- or C-terminus of the proteins; however, the potential of such labels for monitoring conformational rearrangements along the reaction coordinate has been poorly explored.

RNA can be labeled in a number of ways, e.g., at the 5′ or 3′ ends or at internal positions using fluorescent nucleotide derivatives, many of which are commercially available. In the growing field of RNA chemistry, new methods emerge rapidly. An exciting new possibility is to introduce small tags that specifically bind fluorophores, such as the tetracysteine tag (Lumio tag). For labeling natural tRNA, modifications provide a number of useful reactive groups, two of which, thioU at position 8 (Johnson et al. 1982) and dihydroU in the D loop of tRNA (Wintermeyer and Zachau 1974), are commonly found in bacterial tRNAs; an asp^{3}U modification at position 47 of some tRNAs is often used as well. Ribosomes can be labeled in a variety of ways. In a relatively simple approach, the ribosomes are labeled by supplying a fluorescence-labeled ribosomal protein to ribosomes that lack the respective protein because it was deleted from the chromosome in the bacterial strain used as a source of ribosomes. So far, this approach has been successfully applied to proteins L1, L11, and L29 (Munro et al. 2010; Seo et al. 2006). Ribosomal protein L12 can be easily removed from the ribosomes by ethanol/salt treatment and replaced by the purified fluorescence-labeled protein. More complicated labeling protocols suggest full reconstitution of ribosomal subunits using purified components (Hickerson et al. 2005), one of which is labeled, or use labeled oligonucleotides base-paired to engineered extensions in rRNA (Dorywalska et al. 2005).

Methods for fluorescence labeling are well-established and often can be downloaded from the web-sites of the companies from which the dye is purchased. However, an optimization of labeling conditions for each particular protein/RNA-dye pair is recommended, as it may improve both yield and purity of labeled product. Because labeling may alter the properties of the target macromolecules, the functional activity of the labeled components should be tested and compared to the unlabeled species; only those fluorescence derivatives should be used which have properties that are sufficiently similar to the unmodified molecules.

7.4 Measurements

7.4.1 Stopped-Flow

Several observables can be used when working with a stopped-flow apparatus, although fluorescence provides the most sensitive signal, allowing for the use of low reactant concentrations. Fluorescence changes may directly report binding or conformational rearrangements of fluorescence-labeled components. Alternatively, two fluorescence-labeled components can be used which serve as fluorescence donor and acceptor in a FRET measurement, and changes of the FRET efficiency may report complex formation and/or conformational changes. The FRET pair should be chosen in such a way that the emission spectrum of the donor overlaps with the excitation spectrum of acceptor, and the expected distance between the dyes in the complex is close to the critical distance, R_0, of the dye pair which equals the distance at which the FRET efficiency is 50 %. The donor is excited at a wavelength which is optimal for maximum donor excitation and minimum acceptor excitation, and fluorescence emission is measured at the maximum emission of the acceptor in most cases, although donor emission can also be measured. Complications introduced by spectral overlap between donor and acceptor can be avoided by using a nonfluorescent acceptor and monitoring changes of donor fluorescence due to FRET. The FRET method is particularly suitable to follow the kinetics of ligand binding, although subsequent conformational rearrangements are often observed as well. Light scattering is another observable that allows monitoring the association and dissociation of larger macromolecular assemblies. It has proven useful in studies of translation initiation and ribosome recycling (Antoun et al. 2006; Grunberg-Manago et al. 1975; Savelsbergh et al. 2009). Light scattering can be conveniently monitored in a standard stopped-flow device.

7.4.2 Quench Flow

The chemistry steps of protein synthesis, i.e., GTP hydrolysis by translational GTPases or the reactions at the peptidyl transferase center, can be followed in a quench-flow apparatus using radioactively or fluorescence-labeled GTP or amino acids attached to peptidyl-tRNA (pept-tRNA) or aminoacyl-tRNA (aa-tRNA). Reactions are initiated by rapidly mixing the reactants and stopped by the rapid addition of an appropriate quencher, such as a strong acid or base, or EDTA at high concentration. Subsequently, the reaction products are analyzed by HPLC, TLC, or other analytical techniques.

7.4.3 Design and Evaluation of Experiments

The design of a transient kinetics experiment for a multicomponent reaction can be quite complex. Generally, the experiment should be planned in such a way that pseudo-first-order reaction conditions are fulfilled, which greatly simplifies the calculation of rate constants from observed rates. This is accomplished by taking limiting concentrations of the labeled component and adding all other reaction components in large excess, such that the change in concentration of unlabeled components during the reaction is negligible. Apparent rate constants of the reaction (k_{app} or k_{obs}) are measured at constant concentration of the labeled component and increasing excess concentrations of its ligands, and the rate constants of the reaction under study are derived from the analysis of the concentration dependence of the apparent rate constants. In recent years, the complexities of data analysis have been overcome by advances in computational methods to analyze reaction time courses by numerical integration using Matlab or specialized programs (Scientist, Berkeley Madonna, Dynafit, Kinsim, and numerous others). This allows using equimolar concentrations of reagents and determining rate constants by global fitting; however, measuring time courses at a number of different concentrations of reagents is even more important in that case. Data from different approaches should be combined for global analysis of a complex reaction.

7.5 Insights into the Catalytic Center of the Ribosome

The catalytic center of the ribosome has to perform two functions: the catalysis of peptide bond formation during elongation and the hydrolysis of the ester bond in pept-tRNA in the termination phase of protein synthesis. In addition, ribosomes are able to catalyze a number of other, unnatural chemical reactions, e.g., thioester, thioamide, or phosphinoamide formation (Rodnina et al. 2006). Thus, the question arises of how the same active site can be utilized to catalyze such a variety of different chemical reactions. Are the catalytic mechanisms the same or different for peptide bond formation and peptidyl-tRNA hydrolysis? What is the contribution of the ribosome to catalysis?

The recent history of studies on the catalytic mechanisms of the ribosome goes back to 2000 when the first crystal structures of the 50S ribosomal subunit from *Haloarcula marismortui* were obtained in the complex with a transition state analog (Nissen et al. 2000). The structural models have established that the catalytic center consists of rRNA and there are no proteins in the vicinity that could donate functional groups to take part in the reaction. The crystal structures provided an excellent framework to formulate and test hypotheses on the mechanism of catalysis. Because the reactions in vivo must be rapid to account for the measured speed of protein synthesis (about 10 s^{-1}), using rapid kinetic methods is crucial to address these questions. In the following, we summarize those approaches and the outcome of those studies in conjunction to further structural, genetic, and chemical work.

7.5.1   Peptide Bond Formation

On the ribosome, peptide bond formation takes place between two substrates, pept-tRNA in the P site and aa-tRNA in the A site (Fig. 7.1a). The reaction proceeds through the nucleophilic attack of the α-amino group of aa-tRNA on the carbonyl carbon of the peptidyl-tRNA (Fig. 7.1b); from a chemical point of view, the reaction is the aminolysis of an ester. Based on the classical studies with model substrates in solution, the reaction is expected to proceed through two intermediates (Fig. 7.1c). First, a zwitterionic tetrahedral intermediate is formed ($T^{\pm}$). Deprotonation of the positively charged amino nitrogen forms the second intermediate (T^-). In the final step, T^- decomposes to form the reaction products (Satterthwait and Jencks 1974); for peptide bond formation these are deacylated tRNA in the P site of the ribosome and a new pept-tRNA, longer by one amino acid, in the A site.

The ribosome could contribute to catalysis in a number of ways, e.g., by stabilizing charges developing in the transition state (TS) or by acid–base catalysis involving ionizing groups of the ribosome at the active center, analogous to protein enzymes that employ amino acid side chains to shuffle protons in the reactions they catalyze. If the latter were true for the ribosome, the rate of catalysis should depend on pH in a way that reflects the pK_a values of the ionizing group(s) taking part in catalysis. Furthermore, the pH/rate profiles should be sensitive to substitution (mutations) of the presumed catalytic residues in the ribosome's active site. As the catalytic center of the ribosome does not contain proteins, the large repertoire of chemically active groups of amino acids that can act as efficient acid/base catalysts is not available. Instead, rRNA has a very limited choice of groups that would be useful for catalysis at physiological pH, as their pK_a values are far from neutrality. This renders rRNA bases inefficient as acid/base catalysts, unless their pK_a values were shifted towards neutrality in the environment of the ribosome. These considerations provide testable predictions which can be addressed by kinetic studies following the rate-pH-dependence of the peptidyl transfer reaction. These experiments illustrate the use of the quench-flow technique for addressing a reaction mechanism. The setup for these experiments was quite simple: Pept-tRNA and aa-tRNA were allowed to react for a defined time at a given pH, the reaction was stopped, and the concentrations of reactants and products were analyzed for each time point. Using radioactively labeled substrates, e.g., [^{14}C] and [^{3}H]-labeled amino acids, made the analysis straightforward and accurate (Fig. 7.1d) (Katunin et al. 2002a; Wohlgemuth et al. 2008).

One important issue for mechanistic studies of peptide bond formation has been the optimum choice of the reaction substrates. Before peptide bond formation takes place, both substrates, the pept-tRNA and aa-tRNA, should be correctly bound at active site. The simplest biologically relevant P-site substrate is fMet-tRNA^{fMet}, which is placed into the P site with the help of the initiation factors. More complex setups allow the formation of di-, tri-, or oligopeptides in the P site using reconstituted translation mixtures containing the desired set of aa-tRNAs (Katunin et al. 2002a; Wohlgemuth et al. 2008) (Fig. 7.1d). Using the natural A-site substrate is

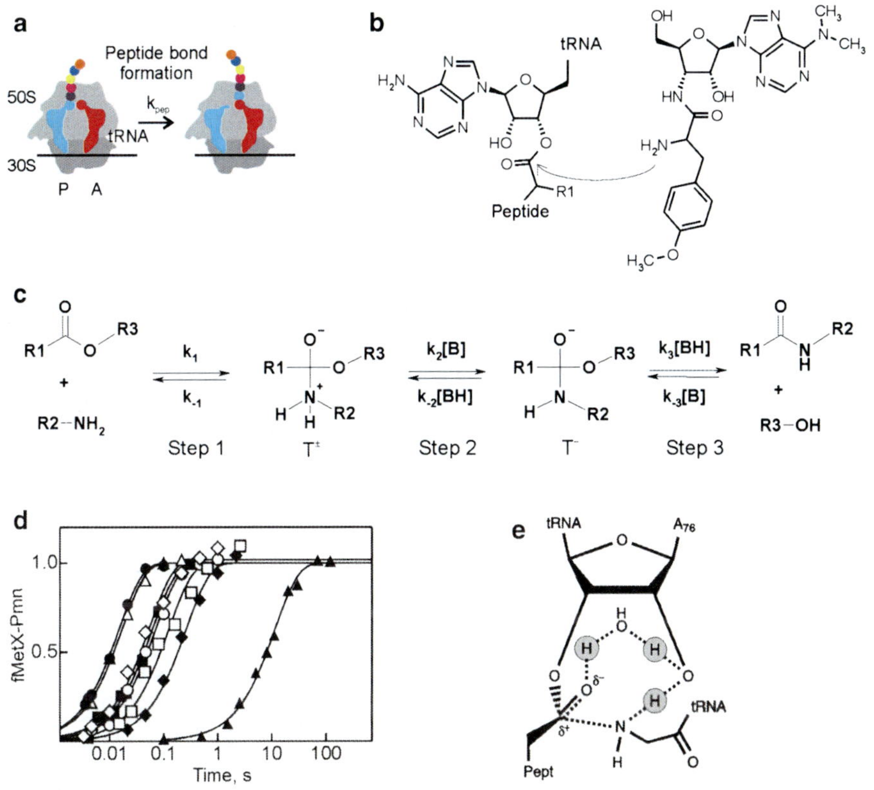

Fig. 7.1 Peptide bond formation. (**a**) Schematic of the reaction. The ribosomal subunits are depicted in *dark gray* (30S) and *light gray* (50S); pept-tRNA is *cyan*, aa-tRNA *red*. (**b**) Reaction substrates, pept-tRNA (*left*) and Pmn (*right*). (**c**) Scheme of the aminolysis reaction mechanism. Proton transfers can be catalyzed by base (B) or acid (BH). R1, R2, R3, substituents. (**d**) Examples of time courses of the reaction between Pmn and fMetX-tRNA, where X is Ala (*closed squares*), Arg (*closed circles*), Asp (*closed diamonds*), Lys (*open triangles*), Phe (*open circles*), Pro (*closed triangles*), Ser (*open diamonds*), or Val (*open squares*). Apparent rate constants were obtained by single-exponential fitting (*continuous lines*) (Wohlgemuth et al. 2008). (**e**) TS of peptide bond formation with an eight-membered proton shuttle. Protons involved in the concerted proton shuttle are *encircled* (Kuhlenkoetter et al. 2011)

more challenging, because aa-tRNA is delivered to the A site in a multistep process involving elongation factor Tu and GTP hydrolysis (Rodnina and Wintermeyer 2001). Some steps on this pathway appear to be partially or fully rate-limiting for peptide bond formation, which precluded studies on the mechanism of the chemistry step. The difficulty could be circumvented by using a small analog of the aminoacyl end of the aa-tRNA, puromycin (Pmn), which binds to the A site of the 50S subunit at a rate that is much higher than the rate of the following chemistry step (Sievers et al. 2004). The reaction between the dipeptidyl-tRNA fMetPhe-tRNA[Phe] in the P site and Pmn in the A site was pH-dependent. At low pH, the rate increased with pH in such a way that the slope of the $\log(k_{pep})$ vs. pH plot was larger than one, suggesting the participation of more than one ionizing group in catalysis. At high pH the rate was pH-independent and the analysis suggested that the pK_a values of the ionizing groups involved in catalysis were in the range of 7–7.5 (Katunin et al. 2002a).

While the initial result seemed encouragingly straightforward, the assignment of the groups that were responsible for the pH-dependence appeared quite difficult. As expected, a large part of the pH-dependence was due to ionization of the α-amino group acting as the nucleophile, which is active only in its deprotonated form (Katunin et al. 2002a). Initially, also the ribosome seemed to contribute to the pH-dependence. However, careful further analysis indicated that the ribosome does not provide ionizing groups for catalysis. Rather, the contribution to pH-dependence was due to a conformational rearrangement of the ribosome outside the active center (Beringer et al. 2005). This rearrangement was observed only when Pmn was used as A-site substrate; when a slightly longer mimic of the tRNA 3′ end, CC-Pmn, was used, only the ionization of the α-amino group was observed to affect the reaction (Beringer and Rodnina 2007; Brunelle et al. 2006). Likewise, no ionizing group(s) other than that of the nucleophile was found to influence the reaction with full-length aa-tRNA as A-site substrate (Bieling et al. 2006). An extensive mutational analysis of the rRNA bases at the core of the catalytic center suggested that replacements of these groups did not impair peptide bond formation with aa-tRNA (Beringer et al. 2003, 2005; Hesslein et al. 2004; Youngman et al. 2004). Taken together, these data strongly suggested that the ribosome does not catalyze peptide bond formation by donating groups acting as general acids/bases.

If the ribosome does not donate active groups, then how does it work? An important step in answering this question was made when activation parameters of the reaction were compared for the ribosome-catalyzed and the uncatalyzed reaction. In contrast to most proteins that catalyze their reactions by providing active residues, which is manifested as a favorable change in the activation enthalpy of the reaction on the enzyme, the 10^7-fold rate enhancement of peptide bond formation on the ribosome is entirely due to a change of the activation entropy of the reaction (Sievers et al. 2004). The physical sources of the favorable entropy change are not easy to understand, because the activation entropy is a macroscopic quantity that can include contributions of desolvation, molecular conformations, and phase-space configuration volumes, of which in many cases solvation effects are dominant (Sharma et al. 2005). Computer simulations suggested that the favorable entropy change was mainly associated with lowering the free energy of solvent reorganization, with a moderate contribution of substrate alignment and proximity at the active site (Sharma et al. 2005; Trobro and Åqvist 2006; Wallin and Aqvist 2010).

Replacing the 2′ OH of A76 of the peptidyl-tRNA in the P site strongly inhibited the reaction on the ribosome (Weinger et al. 2004), but did not affect the uncatalyzed reaction (Sharma et al. 2005; Sievers et al. 2004). These data suggested an important contribution of substrate-assisted catalysis. Although recent data suggest that the magnitude of the effect is not as large as originally reported (Weinger et al. 2004; Zaher et al. 2011), the 2′ OH is important for the transfer of protons in the TS.

A plausible model which was supported by structural and computational analysis suggested a shuttle of protons between the attacking nucleophile, the 2′ OH and 3′ OH groups of A76; in this model, the ribosome would provide a favorable electrostatic environment that stabilizes the substrates, products, and the TS (Wallin and Aqvist 2010). However, the exact pathway of proton transfer and the number of protons actually involved remained unclear.

The next important step in unraveling the mechanism of peptide bond formation was accomplished by chemical synthesis of substrate derivatives and kinetic studies. The detailed analysis of heavy-atom isotope effects using a set of pept-tRNA derivatives with substitutions of essentially every atom taking part in catalysis has demonstrated that the ribosome contributes to chemical catalysis by changing the rate-limiting TS (Hiller et al. 2011; Kingery et al. 2008). In solution, the reaction is rate-limited by the deprotonation of the $T^{\pm}$ intermediate at low pH and by the decomposition of the T^{-} intermediate at high pH (Satterthwait and Jencks 1974). In contrast, on the ribosome the formation of the tetrahedral intermediate and proton transfer from the nucleophilic nitrogen both take place during the rate-limiting step; the breakdown of the tetrahedral intermediate occurs in a separate fast step (Hiller et al. 2011). This result would argue against a fully concerted shuttle where the formation of the 3′-leaving group occurs at the same time as a proton from the α-amino group is received by the 2′ OH of A76 of pept-tRNA. This is important, as the 2′ OH group can receive a proton only if it simultaneously donates a proton to some other group. However, model studies suggested that the carbonyl oxygen of the pept-tRNA, rather than the 3′ OH, accepts the proton from the 2′ OH (Huang et al. 2008; Rangelov et al. 2006). Such a shuttle mechanism may be six-membered, with two protons simultaneously changing their positions in the TS, or eight-membered with three protons "in flight" in the TS; the latter TS includes a water molecule which is found in the right position in the crystal structures of the 50S subunits in the complex with TS analogs (Schmeing et al. 2005a; Wallin and Aqvist 2010). The analysis of kinetic solvent isotope effects showed that three protons move in the rate-limiting TS and that the movement is concerted (Kuhlenkoetter et al. 2011). This not only supports the existence of a concerted proton shuttle, but also favors an eight-membered shuttle with a water molecule serving as intermediary. The results of the kinetic analysis of pH profiles, mutational and chemical substitutions, kinetic isotope effects, in combination with the structural and computational work suggest the following mechanism of peptide bond formation (Fig. 7.1e). In the rate-limiting TS, the attack of the α-amino group on the ester carbon results in an eight-membered transition state, in which a proton from the α-amino group is received by the 2′ OH group of A76, which at the same time donates its proton to the carbonyl oxygen through an adjacent water molecule. Such a scenario would not require a pK_a shift of the 2′ OH group, due to the concerted nature of the bond-forming and bond-breaking events. Protonation of the 3′ OH then would be an independent rapid step. The ribosome catalyzes the reaction by providing a network of interactions that change the rate-limiting TS state and lower the entropy of activation.

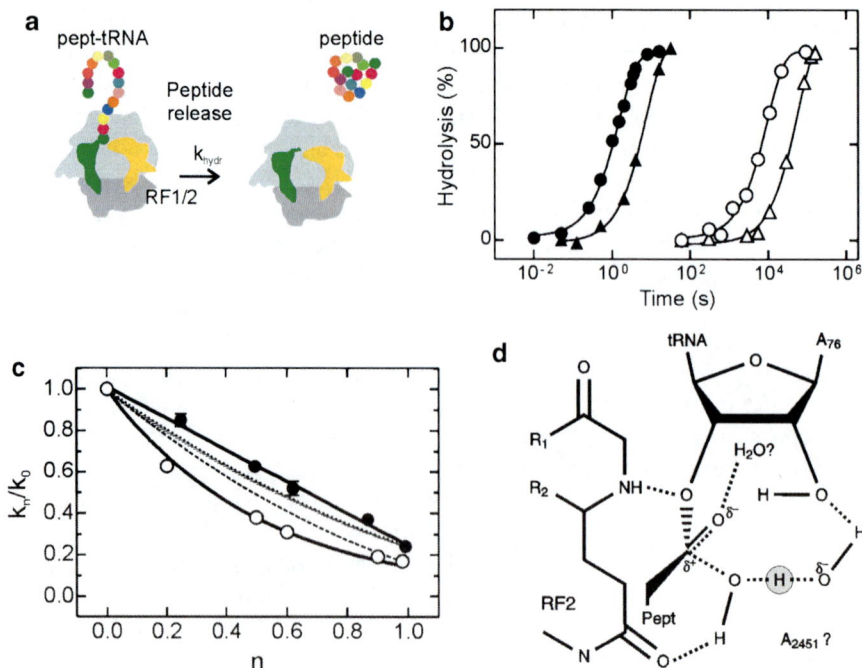

Fig. 7.2 Pept-tRNA hydrolysis in termination. (**a**) Schematic of the reaction. Pept-tRNA is depicted in *green*, RF1/2 in *yellow*. (**b**) Time courses in H_2O and D_2O-containing buffers. Catalyzed (*closed symbols*) and uncatalyzed (*open symbols*) reactions were measured in buffer with either H_2O (*filled circles, open circles*) or D_2O (*filled triangles, open triangles*). (**c**) Proton inventories. *n*, mole fractions of D_2O. For the ribosome-catalyzed reaction (*filled circles*), the linear fit yields an overall KSIE of 4.1; the *dotted line* simulates the rate-limiting simultaneous transfer of two protons in the TS with a fractionation factor of 0.49 each. In the uncatalyzed reaction (*open circles*), many protons contribute to the overall KSIE of 7.0; the *dashed line* simulates the rate-limiting transfer of three protons. (**d**) TS of the catalyzed reaction. The single low-barrier hydrogen bond that determines the reaction rate is *encircled*. The possibility of the involvement of another water molecule and/or residue A2451 of 23S rRNA is indicated

7.5.2 *Peptidyl-tRNA Hydrolysis*

The second reaction catalyzed by the ribosome is the hydrolysis of pept-tRNA, which requires the help of the termination (release) factors, RF1 and RF2 in *Escherichia coli* (Fig. 7.2a). The reaction takes place when the ribosome arrives at a stop codon in the mRNA. RF1/2 are recruited to the ribosome and, upon stop codon recognition, undergo a conformational change which places a conserved GGQ sequence motif of the factors into the catalytic center of the ribosome (Jin et al. 2010; Korostelev et al. 2008; Laurberg et al. 2008; Weixlbaumer et al. 2008). This contact presumably induces a rearrangement of 23S rRNA that allows for the correct positioning of the hydrolytic water molecule for the attack on the ester bond (Schmeing et al. 2005b). Unlike peptidyl transfer, peptide release is impaired by mutations in 23S rRNA of the active site (Polacek et al. 2003; Youngman et al. 2004), although the mechanism of inhibition is not entirely clear. Crystal structures provided information on the mechanism of stop codon recognition and the details of the ribosome's catalytic site augmented by RF1/2 (Jin et al. 2010; Korostelev et al. 2008; Laurberg et al. 2008; Weixlbaumer et al. 2008). However, the mechanism of pept-tRNA cleavage remained unclear. Again, the kinetic analysis utilizing the quench-flow technique turned out to be instrumental. Peptide release takes place with a rate of 0.1–10 s^{-1}, depending on the type of stop codon, the chain length of the peptidyl-tRNA, and the presence of a natural modification of RF1/2, i.e., methylation of the glutamine in the GGQ motif (Brunelle et al. 2008; Dincbas-Renqvist et al. 2000; Kuhlenkoetter et al. 2011; Mora et al. 2003a, b; Shaw and Green 2007). To measure the rate of the release reaction, ribosome complexes with pept-tRNA in the P site were rapidly mixed with excess RF1/2 in the quench-flow apparatus. The reaction was terminated by rapid quenching, and the ratio of intact pept-tRNA relative to released peptide was determined, based on the radioactivity label in the peptidyl moiety.

The first important question was whether one of the residues in the conserved GGQ motif of RF1/2 takes part in catalysis directly. Mutations of the two glycine residues or the glutamine residue in the GGQ motif of RF1/2 did not abolish catalysis, except for the glutamine substitution to proline (which most likely leads to major changes in the protein's structure) (Korostelev et al. 2008; Mora et al. 2003a; Shaw and Green 2007), suggesting that the residues of the GGQ motif do not contribute to catalysis directly. Rather, glutamine appears to control the specificity of the release reaction for water by excluding nucleophiles larger than water from entering the active site (Shaw and Green 2007). Methylation of the glutamine residue of the GGQ motif enhances the reaction up to fivefold, compared to the 10^4-fold rate acceleration brought about by the ribosome (Kuhlenkoetter et al. 2011), i.e., the methyl group is not essential either. Thus, residues of the factor do not seem to take part in the catalysis of peptide release. As a consequence, the ribosome must play a crucial role in catalysis.

The next question was whether ionizing groups take part in the reaction. The rate of pept-tRNA hydrolysis increases linearly up to pH 9, the highest pH attainable experimentally (Kuhlenkoetter et al. 2011). The slope of the pH/rate dependence for both ribosome-catalyzed and uncatalyzed reactions turned out to be close to one. These results suggested that a single ionizing group, with a pK_a >9, takes part in catalysis. For the uncatalyzed reaction, this group must be a water molecule or the 2' OH of A76 of pept-tRNA; water is the more likely candidate, as replacements of the 2' OH did not affect the uncatalyzed reaction. Water is directly involved in the reaction on the ribosome as well and is therefore likely to contribute to the pH-dependence of the ribosome-catalyzed reaction. If the ribosome or RF2 were donating additional ionizing groups, in the simplest case the slope of the pH-dependence would change, which is not observed. Thus, the critical 23S rRNA residues in the active site presumably have a role in positioning the peptidyl-tRNA and/or the release factor in the active site, rather than in chemistry itself (Kuhlenkoetter et al. 2011).

The results described above seemed to suggest similar mechanisms for peptide bond formation and peptide release: the ribosome provides an electrostatic network of interactions that facilitate reaction, consistent with the notion that the ribosome stabilized the substrates and products of the reaction. However, further kinetic analysis demonstrated that the chemical mechanisms of the two reactions are rather different. The ribosome contributes to catalysis of the peptide release by reducing the activation enthalpy (Kuhlenkoetter et al. 2011), in contrast to peptide bond formation which is driven by a favorable change in the activation entropy (Sievers et al. 2004). Moreover, quench-flow measurements of the kinetic solvent isotope effects (Fig. 7.2b) and proton inventories of the peptide release reaction (Fig. 7.2c) suggested that a single proton moved in the TS, in contrast to the TS of peptidyl transfer, where three protons "in flight" were observed (Kuhlenkoetter et al. 2011).

These data exclude a proton shuttle mechanism for pept-tRNA hydrolysis, but would be consistent with several other reaction pathways, the choice of which depends on the assignment of the ionizing group and assumptions concerning the rate-limiting steps of the reaction. According to the model that we consider the most likely (Fig. 7.2d), the attacking water molecule in the TS is engaged in a strong hydrogen bond with a hydroxide ion that facilitates proton transfer and the nucleophilic attack. Density for a water molecule in the active site was observed in the crystal structure of a complex representing the reactant state (Jin et al. 2010). The effect of replacing the 2′ OH group of A76 then may be attributed to a change of the rate-limiting step from an early to a late TS, such that proton transfer to the leaving group becomes rate-limiting. Alternative models for the TS of peptide release are possible (Kuhlenkoetter et al. 2011). In any case, the mechanistic details of peptide release and peptide bond formation are dissimilar beyond the difference in the nature of the attacking nucleophile.

This work demonstrates that one active site, with a small difference introduced by a few amino acids of the release factor, can support rather different reaction mechanisms. Such versatility is unique among ribozymes and rare among protein enzymes. One reason why the active site of the ribosome could retain, or gain, this versatility is probably the fact that ribosomal residues, and residues of the release factor, do not take part in chemistry, but rather provide a network of electrostatic and hydrogen-bonding interactions that help in orienting the substrates and in stabilizing the respective transition state, more or less independent of the chemical nature of the substrates. In the early days of the RNA world, the primordial ribosome, which most likely comprised only the A and P parts of the peptidyl transferase center (Bokov and Steinberg 2009), might have evolved towards accepting a multitude of chemically versatile substrates, thereby providing potential building blocks of polymers for the evolution of life. While the modern ribosome is optimized towards high speed and accuracy of translation, the flexibility of the ribosome's active site presumably has contributed to its stability throughout evolution over four billion years.

7.6   Elongation Factor G in Elongation and Recycling

Among the translation factors EF-G is the only one that has two distinct functions in different phases of protein synthesis. During the elongation phase, EF-G promotes translocation—the movement of the tRNAs together with the mRNA through the ribosome after a peptide bond has been formed. During translocation, EF-G converts the pre-translocation complex to a post-translocation complex in which P- and A-site tRNAs move to the E and P sites, respectively, in a step-wise fashion. Translocation entails movement of tRNA through hybrid states accompanied by the rotation of the ribosomal subunits relative to one another (Agirrezabala et al. 2008; Frank and Agrawal 2000; Julian et al. 2008; Moazed and Noller 1989). Following termination, EF-G, together with ribosome recycling factor, RRF, brings about the rapid dissociation of the post-termination ribosome into subunits (Hirashima and Kaji 1973). In both reactions, conformational rearrangements and movements of components on the ribosome must take place, and these dynamics can be followed by the stopped-flow technique.

A number of specific tools have been developed to study EF-G action. To monitor the movement of the mRNA relative to the ribosome, fluorescence reporter groups attached to the 3' end of the mRNAs were used (Feinberg and Joseph 2001; Peske et al. 2004); movements of the tRNA were monitored using dyes attached to either the A-site peptidyl-tRNA or the deacylated tRNA in the P site (Pan et al. 2007; Robertson et al. 1986; Robertson and Wintermeyer 1981; Rodnina et al. 1997) (Fig. 7.3a). Release of the inorganic phosphate (Pi) can be conveniently followed by the fluorescence change of labeled phosphate-binding protein, which serves as an indicator of free Pi (Brune et al. 1998). Conformational changes at the nucleotide-binding pocket of EF-G can be visualized using fluorescence changes of a methyl-lanthraniloyl (mant) group attached to the ribose moiety of GTP or GDP (Wilden et al. 2006). The key step of ribosome recycling is the dissociation of the post-termination ribosome into 30S and 50S subunits; this could be monitored by changes in light scattering or FRET between donor and acceptor-labeled subunits (Peske et al. 2005; Savelsbergh et al. 2009). Finally, the chemical reaction, GTP hydrolysis by EF-G, was followed by quench flow (Rodnina et al. 1997; Seo et al. 2006). Mixing the pre-translocation complex (with peptidyl-tRNA in the A site and deacylated tRNA in the P site) with EF-G·GTP resulted in multistep kinetics, reflecting several elemental steps of translocation. Likewise, mixing post-termination complexes (with deacylated tRNA in the P site) with EF-G·GTP and RRF allowed monitoring the coordinated disassembly process. Careful analysis of time courses obtained at different EF-G concentrations revealed the sequence of events and provided a framework for the analysis of effects of mutations and of the interference by antibiotics, which ultimately lead to a complex picture of how EF-G acts in translation and recycling.

7.6.1 tRNA-mRNA Translocation

7.6.1.1 Sequence of Events

EF-G binding to the ribosome is rapid and the resulting initial complex is labile (association and dissociation rate constants 10^8 $M^{-1}s^{-1}$ and 100 s^{-1}, respectively)

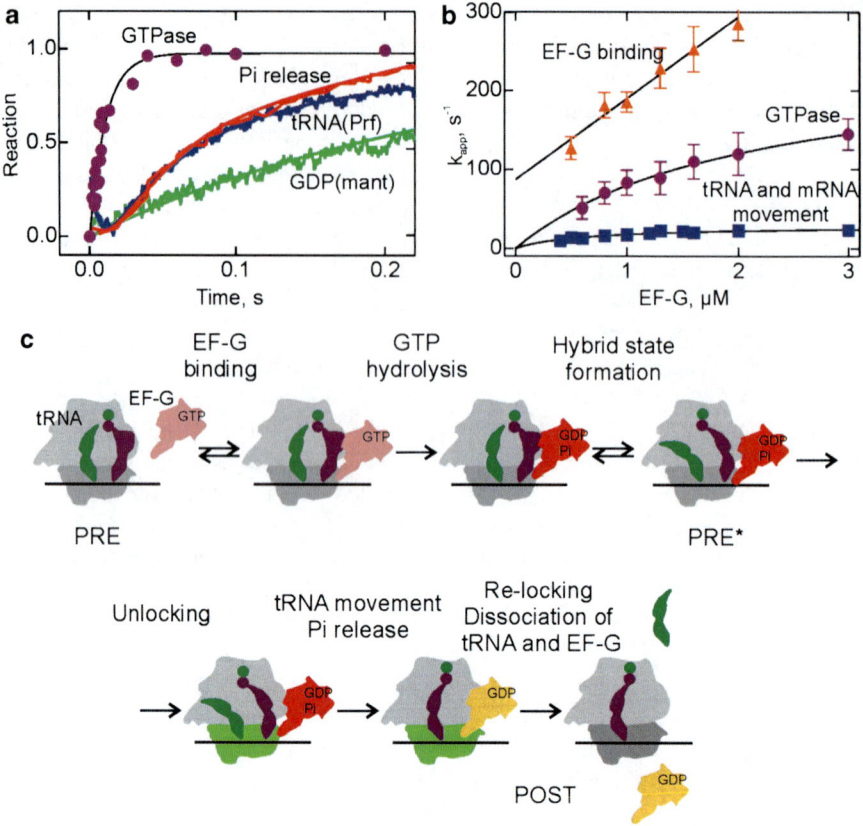

Fig. 7.3 EF-G-catalyzed tRNA-mRNA translocation. (**a**) Examples of time courses measured by fluorescence stopped flow or quench flow of GTP hydrolysis by EF-G (*purple circles*), movements of tRNA (labeled with proflavin; *blue curve*), Pi release (*red line*), as well as conformational changes in the GTP pocket of EF-G (monitored using mant-GTP; *green line*). (**b**) Concentration dependencies of the apparent rate constants (k_{app}) of EF-G binding (*orange triangles*), GTP hydrolysis (*purple circles*), and tRNA-mRNA translocation (*blue squares*) (Savelsbergh et al. 2003). (**c**) Sequence of steps during EF-G-catalyzed translocation. The ribosomal subunits are depicted in *light gray* (50S) and *dark gray* or *green* (30S) to indicate different conformations. EF-G is shown in *pink*, *red*, and *yellow* in the GTP-, GDP·Pi- and GDP-bound forms, respectively. Deacylated tRNA is in *green*, pept-tRNA in *purple*

(Fig. 7.3b). The timing of EF-G binding in relation to the movement of the ribosome into the rotated/hybrid state is controversial. Bulk kinetic analysis suggested that EF-G can bind to the ribosome regardless of the tRNA positions and accelerates the movement of the tRNAs into their hybrid position (Walker et al. 2008; Wintermeyer et al. 2011). In contrast, the results of single-molecule FRET experiments seemed to indicate that EF-G binding and ribosome/tRNA rearrangements are correlated and that spontaneous hybrid state formation is rate-limiting for translocation (Munro et al. 2010). We note that if the spontaneous transition towards the rotated/hybrid state (at a rate around 1 s^{-1} at 25 °C) were rate-limiting for the following steps, EF-G binding would be slow and independent of concentration, which is inconsistent with the experimental evidence (Katunin et al. 2002b; Rodnina et al. 1997; Seo et al. 2006; Walker et al. 2008). In any case, EF-G binding stabilizes the rotated conformation of the ribosome with the tRNAs in hybrid states (Munro et al. 2010; Valle et al. 2003).

Binding to the ribosome induces closing of the nucleotide-binding pocket of EF-G, resulting in a 30,000-fold stabilization of GTP binding (Wilden et al. 2006). GTP hydrolysis by EF-G takes place almost immediately after binding and drives a rearrangement of the ribosome that accelerates the tRNA-mRNA movement on the 30S subunit (Savelsbergh et al. 2003) (Fig. 7.3c). The movement of tRNAs and mRNA can take place spontaneously (Konevega et al. 2007; Shoji et al. 2006) and probably proceeds through a quasi-continuous landscape of intermediate states (Fischer et al. 2010). The release of Pi from EF-G, which takes place in parallel to the tRNA movement on the 30S subunit, induces another conformational change in EF-G that is required for EF-G to dissociate from the ribosome and for the ribosome to return to the (non-rotated) ground state (Savelsbergh et al. 2003, 2005).

7.6.1.2 Rearrangements That Lead to Translocation on the 30S Subunit

The crucial step in tRNA translocation on the 30S subunit is the rearrangement which we termed "unlocking" (Savelsbergh et al. 2003). We note that, in parallel, the term was used to describe the spontaneous rearrangement of the pre-translocation complex into the rotated/hybrid state (Valle et al. 2003). In the following, we use the term "unlocking" to describe the conformational change of the 30S subunit that precedes and drives tRNA translocation (Savelsbergh et al. 2003). The structural mechanism of unlocking is not known; it is probably a composite step itself, and EF-G, driven by GTP hydrolysis, has an important role in promoting it. After the rearrangement step which affects both tRNA movement and Pi release, additional changes occur that affect the two reactions in different ways. Structural evidence suggests that movements of the 30S subunit head may play a crucial role (Agirrezabala and Frank 2009; Dunkle and Cate 2010; Frank and Gonzalez 2010); EF-G could facilitate the swiveling movement of the 30S head (Dunkle et al. 2011). The H583K mutation at the tip of domain 4 of EF-G, which in the complex with the ribosome points towards the decoding center, strongly reduces the rate of translocation; this indicates that the contact of domain 4 of EF-G with the ribosome is involved in opening up the decoding region (Gao et al. 2009; Savelsbergh et al. 2000; Stark et al. 2000; Taylor et al. 2007). Antibiotics that bind to the decoding region selectively inhibit tRNA movement and have no effect on Pi release, suggesting that conformational changes at the decoding site following unlocking are required for movement (Peske et al. 2004). In keeping with these results, a specific cleavage of 16S rRNA at the decoding center by colicin E3, which presumably increases the conformational flexibility of the decoding site, leads to an acceleration of tRNA movement in translocation (Lancaster et al. 2008).

In the presence of EF-G, the directionality of tRNA movement is biased in such a way that in the post-translocation state domain 4 of EF-G—which is crucial for translocation (Rodnina et al. 1997; Savelsbergh et al. 2000)—occupies the 30S A site (Agrawal et al. 1998; Frank and Agrawal 2000; Stark et al. 2000), thus effectively preventing back movement of peptidyl-tRNA while the unlocked state prevails. On EF-G, an insertion in the GTP binding domain, the G' subdomain, appears to be important for EF-G binding to the ribosome and the conformational coupling between GTP hydrolysis, retention of Pi, and unlocking (Agrawal et al. 2001; Nechifor et al. 2007; Ticu et al. 2009).

7.6.2   Ribosome Recycling

The dissociation of the ribosome into subunits following termination (ribosome recycling) is promoted by the RRF and EF-G. The reaction provides an example where the reaction was monitored in stopped-flow utilizing the decrease of light scattering intensity upon dissociation of the 70S ribosome into 30S and 50S subunits (Fig. 7.4a).

7.6.2.1   Sequence of Events

RRF and EF-G·GTP are recruited to the post-termination complex with the deacylated tRNA bound to the hybrid P/E state (Lancaster et al. 2002; Peske et al. 2005). Interactions between RRF and EF-G on the ribosome are crucial for the catalysis of ribosome splitting (Gao et al. 2007; Ishino et al. 2000). GTP hydrolysis by EF-G in the recycling reaction is rapid, comparable to that during translocation (Savelsbergh et al. 2009). However, the presence of RRF delays Pi release by about 30 ms (Fig. 7.4b), suggesting that the complex undergoes a rearrangement that controls the following Pi release. In turn, Pi release appears to drive structural changes that lead to subunit dissociation. The reaction is inhibited by vanadate, an analog of Pi, indicating that Pi release, and the rearrangements coupled to it, are essential for subunit dissociation (Savelsbergh et al. 2009) (Fig. 7.4c). After the dissociation of the 50S subunit, deacylated tRNA as well as mRNA remain bound to the 30S subunit and dissociate rather slowly (Karimi et al. 1999; Peske et al. 2005). IF3 stimulates tRNA dissociation from the 30S subunit, followed by the release of mRNA, which renders the ribosome free for the next round of translation (Karimi et al. 1999; Peske et al. 2005).

7.6.2.2   Distinct Functions of EF-G in tRNA Translocation
          and Ribosome Recycling

Several aspects of EF-G action are similar in translocation and ribosome disassembly. In both reactions, EF-G hydrolyzes GTP rapidly (Rodnina et al. 1997;

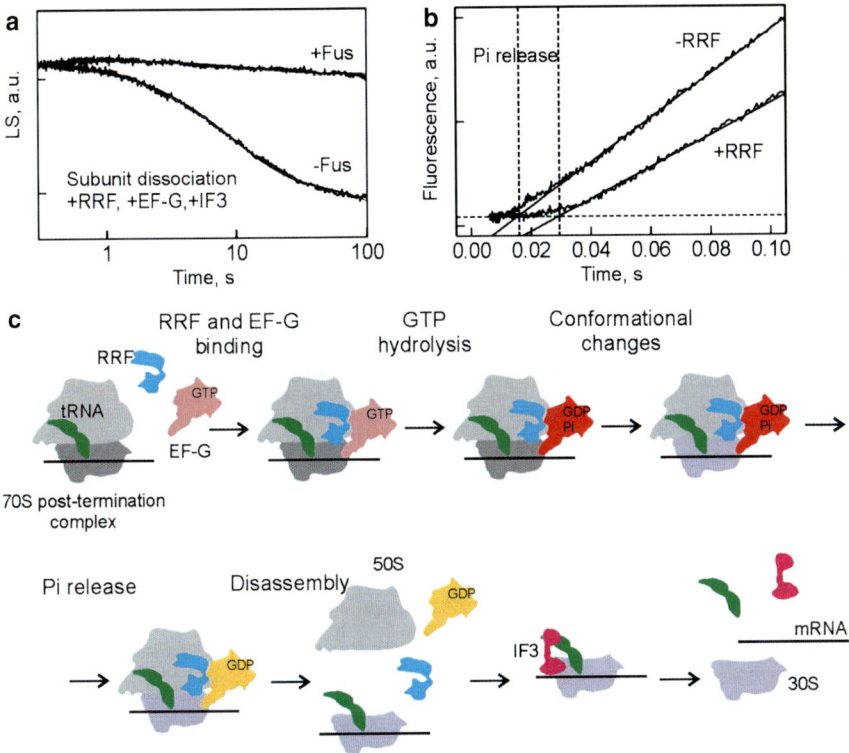

Fig. 7.4 Ribosome recycling. (**a**) Subunit dissociation by RRF, EF-G, and IF3 in the presence and absence of fusidic acid. Changes in light scattering at 435 nm were monitored (arbitrary units, a.u.) (Savelsbergh et al. 2009). (**b**) RRF-mediated delay of Pi release from EF-G (Savelsbergh et al. 2009). (**c**) Sequence of steps in post-termination complex disassembly. The ribosomal subunits are depicted in *light gray* (50S) and *dark gray* or *pale lilac* (30S) depending on the conformation. EF-G is shown in *pink*, *red*, and *yellow* in the GTP-, GDP·Pi- and GDP-bound forms, respectively. RRF is in *cyan*, IF3 in *magenta*, and tRNA in *green*

Savelsbergh et al. 2005, 2009); in fact, GTP hydrolysis by EF-G is independent of the functional state of the ribosome and seems to be required to stabilize EF-G binding to the ribosome (Wilden et al. 2006). Pi release is delayed, albeit to a somewhat different extent during translocation (about 15 ms) compared to recycling (30 ms). However, there are also important differences between the two reactions, indicating that the coupling of the energy of EF-G binding and GTP hydrolysis is different.

When GTP hydrolysis is prevented by replacing GTP with non-hydrolyzable analogs, translocation still takes place, albeit slowly (Katunin et al. 2002b; Rodnina et al. 1997), whereas ribosome disassembly by EF-G/RRF is completely blocked (Karimi et al. 1999; Peske et al. 2005; Zavialov et al. 2005). Vanadate, an analog of Pi, strongly impairs recycling, but has no effect on translocation (Savelsbergh et al. 2009). Fusidic acid, which is a known inhibitor of EF-G function, stalls a particular EF-G conformation which in the absence of the antibiotic is short-lived. Upon translocation, this

EF-G conformation is formed late in the pathway, as fusidic acid does not impair translocation or GTP hydrolysis, but inhibits EF-G turnover, i.e., the dissociation of EF-G following translocation and Pi release (Savelsbergh et al. 2009). In contrast, during recycling a much earlier state must be affected, as fusidic acid completely abrogates ribosome disassembly. The inhibition is complete at a concentration of fusidic acid that is about 1,000-fold lower than the concentration required to inhibit EF-G turnover. This indicates that the inhibition of ribosome recycling is the primary target of fusidic acid, in keeping with submicromolar MIC values observed in vivo. Notably, EF-G action in recycling crucially depends on the presence of RRF. Thus, although translocation and ribosome disassembly are both driven by EF-G, RRF appears to alter the reaction pathway and to change the conformational dynamics of EF-G for the specific purposes of ribosome disassembly into subunits.

Acknowledgments The work in our labs was supported by the Deutsche Forschungsgemeinschaft and the Max Planck Society.

References

Agirrezabala X, Frank J (2009) Elongation in translation as a dynamic interaction among the ribosome, tRNA, and elongation factors EF-G and EF-Tu. Q Rev Biophys 42:159–200

Agirrezabala X, Lei J, Brunelle JL et al (2008) Visualization of the hybrid state of tRNA binding promoted by spontaneous ratcheting of the ribosome. Mol Cell 32:190–197

Agrawal RK, Penczek P, Grassucci RA et al (1998) Visualization of elongation factor G on the *Escherichia coli* 70S ribosome: the mechanism of translocation. Proc Natl Acad Sci U S A 95:6134–6138

Agrawal RK, Linde J, Sengupta J et al (2001) Localization of L11 protein on the ribosome and elucidation of its involvement in EF-G-dependent translocation. J Mol Biol 311:777–787

Antoun A, Pavlov MY, Lovmar M et al (2006) How initiation factors tune the rate of initiation of protein synthesis in bacteria. EMBO J 25:2539–2550

Beringer M, Rodnina MV (2007) Importance of tRNA interactions with 23S rRNA for peptide bond formation on the ribosome: studies with substrate analogs. Biol Chem 388:687–691

Beringer M, Adio S, Wintermeyer W et al (2003) The G2447A mutation does not affect ionization of a ribosomal group taking part in peptide bond formation. RNA 9:919–922

Beringer M, Bruell C, Xiong L et al (2005) Essential mechanisms in the catalysis of peptide bond formation on the ribosome. J Biol Chem 280:36065–36072

Bieling P, Beringer M, Adio S et al (2006) Peptide bond formation does not involve acid–base catalysis by ribosomal residues. Nat Struct Mol Biol 13:423–428

Bokov K, Steinberg SV (2009) A hierarchical model for evolution of 23S ribosomal RNA. Nature 457:977–980

Brune M, Hunter JL, Howell SA et al (1998) Mechanism of inorganic phosphate interaction with phosphate binding protein from Escherichia coli. Biochemistry 37:10370–10380

Brunelle JL, Youngman EM, Sharma D et al (2006) The interaction between C75 of tRNA and the A loop of the ribosome stimulates peptidyl transferase activity. RNA 12:33–39

Brunelle JL, Shaw JJ, Youngman EM et al (2008) Peptide release on the ribosome depends critically on the 2′ OH of the peptidyl-tRNA substrate. RNA 14:1526–1531

Dincbas-Renqvist V, Engstrom A, Mora L et al (2000) A post-translational modification in the GGQ motif of RF2 from Escherichia coli stimulates termination of translation. EMBO J 19:6900–6907

Dorywalska M, Blanchard SC, Gonzalez RL et al (2005) Site-specific labeling of the ribosome for single-molecule spectroscopy. Nucleic Acids Res 33:182–189

Dunkle JA, Cate JH (2010) Ribosome structure and dynamics during translocation and termination. Annu Rev Biophys 39:227–244

Dunkle JA, Wang L, Feldman MB et al (2011) Structures of the bacterial ribosome in classical and hybrid states of tRNA binding. Science 332:981–984

Fabbretti A, Milon P, Giuliodori AM et al (2007) Real-time dynamics of ribosome-ligand interaction by time-resolved chemical probing methods. Methods Enzymol 430:45–58

Feinberg JS, Joseph S (2001) Identification of molecular interactions between P-site tRNA and the ribosome essential for translocation. Proc Natl Acad Sci U S A 98:11120–11125

Fischer N, Konevega AL, Wintermeyer W et al (2010) Ribosome dynamics and tRNA movement by time-resolved electron cryomicroscopy. Nature 466:329–333

Frank J, Agrawal RK (2000) A ratchet-like inter-subunit reorganization of the ribosome during translocation. Nature 406:318–322

Frank J, Gonzalez RL Jr (2010) Structure and dynamics of a processive Brownian motor: the translating ribosome. Annu Rev Biochem 79:381–412

Gao N, Zavialov AV, Ehrenberg M et al (2007) Specific interaction between EF-G and RRF and its implication for GTP-dependent ribosome splitting into subunits. J Mol Biol 374:1345–1358

Gao YG, Selmer M, Dunham CM et al (2009) The structure of the ribosome with elongation factor G trapped in the posttranslocational state. Science 326:694–699

Grunberg-Manago M, Dessen P, Pantaloni D et al (1975) Light-scattering studies showing the effect of initiation factors on the reversible dissociation of Escherichia coli ribosomes. J Mol Biol 94:461–478

Hesslein AE, Katunin VI, Beringer M et al (2004) Exploration of the conserved A + C wobble pair within the ribosomal peptidyl transferase center using affinity purified mutant ribosomes. Nucleic Acids Res 32:3760–3770

Hickerson R, Majumdar ZK, Baucom A et al (2005) Measurement of internal movements within the 30S ribosomal subunit using Forster resonance energy transfer. J Mol Biol 354:459–472

Hiller DA, Singh V, Zhong M et al (2011) A two-step chemical mechanism for ribosome-catalysed peptide bond formation. Nature 476:236–239

Hirashima A, Kaji A (1973) Role of elongation factor G and a protein factor on the release of ribosomes from messenger ribonucleic acid. J Biol Chem 248:7580–7587

Huang KS, Carrasco N, Pfund E et al (2008) Transition state chirality and role of the vicinal hydroxyl in the ribosomal peptidyl transferase reaction. Biochemistry 47:8822–8827

Ishino T, Atarashi K, Uchiyama S et al (2000) Interaction of ribosome recycling factor and elongation factor EF-G with E. coli ribosomes studied by the surface plasmon resonance technique. Genes Cells 5:953–963

Jin H, Kelley AC, Loakes D et al (2010) Structure of the 70S ribosome bound to release factor 2 and a substrate analog provides insights into catalysis of peptide release. Proc Natl Acad Sci U S A 107:8593–8598

Johnson AE (2003) Kinetic analysis of macromolecules. Oxford University Press, Oxford

Johnson AE, Adkins HJ, Matthews EA et al (1982) Distance moved by transfer RNA during translocation from the A site to the P site on the ribosome. J Mol Biol 156:113–140

Julian P, Konevega AL, Scheres SH et al (2008) Structure of ratcheted ribosomes with tRNAs in hybrid states. Proc Natl Acad Sci U S A 105:16924–16927

Karimi R, Pavlov MY, Buckingham RH et al (1999) Novel roles for classical factors at the interface between translation termination and initiation. Mol Cell 3:601–609

Katunin VI, Muth GW, Strobel SA et al (2002a) Important contribution to catalysis of peptide bond formation by a single ionizing group within the ribosome. Mol Cell 10:339–346

Katunin VI, Savelsbergh A, Rodnina MV et al (2002b) Coupling of GTP hydrolysis by elongation factor G to translocation and factor recycling on the ribosome. Biochemistry 41:12806–12812

Kingery DA, Pfund E, Voorhees RM et al (2008) An uncharged amine in the transition state of the ribosomal peptidyl transfer reaction. Chem Biol 15:493–500

Konevega AL, Fischer N, Semenkov YP et al (2007) Spontaneous reverse movement of mRNA-bound tRNA through the ribosome. Nat Struct Mol Biol 14:318–324

Korostelev A, Asahara H, Lancaster L et al (2008) Crystal structure of a translation termination complex formed with release factor RF2. Proc Natl Acad Sci U S A 105:19684–19689

Kuhlenkoetter S, Wintermeyer W, Rodnina MV (2011) Different substrate-dependent transition states in the active site of the ribosome. Nature 476:351–354

Lancaster L, Kiel MC, Kaji A et al (2002) Orientation of ribosome recycling factor in the ribosome from directed hydroxyl radical probing. Cell 111:129–140

Lancaster LE, Savelsbergh A, Kleanthous C et al (2008) Colicin E3 cleavage of 16S rRNA impairs decoding and accelerates tRNA translocation on *Escherichia coli* ribosomes. Mol Microbiol 69:390–401

Laurberg M, Asahara H, Korostelev A et al (2008) Structural basis for translation termination on the 70S ribosome. Nature 454:852–857

Moazed D, Noller HF (1989) Intermediate states in the movement of transfer RNA in the ribosome. Nature 342:142–148

Mora L, Heurgue-Hamard V, Champ S et al (2003a) The essential role of the invariant GGQ motif in the function and stability in vivo of bacterial release factors RF1 and RF2. Mol Microbiol 47:267–275

Mora L, Zavialov A, Ehrenberg M et al (2003b) Stop codon recognition and interactions with peptide release factor RF3 of truncated and chimeric RF1 and RF2 from Escherichia coli. Mol Microbiol 50:1467–1476

Munro JB, Wasserman MR, Altman RB et al (2010) Correlated conformational events in EF-G and the ribosome regulate translocation. Nat Struct Mol Biol 17:1470–1477

Nechifor R, Murataliev M, Wilson KS (2007) Functional interactions between the G′ subdomain of bacterial translation factor EF-G and ribosomal protein L7/L12. J Biol Chem 282:36998–37005

Nissen P, Hansen J, Ban N et al (2000) The structural basis of ribosome activity in peptide bond synthesis. Science 289:920–930

Pan D, Kirillov SV, Cooperman BS (2007) Kinetically competent intermediates in the translocation step of protein synthesis. Mol Cell 25:519–529

Peske F, Savelsbergh A, Katunin VI et al (2004) Conformational changes of the small ribosomal subunit during elongation factor G-dependent tRNA-mRNA translocation. J Mol Biol 343:1183–1194

Peske F, Rodnina MV, Wintermeyer W (2005) Sequence of steps in ribosome recycling as defined by kinetic analysis. Mol Cell 18:403–412

Polacek N, Gomez MJ, Ito K et al (2003) The critical role of the universally conserved A2602 of 23S ribosomal RNA in the release of the nascent peptide during translation termination. Mol Cell 11:103–112

Rangelov MA, Vayssilov GN, Yomtova VM et al (2006) The syn-oriented 2-OH provides a favorable proton transfer geometry in 1,2-diol monoester aminolysis: implications for the ribosome mechanism. J Am Chem Soc 128:4964–4965

Robertson JM, Wintermeyer W (1981) Effect of translocation on topology and conformation of anticodon and D loops of tRNAPhe. J Mol Biol 151:57–59

Robertson JM, Paulsen H, Wintermeyer W (1986) Pre-steady-state kinetics of ribosomal translocation. J Mol Biol 192:351–360

Rodnina MV, Wintermeyer W (2001) Fidelity of aminoacyl-tRNA selection on the ribosome: kinetic and structural mechanisms. Annu Rev Biochem 70:415–435

Rodnina MV, Savelsbergh A, Katunin VI et al (1997) Hydrolysis of GTP by elongation factor G drives tRNA movement on the ribosome. Nature 385:37–41

Rodnina MV, Beringer M, Wintermeyer W (2006) Mechanism of peptide bond formation on the ribosome. Q Rev Biophys 39:203–225

Satterthwait AC, Jencks WP (1974) The mechanism of the aminolysis of acetate esters. J Am Chem Soc 96:7018–7031

Savelsbergh A, Matassova NB, Rodnina MV et al (2000) Role of domains 4 and 5 in elongation factor G functions on the ribosome. J Mol Biol 300:951–961

Savelsbergh A, Katunin VI, Mohr D et al (2003) An elongation factor G-induced ribosome rearrangement precedes tRNA-mRNA translocation. Mol Cell 11:1517–1523

Savelsbergh A, Mohr D, Kothe U et al (2005) Control of phosphate release from elongation factor G by ribosomal protein L7/12. EMBO J 24:4316–4323

Savelsbergh A, Rodnina MV, Wintermeyer W (2009) Distinct functions of elongation factor G in ribosome recycling and translocation. RNA 15:772–780

Schmeing TM, Huang KS, Kitchen DE et al (2005a) Structural insights into the roles of water and the 2′ hydroxyl of the P site tRNA in the peptidyl transferase reaction. Mol Cell 20:437–448

Schmeing TM, Huang KS, Strobel SA et al (2005b) An induced-fit mechanism to promote peptide bond formation and exclude hydrolysis of peptidyl-tRNA. Nature 438:520–524

Seo HS, Abedin S, Kamp D et al (2006) EF-G-dependent GTPase on the ribosome. conformational change and fusidic acid inhibition. Biochemistry 45:2504–2514

Sharma PK, Xiang Y, Kato M et al (2005) What are the roles of substrate-assisted catalysis and proximity effects in peptide bond formation by the ribosome? Biochemistry 44:11307–11314

Shaw JJ, Green R (2007) Two distinct components of release factor function uncovered by nucleophile partitioning analysis. Mol Cell 28:458–467

Shoji S, Walker SE, Fredrick K (2006) Reverse translocation of tRNA in the ribosome. Mol Cell 24:931–942

Sievers A, Beringer M, Rodnina MV et al (2004) The ribosome as an entropy trap. Proc Natl Acad Sci U S A 101:7897–7901

Stark H, Rodnina MV, Wieden H-J et al (2000) Large-scale movement of elongation factor G and extensive conformational change of the ribosome during translocation. Cell 100:301–309

Taylor DJ, Nilsson J, Merrill AR et al (2007) Structures of modified eEF2.80S ribosome complexes reveal the role of GTP hydrolysis in translocation. EMBO J 26:2421–2431

Ticu C, Nechifor R, Nguyen B et al (2009) Conformational changes in switch I of EF-G drive its directional cycling on and off the ribosome. EMBO J 28:2053–2065

Trobro S, Åqvist J (2006) Analysis of predictions for the catalytic mechanism of ribosomal peptidyl transfer. Biochemistry 45:7049–7056

Valle M, Zavialov A, Sengupta J et al (2003) Locking and unlocking of ribosomal motions. Cell 114:123–134

Walker SE, Shoji S, Pan D et al (2008) Role of hybrid tRNA-binding states in ribosomal translocation. Proc Natl Acad Sci U S A 105:9192–9197

Wallin G, Aqvist J (2010) The transition state for peptide bond formation reveals the ribosome as a water trap. Proc Natl Acad Sci U S A 107:1888–1893

Weinger JS, Parnell KM, Dorner S et al (2004) Substrate-assisted catalysis of peptide bond formation by the ribosome. Nat Struct Mol Biol 11:1101–1106

Weixlbaumer A, Jin H, Neubauer C et al (2008) Insights into translational termination from the structure of RF2 bound to the ribosome. Science 322:953–956

Wilden B, Savelsbergh A, Rodnina MV et al (2006) Role and timing of GTP binding and hydrolysis during EF-G-dependent tRNA translocation on the ribosome. Proc Natl Acad Sci U S A 103:13670–13675

Wintermeyer W, Zachau HG (1974) Replacement of odd bases in tRNA by fluorescent dyes. Methods Enzymol 29:667–673

Wintermeyer W, Savelsbergh A, Konevega AL et al (2011) Function of elongation factor G in translocation and ribosome recycling. In: Rodnina MV, Wintermeyer W, Green R (eds) Ribosomes structure, function, and dynamics. Springer, New York, pp 329–338

Wohlgemuth I, Brenner S, Beringer M et al (2008) Modulation of the rate of peptidyl transfer on the ribosome by the nature of substrates. J Biol Chem 283:32229–32235

Youngman EM, Brunelle JL, Kochaniak AB et al (2004) The active site of the ribosome is composed of two layers of conserved nucleotides with distinct roles in peptide bond formation and peptide release. Cell 117:589–599

Zaher HS, Shaw JJ, Strobel SA et al (2011) The 2′-OH group of the peptidyl-tRNA stabilizes an active conformation of the ribosomal PTC. EMBO J 30:2445–2453

Zavialov AV, Hauryliuk VV, Ehrenberg M (2005) Splitting of the posttermination ribosome into subunits by the concerted action of RRF and EF-G. Mol Cell 18:675–686

Chapter 8
Investigating RNAs Involved in Translational Control by NMR and SAXS

Kathryn D. Mouzakis, Jordan E. Burke, and Samuel E. Butcher

8.1   Introduction

Translational control, broadly defined, is the posttranscriptional regulation of gene expression within a cell. The majority of translational control is thought to rely on the interaction of regulatory proteins with messenger RNA (mRNA) (Keene 2007). However, a growing body of evidence indicates that mRNA structures play an important role in translational control. For example, certain classes of riboswitches form dynamic RNA structures that influence translation (Lemay et al. 2011). Additionally, RNA–RNA interactions can broadly control translation, as in the case of microRNA (Bartel 2009). Therefore, it is important to consider the role of RNA structure and intrinsic dynamic motions that can modulate RNA function during translation (Talini et al. 2011; Bailor et al. 2010).

Historically, the most informative structural information related to RNA structure and its role in translation has been derived from X-ray crystallography, with the most striking example of this being the structure of the ribosome (Yusupov et al. 2001; Ban et al. 2000; Carter et al. 2000; Selmer et al. 2006; Gluehmann et al. 2001; Pioletti et al. 2001; Zhao 2011). However, obtaining diffraction quality crystals of RNA can be tremendously challenging. Alternatives to X-ray crystallography are nuclear magnetic resonance (NMR) and small angle X-ray scattering (SAXS), which are solution-based methods. In addition to providing structural information, NMR can be used to investigate dynamic motions. In this chapter, we review methods for investigating RNA structure and dynamics in solution. In Sect. 8.1, we focus predominately on NMR spectroscopy and SAXS as complementary techniques for analyzing RNA structure and dynamics in solution. In Sect. 8.2, specific examples of RNAs involved in translational control are discussed.

K.D. Mouzakis • J.E. Burke • S.E. Butcher (✉)
Department of Biochemistry, University of Wisconsin,
Madison, WI, USA
email: sebutcher@wisc.edu

J.D. Dinman (ed.), *Biophysical Approaches to Translational Control of Gene Expression*,   141
Biophysics for the Life Sciences 1, DOI 10.1007/978-1-4614-3991-2_8,
© Springer Science+Business Media New York 2013

8.1.1   From RNA Secondary Structure Prediction to Biophysical Studies

The first step to understanding RNA structure using any solution technique is identification of the RNA sequence of interest and determination of its secondary structure. Prediction of RNA secondary structure is routinely accomplished with Mfold (Zuker 2003). Mfold predicts secondary structure of RNA using free energy minimization based on thermodynamic nearest-neighbor parameters. Mfold is approximately 73% accurate for RNAs less than 700 nt in length (Mathews et al. 1999), and its predictive power greatly improves for smaller RNAs. While the Mfold algorithm alone cannot predict base-pairs involved in pseudoknot formation, several alternative programs have been developed that can (Theis et al. 2008; Andronescu et al. 2005; Hart et al. 2008). The accuracy of secondary structure prediction by Mfold can be improved by incorporating experimental data from sources such as chemical probing (Mathews et al. 2004), NMR (Hart et al. 2008), and/or microarray (Kierzek et al. 2006) experiments. Of these, NMR provides the most direct and rigorous experimental verification of secondary structure, since it is the only method that can directly detect the hydrogen-bonded protons involved in base-pair formation. For larger RNA sequences, accurate prediction of secondary structure is more challenging due to increased base-pairing possibilities leading to alternative secondary structures with similar overall free energies. Prediction of these large secondary structures can be assisted by the use of phylogenetic analysis (Juan and Wilson 1999; Mathews et al. 2010) and chemical probing (Mathews et al. 2004; Weeks 2010; Weeks and Mauger 2011). If the sequence of interest is derived from a larger RNA, it is critical to verify that the secondary structure of the RNA was not perturbed due to the truncation. This can be demonstrated with NMR, by confirming that the chemical shifts for each subdomain are consistent with those observed in the context of the larger RNA. This approach was recently used to examine the secondary structure of a 356 nt HIV-1 genome packaging 5′ leader RNA (Lu et al. 2011).

In vitro transcription with T7 RNA polymerase is standard for production of the milligram quantities of RNA required for NMR (Milligan and Uhlenbeck 1989; Hennig et al. 2001; Scott and Hennig 2008). The majority of the DNA template for transcription can be single-stranded, but the 20 base-pair T7 RNA polymerase promoter region must be double stranded, as described (Milligan and Uhlenbeck 1989). Chemical synthesis of the DNA template strand with phosphoramidite chemistry is customary for synthesis of oligonucleotides 90 nts in length or smaller. For DNA constructs longer than 90 nts, standard cloning techniques (Sambrook and Russell 2001) are used to insert the DNA duplex into a DNA plasmid. Once milligram quantities of plasmid DNA are produced, the plasmid is linearized with a restriction enzyme that cuts immediately following the end of the DNA template sequence (Hart et al. 2008). This linearized product is then used for in vitro transcription. Additionally, T7 RNA polymerase requires at least one but preferably two guanosine residues at the 5′ end of the RNA construct for initiation of transcription (Milligan and Uhlenbeck 1989).

Purification techniques for milligram quantities of RNA have historically relied on a denaturing purification process followed by a refolding step (Hart et al. 2008). These techniques have been explained in great detail elsewhere (Hart et al. 2008; Wyatt et al. 1991). Generally NMR data acquisition requires low monovalent salt concentrations, 100 mM or less, because higher salt concentrations decrease signal to noise (Voehler et al. 2006). Excess salt from the purification process can be removed with size exclusion chromatography, buffer exchange using dialysis, or centrifugal filtration (Lukavsky and Puglisi 2004). Once the RNA sample is in low ionic strength buffer, it can be subjected to a refolding procedure by diluting to ~10 µM, heating to 90–95 °C for 1–2 min, followed by rapid cooling on ice. This step promotes homogenous folding of the RNA and disrupts intermolecular associations (nonnative base-pair formation between molecules) that can occur at higher concentrations. The chemical integrity of the RNA and conformational homogeneity should be analyzed using denaturing and non-denaturing PAGE (Woodson and Koculi 2009), respectively.

A handful of approaches for native purification of RNA have also been published (Lukavsky and Puglisi 2004; Batey and Kieft 2007; Pereira et al. 2010; Luo et al. 2011). These studies employ size exclusion (Lukavsky and Puglisi 2004) and affinity chromatography (Batey and Kieft 2007; Pereira et al. 2010; Luo et al. 2011). Native purification is advantageous because it eliminates (1) RNA aggregates that can form during the ethanol precipitation step after transcription (Lukavsky and Puglisi 2004), and (2) RNA misfolding (Batey and Kieft 2007; Pereira et al. 2010; Luo et al. 2011). RNA aggregation and misfolding can be particularly problematic for RNAs as small as 50 nucleotides. Furthermore, for some RNAs co-transcriptional folding is critical for their homogeneity (Pereira et al. 2010) and catalytic activity (Luo et al. 2011). As is discussed in Sect. 8.1.3, sample homogeneity is required for high quality SAXS data acquisition.

8.1.2 Nuclear Magnetic Resonance Spectroscopy

NMR spectroscopy is a powerful tool for investigating the solution structures of small molecules, proteins, and nucleic acids. Its power stems from direct detection of atoms (typically ^1H, ^{13}C, or ^{15}N) that report structural and dynamic information in solution. RNA secondary, tertiary, and quaternary structure, as well as dynamics, can be probed using NMR. There are numerous examples of RNA solution structures solved using NMR; 489 NMR structures with RNA have been deposited in the Protein Data Bank (PDB) (www.pdb.org).

NMR has several limitations. The first limiting factor is molecular size. The signal-to-noise ratio in an NMR experiment is dependent upon spectrometer sensitivity, the sample concentration, and the NMR relaxation rate. As molecular weight increases, the NMR relaxation rate also increases, which results in line broadening and rapid loss of signal. Transverse relaxation optimized spectroscopy (TROSY) has greatly extended the molecular weight limit for molecules studied with NMR

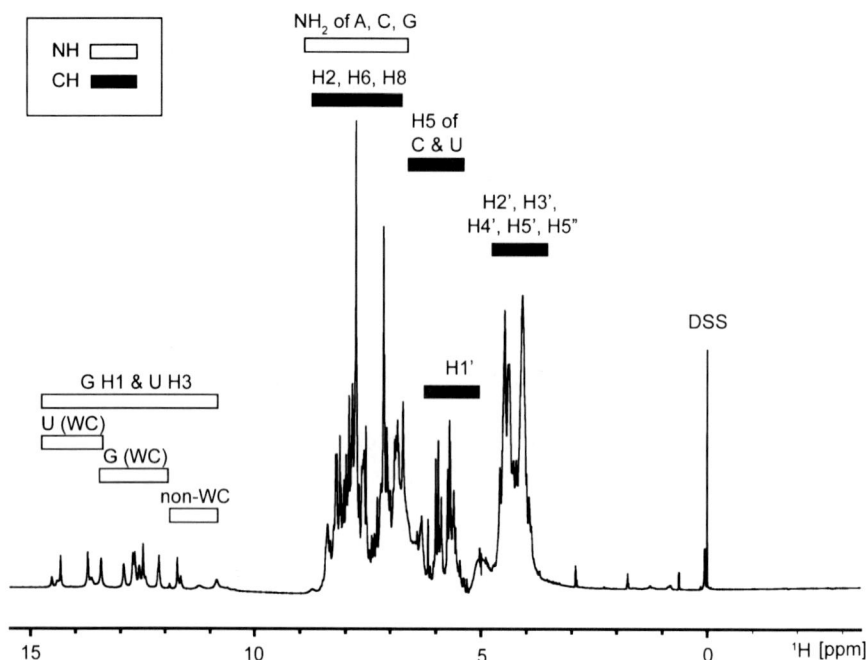

Fig. 8.1 ^{1}H 1D spectrum of a 32 nt RNA (PDB ID 1XHP). Characteristic ^{1}H chemical shift ranges are indicated for nitrogen-bonded protons (*white bars*) and carbon-bonded aromatic and ribose protons (*black bars*). Imino protons involved in Watson-Crick (WC) base-pairing have a chemical shift range of 10–15 ppm. DSS is 4,4-dimethyl-4-silapentane-1-sulfonic acid, a chemical shift calibration standard

(Tzakos et al. 2006; Fernández and Wider 2003). TROSY partially counters the dipole–dipole relaxation between neighboring nuclei using constructive interference from chemical shift anisotropy (Hennig et al. 2001; Pervushin et al. 1997). Currently, the practical size limitation for RNA structure determination by NMR is approximately 40 kDa, or around 100 nucleotides. However, in some favorable cases, NMR can be used to determine secondary structures of much larger RNAs (Lu et al. 2011). Approaches for resolving chemical shift overlap for large RNAs are discussed in greater detail in Sect. 8.1.2.3. The second potentially limiting factor is sample concentration. NMR is an inherently insensitive technique, so the method generally requires high concentrations of sample. For biological macromolecules, this is typically between 0.1 and 2 mM. Even with 1 mM sample concentrations, signal averaging of multiple experiments is typically used to increase the signal-to-noise.

Resolution is the third major challenge in NMR and is particularly problematic for RNA. Unlike proteins, which are composed of 20 distinct amino acids, nucleic acids are made up of only four chemically similar nucleotides, which results in a small degree of chemical shift dispersion (Fig. 8.1). Additionally, the proton density within a nucleotide is concentrated primarily in the ribose ring, and the majority of protons in the ribose ring resonate in a narrow spectral window of ~1 ppm. This high level of chemical shift overlap can make it challenging to obtain complete resonance assignments. Therefore, isotopic ^{13}C and ^{15}N labeling (Nikonowicz et al. 1992; Nikonowicz and Pardi 1993) is required to resolve ^1H chemical shifts into additional dimensions by correlating ^1H shifts to their bonded carbon and/or nitrogens. Such multidimensional heteronuclear NMR experiments are instrumental in resolving overlapped ^1H chemical shifts (Hennig et al. 2001).

8.1.2.1   NMR Analysis of Base-Pairing Interactions and Secondary Structure

The first step in investigating RNA structure using NMR is measuring a 1D ^1H spectrum. The observation of imino proton resonances from G and U nucleotides (note there are no imino protons on A and C), in the 10–15 ppm range, is indicative of hydrogen bond formation and base-pairing (Fig. 8.1). The imino resonances can be assigned to their respective nucleotides using ^1H-^1H 2D nuclear Overhauser effect spectroscopy (NOESY) experiments. The NOESY experiment provides information about the number and type of base-pairs, as well as their sequential neighbors (Heus and Pardi 1991). In a ^1H-^1H 2D NOESY experiment, cross-peaks (NOEs) are observed between protons that are less than 6 Å apart. In a standard A-form helix, an NOE cross-peak is observed between adjacent base-pairs (Hennig et al. 2001) (Fig. 8.2b). Thus, sequential neighbors for base-pairs are identified by "walking-through" these cross-peaks (Fig. 8.2c).

The HNN-COSY experiment (Dingley and Grzesiek 1998) is extremely useful for investigating base-pairing in ^{15}N-labeled RNA, because it directly detects

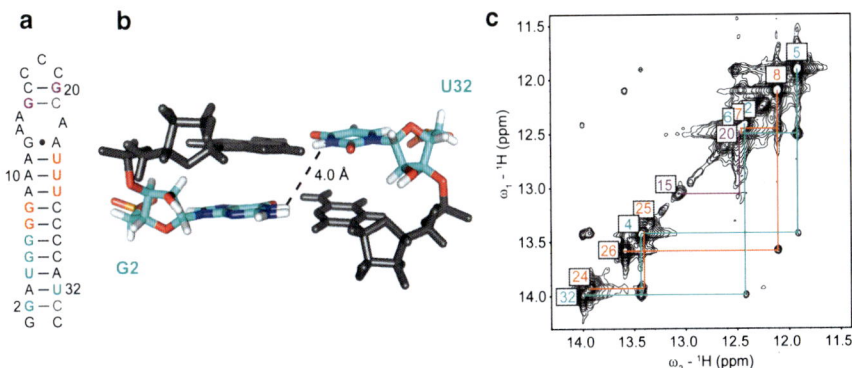

Fig. 8.2 ^1H-^1H 2D NOESY experiment for assignment of imino protons in the SIV frameshift stem-loop RNA structure (Marcheschi et al. 2007). (**a**) RNA secondary structure. (**b**) Base-pairs stacked in A-form helical regions are oriented such that the G and U imino protons are within 5 Å and will exhibit a cross-peak in the ^1H-^1H 2D NOESY. (**c**) Zoomed-in region of 2D NOESY showing imino cross-peaks. A sequential walk is shown. *Colored lines* correspond to (**a**)

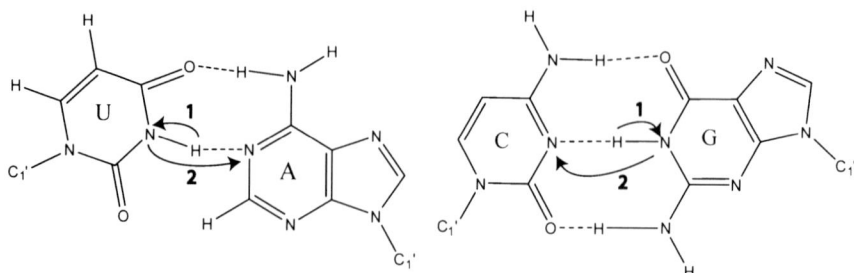

Fig. 8.3 The HNN-COSY experiment directly detects base-pairing by correlating the acceptor and donor nitrogen atoms involved in the hydrogen bond. Magnetization is transferred from the imino proton to the donor nitrogen (1), and then from the donor nitrogen through the hydrogen bond to the acceptor nitrogen (2). After labeling the magnetization, it is transferred back to the imino proton through the reverse process. Adapted from Tzakos et al. (2006)

hydrogen bonding. The simplest application of this experiment correlates the donor and acceptor ^{15}N atoms across N-H⋯N hydrogen bonds in Watson-Crick base-pairs (Tzakos et al. 2006; Dingley and Grzesiek 1998). During the experiment, magnetization is transferred in two steps (Fig. 8.3): first, from the imino proton to its directly bonded donor nitrogen, and second, from the donor nitrogen through the hydrogen bond to the acceptor nitrogen. The magnetization is transferred back to the imino proton through the reverse of the first two steps. These experiments have been extended to detect NH⋯O = C, OH⋯N-, OH⋯O = P and NH$_2$⋯O = P hydrogen bonds (Duchardt-Ferner et al. 2011).

8.1.2.2 Towards Solving NMR Structures

The quality of the ensemble of NMR structures is dependent upon the quality of restraints, which includes the overall number of distance restraints (NOEs) measured (Allain and Varani 1997). RNA structure determination by NMR has historically relied on the measurement of many short range distance restraints. This approach requires unambiguous chemical shift assignment for as many of the protons in the RNA as possible. For RNAs larger than 30–40 nucleotides, uniform ^{13}C, ^{15}N labeling may not be sufficient to resolve proton spectral overlap. To circumvent this problem, RNAs can be labeled using nucleotide-type specific labeling strategies (Hennig et al. 2001; Lu et al. 2010). Selective deuteration (Davis et al. 2005) and perdeuteration (Lu et al. 2010) can also be used to dramatically simplify NMR spectra and resolve chemical shift overlap (discussed in Sect. 8.1.2.3).

Chemical shift assignment is accomplished in two basic ways. The first and most direct is via through-bond experiments that take advantage of scalar couplings between protons and neighboring ^{1}H, ^{13}C, and ^{15}N nuclei (Hart et al. 2008; Hennig et al. 2001; Tzakos et al. 2006; Clos et al. 2011). Because chemical shifts are dominated by local electronic environment, the chemical shift of the proton and its directly bonded and neighboring nuclei can be used to help assign the identity of the proton. Thus, experiments that correlate a proton chemical shift to its directly bonded nucleus, such as 2D heteronuclear single quantum coherence (HSQC), are invaluable for proton resonance assignment (Hennig et al. 2001). Experiments that correlate multiple bonds can provide even more information. For example, triple resonance HCN-type heteronuclear NMR experiments can be used to correlate nucleobase aromatic protons to a particular ribose group by appropriately transferring magnetization through the glycosidic bond (Sklenář et al. 1998).

The second strategy for proton resonance assignment is based on through-space transfer of magnetization, via the NOE. The nucleotide structure contains a number of short (<6 Å) interproton distances that give rise to NOEs. Additionally, RNA structures are dominated by the A-form helical geometry, which gives rise to predictable NOE patterns that can be used to obtain resonance assignments of neighboring nucleotides (Hennig et al. 2001). Because the intensity of the NOE is distance-dependent (r^{-6}), NOEs obtained from these experiments are utilized as distance restraints for use in restrained molecular dynamics simulations to produce an ensemble of structures that satisfy the restraints (Lukavsky and Puglisi 2005).

Torsion angle restraints are also important for structure determination. The sugar-phosphate backbone and ribose ring are characterized by six backbone torsion angles and the ribose sugar pucker, respectively. Ribose sugar puckers in A-form RNA are C3'-endo, but in non-helical regions such as loops the ribose ring may adopt a C2'-endo conformation or undergo dynamic exchange between C3'- and C2'-endo conformations. The sugar pucker can be analyzed using 2D COSY or TOCSY experiments (Hennig et al. 2001). Backbone torsion angle measurements around the phosphodiester bond can be quantitatively measured (Nozinovic et al. 2010a, b). However, experiments designed to quantitatively measure backbone torsion angles are often impractical for RNAs larger than 30 nucleotides, because they rely on the ability to resolve phosphorous chemical shifts, which are usually extensively overlapped.

Ideally, an NMR structure should be constrained by a large number of long-range NOEs between protons far apart in sequence (Allain and Varani 1997). However, the vast majority of NOEs occur between protons within a nucleotide (intraresidue) or between neighboring nucleotides. While these NOEs are important for establishing resonance assignments and defining local structure, they cannot sufficiently define the global RNA structure. RNA NMR structure quality can therefore be significantly improved by inclusion of long-range distance restraints. Such information can be obtained through the measurement of residual dipolar couplings (RDCs) (Prestegard et al. 2000; Bax et al. 2001). RDCs measure the orientation of a bond vector relative to the overall alignment of a molecule within the magnetic field. Therefore, RDCs provide angular restraints that improve both the local and global structural quality (Tzakos et al. 2006) and can also be used to characterize RNA dynamics (discussed in greater detail in Sect. 8.1.2.4).

RDCs are not normally observed in solution under isotropic conditions because the overall tumbling of a molecule causes the angular term in the RDC equation (Fig. 8.4) to average to a value that approaches zero (Tolman 2001). However, by

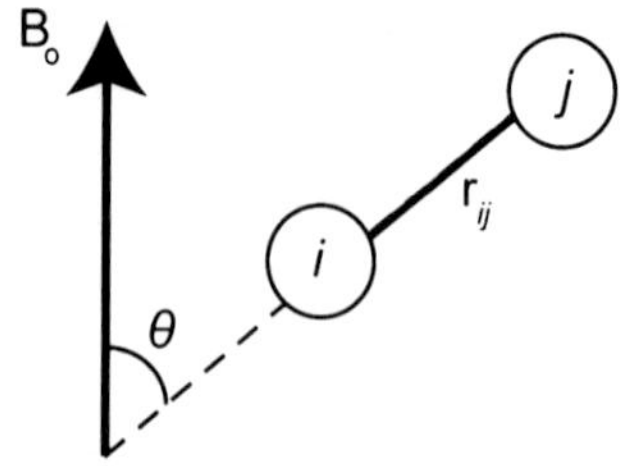

$$D_{ij} = -\frac{\gamma_i \gamma_j \mu_o h}{(2\pi)^3 (r_{ij})^3} \left\langle \frac{3\cos^2(\theta - 1)}{2} \right\rangle$$

Fig. 8.4 RDCs (D_{ij}) between spins i and j provide long-range distance restraints for the average orientation (θ) of the internuclear vector relative to the magnetic field (B_o). Here, γ_i and γ_j are the gyromagnetic ratios of spins i and j, respectively, r_{ij} is their internuclear distance, h is Plank's constant, μ_o is the magnetic permittivity of vacuum, and the angular term is shown in *brackets*. RDCs are not observed in isotropic solution conditions because the angular term averages to nearly zero. Adapted from Getz et al. (2007)

imparting a small degree of alignment to the sample, the dipolar coupling retains information about the angle of a bond vector relative to the alignment tensor. RDC values are determined by subtracting the measured dipolar coupling in isotropic conditions from the coupling in partially aligned conditions. For RNA, alignment is typically achieved by adding filamentous bacteriophage (Pf1) to the RNA (Prestegard et al. 2000; Hansen et al. 1998).

8.1.2.3   Solution Structures of Large RNAs

Solving solution structures of even moderately sized RNAs (50–100 nts) can be extremely challenging due to spectral overlap and fast relaxation rates. Only a handful of RNA structures greater than 75 nucleotides have been solved using NMR. These include the 77 nt hepatitis C viral (HCV) internal ribosome entry site (IRES) domain II (Lukavsky et al. 2003), an 86 nt tetraloop receptor complex (Davis et al. 2005), a 101 nt core encapsidation signal of Moloney murine leukemia virus (MoMuLV) (D'Souza et al. 2004) (see Sect. 8.2), a 102 nt ribosome-binding structural element (RBSE) in Turnip Crinkle Virus (TCV) (Zuo et al. 2010), and the 140 nt dimeric genome packing signal in MoMuLV (Miyazaki et al. 2010). Each of these studies employed a different approach to reduce spectral overlap and improve NMR relaxation rates associated with large macromolecules (Fig. 8.5).

The most common approach for studying large RNAs has been the so-called divide and conquer approach (reviewed in Clos et al. 2011). Here, the RNA is separated into small, thermodynamically stable helical domains. By breaking the RNA into smaller domains, spectral overlap is reduced and assignment of the domain is often feasible. Unfortunately, the structure of the subdomain may change in the context of the larger, more biologically relevant RNA, if it is involved in tertiary interactions (Tzakos et al. 2006). Therefore, it is critical to verify that the chemical shifts for each subdomain are consistent with those observed in the context of the larger RNA.

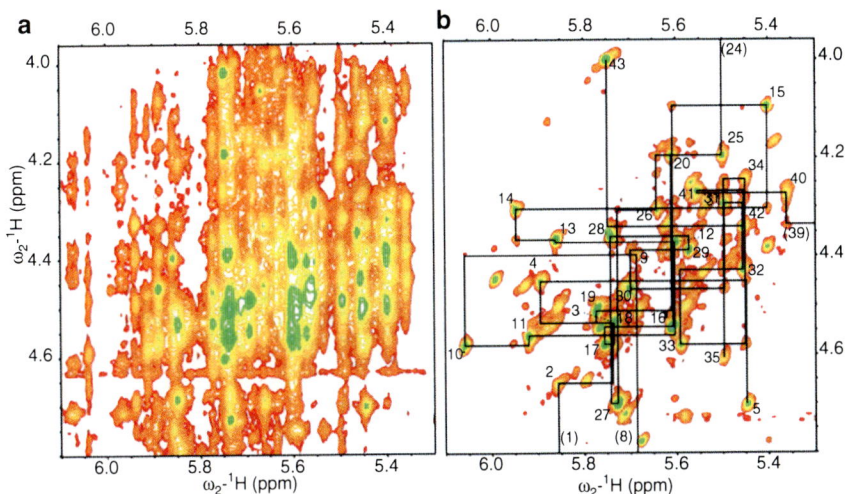

Fig. 8.5 Selective deuteration greatly simplifies NMR spectra (Davis et al. 2005). (**a**) A 900 MHz 2D NOESY spectrum of the inter- and intra-ribose region. (**b**) A 900 MHz 2D NOESY of the same region of the selectively deuterated sample. A sequential internucleotide walk from H2' (*i*) to H1' (*i* + 1) is drawn. *Numbers in parentheses* are for those sequential peaks not in the region displayed. Reprinted from Davis et al. (2005), with permission from Elsevier

Several labeling strategies can be used to facilitate chemical shift assignment. Because ^{2}H nuclei are not visible by ^{1}H NMR, selective deuteration can greatly simplify highly crowded regions of the spectrum, like the ribose region (Fig. 8.5) (Davis et al. 2005). Alternatively, incorporation of a protonated nucleotide-type into an otherwise perdeuterated RNA chain is another successful strategy for greatly simplifying NMR spectra of large RNAs (Lu et al. 2010; D'Souza et al. 2004; Miyazaki et al. 2010). Deuteration also improves relaxation properties for larger RNAs, leading to an improved signal-to-noise ratio (Scott et al. 2000). Deuteration has been useful for several large RNA structures (Davis et al. 2005; D'Souza et al. 2004; Miyazaki et al. 2010).

A second method facilitating chemical shift assignment uses selective nucleotide-type ^{13}C,^{15}N labeling in conjunction with filter-edited NOESY experiments (D'Souza et al. 2004; Peterson et al. 2004). Filter-edited NOESY experiments identify intermolecular NOEs between isotope labeled and unlabeled nucleotides. We found that 2D filter-edited experiments take approximately 24 h and were sufficient for assignment of an 86 nt RNA complex (Davis et al. 2005).

8.1.2.4 RNA Dynamics

It is increasingly clear that RNA can sample a wide range of structural conformations due to inherent dynamics (Bailor et al. 2010). A complete understanding of RNA structure and function requires knowledge of its dynamic motions. Many biophysical

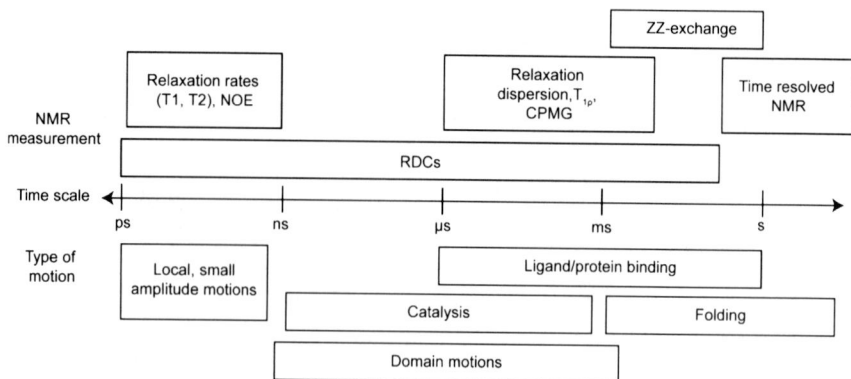

Fig. 8.6 NMR measurements used to study RNA dynamics for motions in the picosecond (ps) to second (s) timescale. Adapted from Shajani and Varani (2007)

approaches have been developed to probe RNA dynamics: molecular dynamics simulations (Hall 2008; Chen 2008; Mackerell and Nilsson 2008), single-molecule fluorescence spectroscopy (Roy et al. 2008; Zhao and Rueda 2009; Walter 2001), time-resolved hydroxyl radical footprinting techniques (Sclavi et al. 1998; Shcherbakova et al. 2006), single-molecule force measurements (Li et al. 2008), and NMR methods (Shajani and Varani 2007; Getz et al. 2007; Furtig et al. 2007; Rinnenthal et al. 2011). In this section, we highlight the different approaches for measuring RNA dynamics using NMR.

NMR provides a unique way to investigate dynamics over time scales encompassing biologically relevant motions (Fig. 8.6) (Shajani and Varani 2007; Getz et al. 2007; Rinnenthal et al. 2011). Dynamic motions can occur at the atomic level or can correspond to entire domains (Getz et al. 2007). RNA motions in the picosecond to nanosecond range reflect local motions (Shajani and Varani 2007). These motions are characterized by measuring the longitudinal (T_1) and transverse (T_2) relaxation rates for ^{13}C or ^{15}N nuclei, and the heteronuclear NOE between these nuclei and their bonded protons. The Lipari-Szabo model free approach (Lipari and Szabo 1982) is the most popular method utilized to quantitatively describe these motional amplitudes and their rates. These types of experiments have proved very useful in characterizing local motions for several RNAs (Shajani and Varani 2005; Duchardt and Schwalbe 2005; Akke et al. 1997; Dayie et al. 2002; Musselman et al. 2010; Hall and Tang 1998). Slower motions in the microsecond to millisecond range are often reflective of more complex dynamic processes such as conformational changes, domain motions, folding, and ligand binding. These motions can be investigated using relaxation dispersion experiments (Shajani and Varani 2005; Kloiber et al. 2011; Johnson and Hoogstraten 2008). Several examples demonstrating the use of these techniques to study RNA are available (Duchardt and Schwalbe 2005; Kloiber et al. 2011; Blad et al. 2005; Dethoff et al. 2008). Slow processes on the millisecond to second timescale can be measured using ZZ-exchange NMR experiments (Latham et al. 2009). Rates for processes that occur on even longer timescales can be measured by time-resolved NMR methods (Furtig et al. 2007; Schanda et al. 2005; Lee et al. 2010).

The study of RNA dynamics by NMR has been expanded by the measurement and analysis of RDCs. RDCs report changes in bond orientation relative to the magnetic field resulting from internal and overall molecular motions (Tolman 2001) (Fig. 8.4). These couplings report on the weighted average of all conformations sampled by a macromolecule over the course of picoseconds to milliseconds. This time range encompasses both local motions as well as large-scale conformation changes associated with RNA domain motions (Fig. 8.6).

A-form helices are the predominate building blocks of RNA structures. Because they are semi-rigid and highly regular, their overall geometries are independent of sequence as long as base-pairing is maintained (Wang et al. 2010). Therefore, a large, complex RNA structure can be thought of as a combination of A-form helical domains, which can be directly linked or separated by single-stranded regions that form loops, turns, bulges, and linkers (Hermann and Patel 2000; Butcher and Pyle 2011). These elements allow RNA to sample a wide range of conformations. NMR provides a unique way to quantitatively characterize domain motion and orientation using RDCs (Getz et al. 2007; Zhang and Al-Hashimi 2009). Domain motion and orientation can be efficiently examined using RDCs by solving the order tensor solution (Bailor et al. 2007), which describes the average orientation of each domain relative to the magnetic field and provides information on direction and amplitude of motion.

Investigation of RNA dynamics using order tensor analysis is efficient because solution structures and complete resonance assignments are not required. Order tensor analysis requires a minimum of five nonparallel RDCs for each helical domain (excluding RDCs from terminal base-pairs, which are dynamic due to fraying). Because helices are highly regular structures, they can be accurately modeled using RNA modeling programs (Bailor et al. 2007). Practically, a set of ≥ 11 RDCs is needed for a well-determined order tensor solution. Consider an RNA molecule with two helices separated by an asymmetrical bulge (Bailor et al. 2007). To determine the interhelical bend angle and dynamics between the two helical domains, order tensor solutions are determined for each helix independently. A ^{13}C, ^{15}N labeled RNA sample is used to measure aromatic (CH), ribose (C1'H1', C2'H2', C3'H3', and C4'H4'), and imino (NH) dipolar couplings in isotropic and partially aligned conditions. The difference in dipolar coupling in the partially aligned and isotropic conditions yields the RDC measurement. The dipolar coupling is manifested as a measurable peak splitting (Hz), which should be measured in duplicate along the $^{13}C/^{15}N$ and ^{1}H dimensions to assess the RDC error. The order tensor solution for each helix can be computed using software such as RAMAH (Hansen and Al-Hashimi 2006). The Euler angles from the order tensor solution are used to rotate model A-form helices into their principle axis system (PAS). In their PAS, the helices are connected according to their covalent bonding and the interhelical bend angle can be calculated using the Euler-RNA software (Bailor et al. 2007). RNA dynamics are encoded by the general degree of order (GDO) and η terms in the order tensor solution (Getz et al. 2007). The GDO describes the structural rigidity and amplitude of helical motions, while η describes the directionality of those motions.

The method described is limited by the 4^{n-1} degeneracy of the order tensor solution (Tolman and Ruan 2006), where n is the number of independent rigid helical domains that need to be oriented to each other. While this degeneracy can sometimes be partially overcome by covalent connectivity, steric clash, or experimental restraints (Bailor et al. 2007), it may be challenging to determine the correct orientation for RNAs with more than two domains (Zuo et al. 2008). Additionally, domain motions can occur on the same timescale as overall rotational motions (Dethoff et al. 2009), which convolutes dynamics analysis. One way to decouple domain motions from the overall reorientation of the molecule is through RNA helical extension (Zhang and Al-Hashimi 2009) such that the domain motions occur on a different timescale from the overall rate of molecular tumbling.

The dynamics of the HIV-1 transactivation response (TAR) RNA (Al-Hashimi et al. 2002; Pitt et al. 2004; Zhang et al. 2006; Casiano-Negroni et al. 2007) has been extensively studied using RDCs. Located in the 5′ end of all HIV-1 pre-mRNA transcripts (Muesing et al. 1987), the TAR hairpin regulates viral replication (Frankel 1992) through its interaction with the trans-activator protein (Tat) (Cullen 1986). Al-Hashimi *et al.* demonstrated that in low salt conditions TAR has an average interhelical bend angle of 47° and moves with isotropic motions sampling all positions within a cone radius angle of 46° (Al-Hashimi et al. 2002). In the presence of magnesium chloride and increasing concentrations of sodium chloride, TAR's dynamic motions are quenched and the RNA linearized (Pitt et al. 2004; Casiano-Negroni et al. 2007).

For the HIV-1 TAR RNA, using elongated-helical domains to study RNA dynamics yielded residue-specific measurements of internal motions in TAR that were not apparent using the smaller, non-elongated RNA (Zhang et al. 2006). Molecular dynamics and order tensor analysis were further combined to trace out the entire helical trajectory for TAR where the helices bend and twist in a correlated manner (Frank et al. 2009). These studies of TAR demonstrate the utility of using NMR to study RNA dynamics.

8.1.3   RNA Structure Analysis by Small Angle X-Ray Scattering

Many RNA structures important for translational control of gene expression are large by NMR standards, making structure determination quite challenging and expensive. Fortunately, SAXS is a complementary technique for extracting global shape information about macromolecules in solution. SAXS also provides a method for monitoring the kinetics of macromolecular folding in real time and can be used to assess and compare the solution state conformation of molecules to structures determined by X-ray crystallography and cryo-electron microscopy. Because SAXS alone provides low resolution structural information, the combination of SAXS with an NMR-based approach is ideal for characterization of large RNA molecules in solution. SAXS experiments can be performed at much lower concentrations than NMR, requiring significantly less sample. Because data collection takes only milliseconds with synchrotron X-ray sources, SAXS is also a valuable method for characterization of folding kinetics and RNA dynamics. While characterization of large macromolecules by NMR is often difficult or intractable, SAXS effectively has no upper size limit. This section addresses sample requirements, analysis of SAXS data, modeling of RNA structures based on SAXS, and implementation of a combined NMR/SAXS approach.

8.1.3.1 Sample Requirements for SAXS

Since all species in solution contribute to X-ray scattering, the sample must be pure. Typically, sample concentrations range from 0.1 to 3 mg/mL. RNA samples at concentrations of 3 mg/mL or higher may exhibit structure factors related to repulsion that can interfere with accurate measurement of molecular size; however, higher salt concentrations can be used to reduce these effects. Additionally, the buffer scattering must be subtracted and buffer matching is crucial and can be achieved by extensive dialysis.

SAXS measures the simultaneous scattering of all orientations of molecules in solution, which is represented by the scattering profile, $I(q)$ (Fig. 8.7a). Species that are present even at low concentrations contribute to X-ray scattering, such that contamination from formation of higher order complexes or misfolded species can greatly interfere with obtaining high quality data. Misfolded species can be eliminated through the use of size exclusion purification immediately before data collection (Rambo and Tainer 2010a).

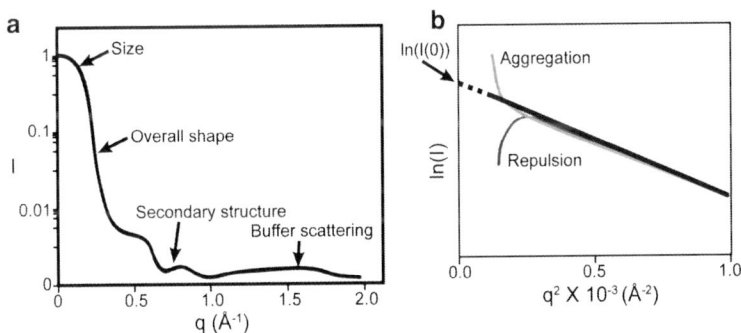

Fig. 8.7 Small angle X-ray scattering (SAXS) data. (**a**) X-ray scattering intensity (I) vs. scattering angle (q). *Arrows* point to regions of plot that correspond to structural information. (**b**) A Guinier plot, $\ln(I)$ vs. q^2, provides information about the size of the molecule and the quality of the sample. This plot should be linear at small values of q ($q_{max} * R_g < 1.2$). A sharp drop-off in the plot is indicative of repulsion while an upward curve is indicative of aggregation. *Dashed line* indicates data extrapolation, based on the Guinier equation, to $q = 0$ (*arrow*)

   Two basic types of X-ray sources are available for collection of SAXS data: synchrotron radiation and bench-top sources. While synchrotron radiation provides a higher intensity of X-rays and thus reduces data collection time, it can also increase the risk of radiation damage to the sample. To minimize radiation damage, the sample can be flowed back and forth across the beam during data collection. Secondary radiation damage can also occur due to X-ray-induced generation of hydroxyl radicals, which can promote RNA hydrolysis. To minimize this, tris buffer (~20 mM) can be included as an effective hydroxyl radical scavenger. Collection of SAXS data using a bench-top instrument requires a smaller sample volume (as little as 30 μL); however, the low intensity of the X-rays requires a much longer data collection time. While it is tempting to overcome this problem by using a higher concentration of RNA, this may result in interparticle repulsion as described above (Putnam et al. 2007).

8.1.3.2 Interpreting SAXS Data

The SAXS profile, $I(q)$, results from observation of X-ray scattering from all orientations of the molecule in solution (Rambo and Tainer 2010b). The resulting 2D scattering pattern can therefore be radially averaged and converted into a 1D scattering curve to maximize the amount of signal obtained from a given experiment. The scattering curve contains valuable information about the dimensions, volume, and fold of an RNA molecule (Fig. 8.7a). The scattering angle is expressed in reciprocal space as a function of q (Koch et al. 2003):

$$q = \frac{4\pi \sin\theta}{\lambda} \tag{8.1}$$

where θ is one-half the scattering angle from the incident beam and λ is the wavelength of the X-ray radiation. The maximum intensity of the scattering curve is dependent on the source and type of detector and is therefore frequently normalized to 1 to allow for comparison of different data sets. The smallest angles provide information about the size of the molecule (Fig. 8.7a). For most molecules, this falls in the range of $q < 0.05$ Å^{-1}. Scattering in the range of $q < 0.3$ Å^{-1} contains information about the shape of the molecule. Peaks observed in the range of $0.3 < q < 1.0$ Å^{-1} arise from internal secondary structure within the molecule (Fig. 8.7a). Unfortunately, due to the fact that the molecules in solution are randomly oriented, high resolution information cannot be extracted from this region for the purposes of structure analysis.

   The radius of gyration (R_g) is the root mean square distance of electron density from the center of mass and provides an accurate measure of size and shape that is useful for comparison between different samples. R_g can be estimated using a Guinier transform which exploits an approximately linear relationship between $\ln(I(q))$ and q^2 at low values of q (Konarev et al. 2003):

$$\ln I(q) = I(0) - \frac{1}{3}R_g q^2 \tag{8.2}$$

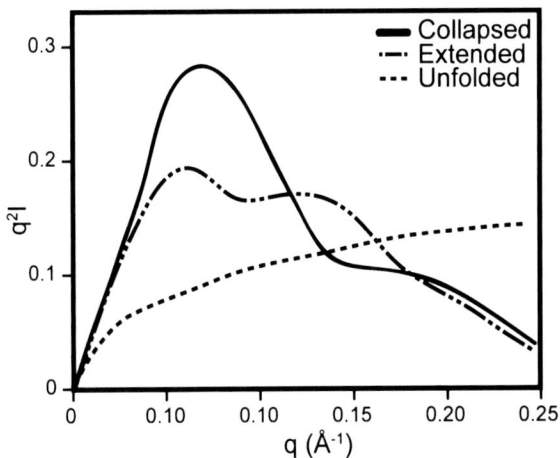

Fig. 8.8 The Kratky plot, q^2I vs. q, indicates the extent of folding within an RNA molecule. Molecules with an extensive tertiary fold (collapsed) result in a different profile than non-globular molecules (extended) or unfolded molecules. Adapted from Bai et al. (2005)

where $I(0)$ is the intensity of scattering at $q=0$. $I(0)$ is related to the molecular weight of the molecule and should not change with sample concentration. The data range for this method should be limited such that $q_{max}*R_g < 1.2$, where q_{max} is the maximum q value. The Guinier plot also provides information about sample quality (Fig. 8.7b). Deviations from linearity indicate possible interparticle interactions such as aggregation or, more commonly with nucleic acids, repulsion (Fig. 8.7b). Changes in R_g have been employed to measure both kinetics of binding and affinity in ribozyme systems (Fang et al. 2000). Frequently, this type of analysis is adequate for studying global tertiary collapse during RNA folding (Pollack 2011) and has been used to characterize distinct stages during folding of RNase P (Roh et al. 2010).

Another transform of the scattering profile, the Kratky plot (Fig. 8.8), is useful for inferring qualitative information about the level of structure in a molecule. Molecules that are completely unfolded exhibit a very different profile from molecules with both extensive secondary and tertiary structure (Fig. 8.8) (Bai et al. 2005). These profiles are distinct from a "random" fold, in which helices are randomly oriented in relationship to one another, or a non-globular or "extended" fold where the RNA helices are rigidly extended away from one another (Fig. 8.8) (Bai et al. 2005). Kratky analysis of time-resolved SAXS experiments has been used to assess the kinetics of folding of RNase P (Roh et al. 2010) and the Tetrahymena ribozyme (Das et al. 2003), as well as the contribution of cation concentration to the extent of collapse within the glycine riboswitch (Lipfert et al. 2010).

The pair distance distribution function (PDDF) or $p(r)$ reflects structural features of the molecule (Fig. 8.9) and is calculated via an indirect Fourier transform that can be performed using the GNOM program (Svergun 1992). The PDDF represents the real space distribution of interatomic vectors within the RNA that can be thought of

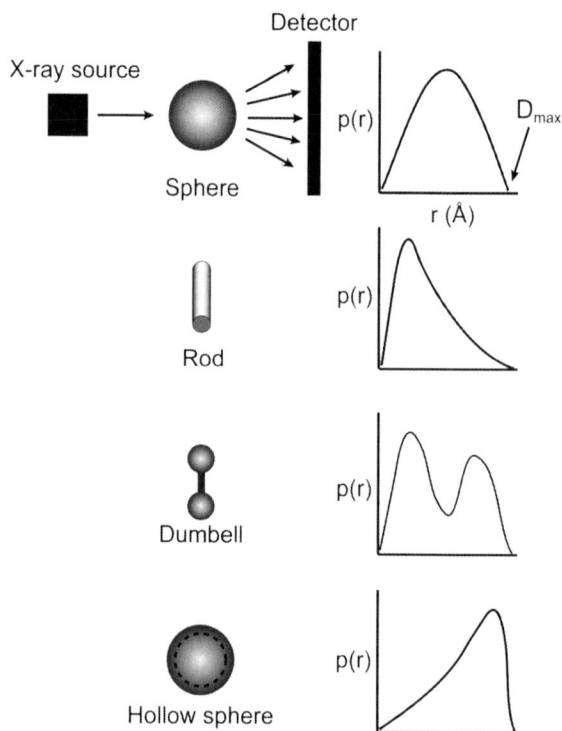

Fig. 8.9 The pair distance distribution function (PDDF) or $p(r)$ reflects the shape of the molecule. The PDDF can be thought of as the probability of finding two atoms in the molecule a given distance (r) apart. Adapted from Koch et al. (2003)

as the probability of finding two atoms a given distance (r) apart. R_g and $I(0)$ are more accurately determined from the PDDF than from the Guinier transform (Svergun 1992). Additionally, the PDDF provides an estimate of the maximum dimension (D_{max}) of the molecule; however, the value of D_{max} cannot be determined with the same level of accuracy as R_g and is therefore not as widely employed.

8.1.3.3 Ab Initio Modeling of Molecular Shapes from SAXS Data

The information provided by the PDDF can be utilized to generate a low resolution envelope of an RNA using ab initio modeling techniques. Currently, DAMMIN (Svergun 1999) and its more recent implementation, DAMMIF (Franke and Svergun 2009), are the best software options for ab initio modeling of nucleic acid molecules. Both of these programs employ a set of dummy atoms or beads to simulate the general shape of the molecule. The predicted scattering amplitude of the dummy atom model is calculated and compared to the experimental scattering data. This process is repeated iteratively until a good fit is achieved. Typically, only data in the range of $q < 0.33$ Å^{-1} are used with this method for the reasons described in Sect. 8.1.3.2.

Due to the inherently low resolution associated with ab initio modeling from SAXS data, the arrangement of domains in a molecule cannot always be unambiguously determined using this method. For RNA, helical extensions have been employed in a variety of studies to assign structural features within a low resolution structure (Ali et al. 2010). Also, because these programs use the output of GNOM, it is essential that the parameters employed to generate the PDDF (e.g., D_{max}) are optimized before calculating an envelope. Typically, several envelope models are generated (>10) and then compared with each other to gauge the quality of the model. This can be achieved using the DAMAVER software package (Volkov and Svergun 2003), which provides a measure of agreement between the molecules, the normalized spatial discrepancy (NSD). The NSD should fall between 0.7 and 0.9 for a unique, well-determined model (Volkov and Svergun 2003).

8.1.3.4 All-Atom Molecular Modeling

Because of the difficulty of inferring structural details about the position of molecular features from a low-resolution envelope, a preferable method for modeling molecular structure employs the use of all-atom models that have been selected for and/or refined against the scattering data (Baird et al. 2005). Detailed structural information cannot be directly derived from the observed scattering profile; therefore, selection of the best models is dependent upon accurate back-prediction of the scattering profile for each model.

If an existing structure is available for a related molecule, the best modeling method is homology modeling. This can be achieved using nonbonded noncrystallographic symmetry (NCS) restraints in XPLOR-NIH, as in the case of the structure of tRNA^{Val} (Grishaev et al. 2008a) (Fig. 8.10). A model of tRNA^{Val} was first

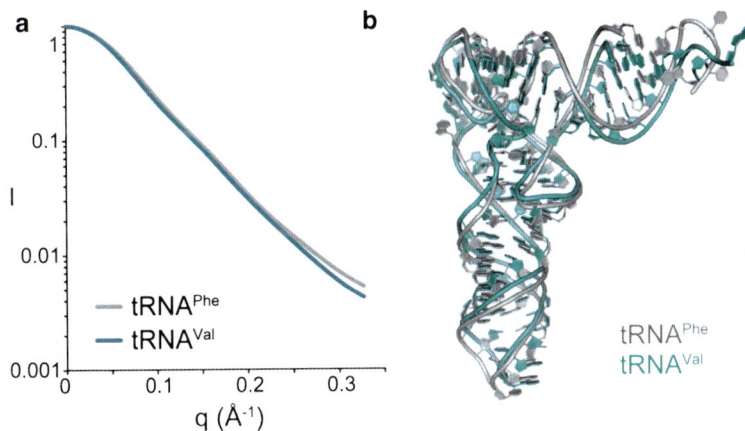

Fig. 8.10 Homology modeling and refinement of tRNA^{Val} using SAXS and RDC measurements. (**a**) Predicted scattering curves of tRNA structures. (**b**) Comparison of the tRNA^{Phe} crystal structure (PDB ID 1TRA) with the refined tRNA^{Val} homology model (PDB ID 2K4C)

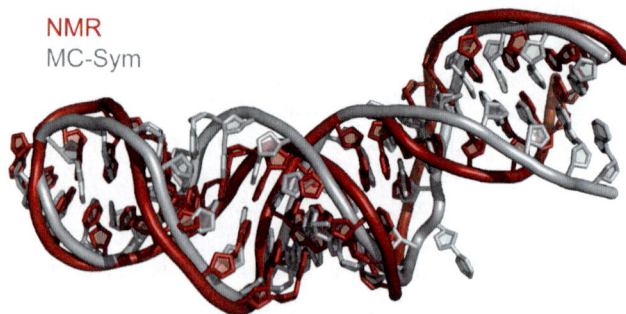

Fig. 8.11 Modeling the structure of an RNA with MC-Sym. Model of a small (43 nucleotide) tetraloop receptor RNA generated with the MC-Fold/MC-Sym pipeline (*gray*), which has a 4.1 Å RMSD from the NMR structure (*red*) (Nikonowicz and Pardi 1993)

built by constraining the model based on the crystal structure of tRNA[Phe], and then refined against SAXS (Fig. 8.10a) and RDC measurements in XPLOR-NIH, resulting in a structural model that provided information about tRNA[val] (Fig. 8.10b).

For RNA molecules, the most straightforward method in the absence of a homology model is the MC-Fold/MC-Sym pipeline (Parisien and Major 2008). Some models predicted using this software can deviate by as little as 2–4 Å from the experimentally determined structure (Parisien and Major 2008). Accuracy of models can be judged in part by observation of base-pairing, either by biochemical methods or NMR (see Sect. 8.1.2.1). MC-Sym employs a library of structures from the PDB to create mosaic models that include small fragments from matching sequences in the library. For small helical RNA structures, this modeling method can be surprisingly accurate (Fig. 8.11) (Davis et al. 2007). This method has been successfully employed with the TPP riboswitch to create all-atom models that are consistent with SAXS measurements (Ali et al. 2010). Another suite of software, the Simbios NAST/C2A package, allows for coarse grain modeling of RNA molecules and then conversion of the coarse grain models to all-atom models (Jonikas et al. 2009).

To determine which molecular models are consistent with the experimental data, the scattering profile for each model must be accurately predicted. Back-calculated scattering amplitudes for all-atom models can be determined using the Debye equation (Debye 1915):

$$I(q) = \sum_{i=1}^{N} \sum_{j=1}^{N} \frac{f_i(q)f_j(q)(\sin(qd_{ij}))}{qd_{ij}} \tag{8.3}$$

where d_{ij} is the distance between atoms i and j and N is the number of atoms in the molecule. Most available software uses some approximation to limit the computational time necessary. CRYSOL (Svergun et al. 1995) and the FOXS Web server (Schneidman-Duhovny et al. 2010) employ coarse grain approximations that create a bead for each residue in the molecule to obtain the χ^2 goodness of fit between the predicted and experimental scattering amplitudes. CRYSOL provides not only the predicted scattering amplitudes, but also the predicted PDDF and R_g. FOXS is a higher throughput method and can calculate predicted scattering based on form factors for each atom type in the PDB file, making it a useful tool for nucleic acids. Fast-SAXS-RNA software (Yang et al. 2010) is fine-tuned specifically for use with nucleic acids and uses two form factor approximations per residue (one for the sugar moiety and one for the base). Model building can also be used in conjunction with the integrative model platform (IMP) (Forster et al. 2008), which refines structures using χ^2 minimization. IMP employs rigid body treatment of user specified domains to continuously adjust the shape of the molecule until agreement with the experimental scattering curve is optimized.

Finally, an exciting new direction is available for analyzing inherently flexible RNA molecules and RNA-protein complexes in solution using SAXS. Two independently developed ensemble modeling methods, Ensemble Optimization Method (EOM) (Bernado et al. 2007) and Minimal Ensemble Search (MES) (Pelikan et al. 2009), employ either a library of conformations or high temperature molecular dynamics simulations, respectively, to sample conformational space. From the generated conformations, a set that best represents the scattering profile is selected and can be subjected to energy minimization. These approaches help to define the conformational space sampled by flexible domains in a given molecule (Bernstein et al. 2009). MES can also be used in conjunction with the FOXS server (Schneidman-Duhovny et al. 2010) if a large variety of structural models have already been generated.

8.1.3.5 Combining SAXS and NMR

Incorporation of both SAXS and NMR data can be very powerful for analyzing the structures of RNA molecules in solution. NMR structures are often underdetermined, in part because the data are often insufficient to accurately define the global molecular shape. SAXS provides low-resolution global information without any detailed information about interatomic distances. Thus, the combination of these two techniques is ideal for analyzing molecular structures in solution (Wang et al. 2009).

SAXS data can be used to further refine the overall size and shape of an already determined NMR structure (Zuo et al. 2008; Grishaev et al. 2008b). We have used this method to solve the structure of an 86 nt RNA/RNA complex in solution (Fig. 8.12). Refinement with SAXS data improved the fit of the structure to both the measured R_g (Zuo et al. 2008) and the calculated ab initio structure (Fig. 8.12) (unpublished data).

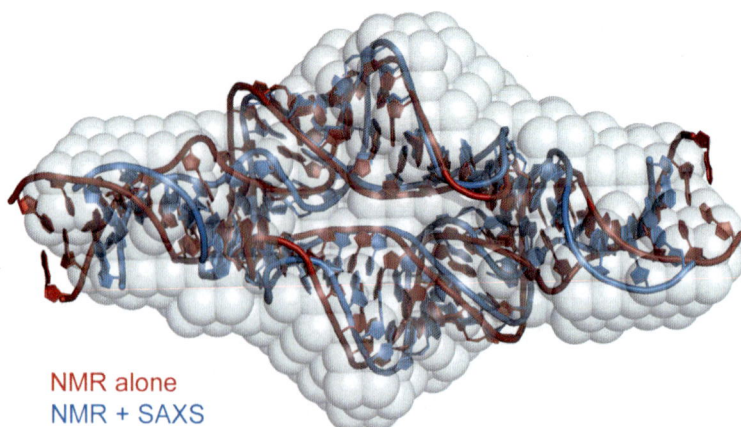

Fig. 8.12 NMR and SAXS refinement of an RNA:RNA complex. Comparison of the ab initio structure (*gray spheres*) with the NMR structure of the complex refined in the absence (*red*) and presence (*blue*) of SAXS restraints. Inclusion of SAXS restraints results in better agreement with measured R_g (Zuo et al. 2008) and a better fit with the ab initio structure (unpublished data)

8.2   Using NMR and SAXS to Study RNAs Involved in Translational Control

RNA structural elements, such as internal ribosomal entry sites (IRESs) (Balvay et al. 2009) and programmed ribosomal frameshift sites (PRFS) (Brierley 1995; Farabaugh 1996), are widely used in biology to regulate translation. IRESs facilitate cap-independent translation (Balvay et al. 2009; Shatsky et al. 2010), range in size from 250 to 500 nucleotides, and are typically composed of several stable domains, including stem-loops and pseudoknots. PRFS, which increase genomic coding capacity by promoting a change in reading frame during translation, are composed of three essential elements: a heptanucleotide "slippery" sequence, a linker region, and a downstream RNA structure, such as a pseudoknot or stem-loop (Giedroc and Cornish 2009). While biochemical and genetic results highlight the functional importance of IRESs and PRFS, structural information is needed to understand the roles of structure and dynamics in the function of these RNAs. The following sections highlight several examples of RNAs involved in translation control that have been characterized using NMR and SAXS.

8.2.1   Programmed Ribosomal Frameshift Sites

NMR has been employed to investigate the downstream RNA structures of numerous frameshift sites (Marcheschi et al. 2007, 2009; Staple and Butcher 2005a; Wang et al. 2002; Gaudin et al. 2005; Liphardt et al. 1999; Pennell et al. 2008; Cornish et al. 2006). The most common type of frameshift is a −1 PRF, which promotes a one nucleotide shift of the translating ribosome in the 5′ direction. The downstream stimulatory structure for −1 PRF sites is most commonly an H-type pseudoknot (Giedroc and Cornish 2009), a stem-loop in which nucleotides from the 3′ strand fold back to base-pair with the loop, creating a second helical stem (Staple and Butcher 2005b). The first structure of a functional pseudoknot was determined with NMR (Shen and Tinoco 1995). A large bend and tertiary interactions in this structure suggested multiple possible mechanisms for frameshift enhancement.

NMR has been further employed to examine the differences in RNA structure and dynamics between frameshift stimulating (S) and non-stimulating (NS) H-type pseudoknots (Wang et al. 2002). Identical stretches of sequence in S and NS pseudoknots have excellent ^{1}H and ^{15}N chemical shift agreement, suggesting they have extremely similar structures. As a result, the characteristic bend observed for these pseudoknots cannot explain frameshift stimulation. Further, NMR relaxation rates for imino ^{15}N nuclei within the helical regions for the S and NS pseudoknots did not reveal complex dynamics or striking differences between the two types.

While most frameshift sites contain a pseudoknot structure, structural studies using NMR showed that the structure in the human immunodeficiency virus type-1 (HIV-1) PRFS is an extended stem-loop (Staple and Butcher 2005a; Gaudin et al. 2005). The HIV-1 PRFS RNA structure has an extremely stable upper stem-loop separated from a meta-stable lower helix by a dynamic three-purine bulge. While the lower helix must be unwound prior to frameshifting, the overall bend adopted by the RNA structure may be important for frameshift stimulation for reasons that are not yet entirely clear (Staple and Butcher 2005a; Dulude et al. 2002). In HIV-1, the frameshifting efficiency controls the ratio of structural to enzymatic proteins that is critical for virion infectivity (Dulude et al. 2006). Therefore, small molecules that target the HIV PRFS stem-loop and perturb its stability or structure may be able to attenuate viral replication.

The conserved three-purine bulge in the HIV-1 PRFS structure served as a target for a high-throughput screen for small molecules (Marcheschi et al. 2009). Compound binding was verified by NMR. Although the compound with the highest affinity only had a modest impact on frameshifting efficiency, this study demonstrates the utility of NMR to screen, validate, and map RNA-small molecule interactions that have the potential to modulate −1 PRF during translation.

NMR revealed that the related simian immunodeficiency virus (SIV) PRFS also contained a stem-loop structure instead of a previously proposed pseudoknot structure (Marcheschi et al. 2007). Three RNA constructs of various lengths were examined by NMR. While two of the three constructs were designed to allow pseudoknot formation, the shortest construct contained only a stem-loop (Fig. 8.2a). Chemical shift mapping indicated that all three constructs contained a stem-loop structure, and that a pseudoknot does not form. NMR structure determination revealed that the SIV frameshift stem-loop forms two G–C base-pairs across the loop, which preclude pseudoknot formation (Marcheschi et al. 2007).

8.2.2 Untranslated Regions

The 5′ and 3′ untranslated regions (UTRs) of mRNA flank the protein coding region and are often highly structured. Many of the structures in these elements play a large role in translational regulation. Furthermore, the RNA structures found in UTRs are highly diverse, and can utilize very different strategies to control translation. This section reviews the application of NMR and SAXS to study four different types of structures found in UTRs: an IRES, a cap-independent translation element, an "RNA thermometer," and an adenine sensing riboswitch.

8.2.2.1 Internal Ribosomal Entry Sites

Due to the inherently large size of internal ribosomal entry sites (IRESs), NMR investigations of their structure and function have focused on individual IRES domains. For example, NMR has been employed to investigate an IRES domain in HCV RNA (Lukavsky et al. 2003; Paulsen et al. 2010). Translation of the HCV mRNA is initiated in the 5′ UTR on a highly structured IRES. While the 100 kDa IRES is made up of multiple domains, the 77 nt domain II is required for IRES activity (Kieft et al. 2001) and contacts the 40S head region of the ribosome (Spahn et al. 2001). Lukavsky et al. demonstrated that the full-length domain II folds independently in the context of the 100 kDa IRES using segmental ^{15}N labeling and chemical shift mapping (Lukavsky et al. 2003). The NMR structure of domain II was solved using a divide and conquer approach, in which domain II of the HCV IRES was divided into two manageable pieces of 34 nt (domain IIb) and 55 nt (domain IIa). Local NOE distance restraints were combined with RDC data from the full-length construct to determine the solution structure of domain II. The structure adopts a bent L-shape stabilized by the internal stacking of 5 nts in the asymmetric bulge found in domain IIa.

The 90° bend in domain IIa of the HCV IRES appears to be critical for IRES function (Paulsen et al. 2010). Several small molecules that bind to domain IIa with low micromolar affinity and inhibit HCV viral replication have been reported (Seth et al. 2005). The compounds have binding sites that localized to the 5 nt bulge. The structure of domain IIa in complex with one such inhibitor, Isis-11, was solved by NMR (Paulsen et al. 2010). Interestingly, compound binding trapped a nearly linear orientation of the RNA, eliminating the 90° bend observed in the free RNA. A structural understanding of this complex at the atomic level may facilitate structure-based drug design for future classes of improved antiviral inhibitors.

8.2.2.2 Ribosome-Binding Structural Elements

The global structure of a 102 nt RBSE found within the 3′ UTR of the TCV genomic RNA was characterized through a combination of NMR and SAXS (Zuo et al. 2010)

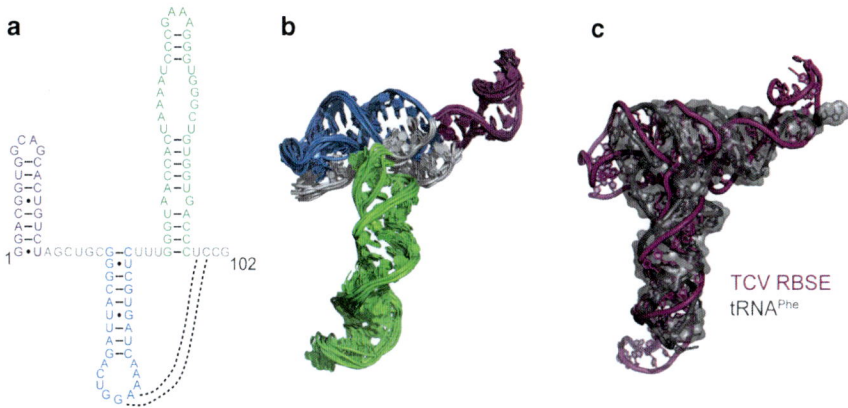

Fig. 8.13 Solution structure of the TCV RBSE (Zuo et al. 2010). (**a**) Secondary structure of the RBSE. Long-range base-pairing interactions are indicated by *dashed lines*. (**b**) The solution structure of the TCV RBSE (PDB ID 1KRL) as determined by a combination of SAXS and NMR reveals a twisted T-shape. (**c**) Despite the lack of similarity in secondary structure, the TCV RBSE (*purple*) resembles a tRNA (*gray*) in its three-dimensional fold

(Fig. 8.13a). The RBSE assists in recruitment of the large ribosomal subunit and contains a large asymmetric loop that is essential for the coordination of translation of the viral proteins with genome replication. This study utilized the G2G software package that first models the individual A-form helices in the RNA, then aligns the helices by fitting them to the RDC data (Wang et al. 2009). Linker sequences are then modeled into the aligned helices to create a complete model. The resulting model was iteratively refined against both SAXS and NMR data in XPLOR-NIH (Fig. 8.13b). Interestingly, the RBSE RNA mimics a tRNA molecule in overall shape (Fig. 8.13c), despite the fact that the RBSE secondary structure is quite distinct from the typical cloverleaf tRNA secondary structure (Fig. 8.13a) (Zuo et al. 2010). Thus, the TCV RBSE likely binds to the ribosome via its ability to mimic a cellular tRNA.

8.2.2.3 RNA Thermometers

RNA thermometers regulate translation initiation through temperature-dependent RNA structure formation. Rinnenthal *et al.* have investigated a stem-loop structure found in the 5′ UTR of the Salmonella fourU RNA that base-pairs to the Shine-Dalgarno sequence at low temperatures, preventing ribosome recruitment (Rinnenthal et al. 2010). At high temperatures, the structure unfolds, exposing the Shine-Dalgarno sequence and allowing translation to occur. To investigate the temperature-dependent dynamics of this structure, NMR was used to measure the thermodynamic stability of individual base-pairs (Rinnenthal et al. 2010). Thermodynamic stability of the base-pairs was quantified by measuring the imino ^{1}H exchange rates along a 5–50 °C temperature gradient. Based on the measured imino ^1H exchange rates, ΔG, ΔH, and ΔS were calculated for each base-pair. This information is consistent with a zipper-type of unfolding mechanism for this RNA thermometer (Rinnenthal et al. 2010). Surprisingly, mutations were found to have long-range effects on base-pair stability over a distance of six base-pairs, an effect which may be transduced via perturbations of the RNA hydration shell (Rinnenthal et al. 2010).

8.2.2.4 Riboswitches

Riboswitches typically reside in the untranslated 5′ UTR regions of bacterial mRNA (Haller et al. 2011). These regulatory RNA elements either increase or decrease expression of the downstream gene as a result of ligand binding (Serganov and Patel 2007). Riboswitches are composed of two structural elements: an aptamer domain, which binds the ligand (usually a metabolite), and an expression platform, which is linked to the aptamer domain and undergoes a conformational change in response to ligand binding. Multiple crystal structures have been solved of various riboswitch aptamer domains bound to their ligands (Edwards et al. 2007; Alexander 2009; Henkin 2008). While the crystal structures reveal how ligands are bound, they do not reveal the dynamic interplay between ligand binding and the resulting conformational change in the expression platform. NMR has proved useful for characterization of purine riboswitch folding (Buck et al. 2010; Ottink et al. 2007; Noeske et al. 2007a, b). Recently, the folding of an adenine sensing riboswitch, which regulates synthesis of enzymes responsible for purine synthesis, was followed for 3 min using time-resolved NMR (Lee et al. 2010). Ultrafast acquisition of ^1H-^{15}N 2D HSQC spectra revealed that after the addition of the adenine metabolite and magnesium to the free RNA solution, the core ligand-bound structure formed in the first 20 s. Within 30 s of core formation, tertiary interactions between two helical loops were stabilized; however, stabilization of all anticipated base-pairs lasted up to 3 min. The sensitivity of SAXS to overall molecular shape also renders this technique appropriate for characterization of riboswitches (Rambo and Tainer 2010a; Kulshina et al. 2009; Stoddard et al. 2010; Baird and Ferré-D'Amaré 2010; Lipfert et al. 2010; Baird et al. 2010), which may have multiple conformations in their unbound states.

8.3 Summary

Only 2% of the human genome is translated into protein, whereas more than 80% is transcribed into RNA (ENCODE Project Consortium 2007). Given the abundance of RNA in human biology, it is clear that RNA structural studies are going to be important for many years. As more RNAs are investigated, more functions will be discovered, including new strategies for regulating translation. Both NMR and

SAXS methods are continuing to improve, and we are now witnessing the integration of these methods to provide a more detailed and comprehensive view of RNA structure, function, and dynamics. In the future, we can expect that NMR and SAXS will play significant roles in advancing our understanding of translational control.

References

Akke M, Fiala R, Jiang F, Patel D, Palmer AG (1997) Base dynamics in a UUCG tetraloop RNA hairpin characterized by 15N spin relaxation: correlations with structure and stability. RNA 3:702–709

Alexander S (2009) The long and the short of riboswitches. Curr Opin Struct Biol 19:251–259

Al-Hashimi HM, Gosser Y, Gorin A, Hu W, Majumdar A, Patel DJ (2002) Concerted motions in HIV-1 TAR RNA may allow access to bound state conformations: RNA dynamics from NMR residual dipolar couplings. J Mol Biol 315:95–102

Ali M, Lipfert J, Seifert S, Herschlag D, Doniach S (2010) The ligand-free state of the TPP riboswitch: a partially folded RNA structure. J Mol Biol 396:153–165

Allain FHT, Varani G (1997) How accurately and precisely can RNA structure be determined by NMR? J Mol Biol 267:338–351

Andronescu M, Zhang ZC, Condon A (2005) Secondary structure prediction of interacting RNA molecules. J Mol Biol 345:987–1001

Bai Y, Das R, Millett IS, Herschlag D, Doniach S (2005) Probing counterion modulated repulsion and attraction between nucleic acid duplexes in solution. Proc Natl Acad Sci U S A 102:1035–1040

Bailor MH, Musselman C, Hansen AL, Gulati K, Patel DJ, Al-Hashimi HM (2007) Characterizing the relative orientation and dynamics of RNA A-form helices using NMR residual dipolar couplings. Nat Protoc 2:1536–1546

Bailor MH, Sun X, Al-Hashimi HM (2010) Topology links RNA secondary structure with global conformation, dynamics, and adaptation. Science 327:202–206

Baird NJ, Ferré-D'Amaré AR (2010) Idiosyncratically tuned switching behavior of riboswitch aptamer domains revealed by comparative small-angle X-ray scattering analysis. RNA 16:598–609

Baird NJ, Westhof E, Qin H, Pan T, Sosnick TR (2005) Structure of a folding intermediate reveals the interplay between core and peripheral elements in RNA folding. J Mol Biol 352:712–722

Baird NJ, Kulshina N, Ferré D'Amaré AR (2010) Riboswitch function: flipping the switch or tuning the dimmer? RNA Biol 7:328–332

Balvay L, Soto Rifo R, Ricci EP, Decimo D, Ohlmann T (2009) Structural and functional diversity of viral IRESes. Biochim Biophys Acta 1789:542–557

Ban N, Nissen P, Hansen J, Moore PB, Steitz TA (2000) The complete atomic structure of the large ribosomal subunit at 2.4 A resolution. Science 289:905–920

Bartel DP (2009) MicroRNAs: target recognition and regulatory functions. Cell 136:215–233

Batey RT, Kieft JS (2007) Improved native affinity purification of RNA. RNA 13:1384–1389

Bax A, Kontaxis G, Tjandra N (2001) Dipolar couplings in macromolecular structure determination. Methods Enzymol 339:127–174

Bernado P, Mylonas E, Petoukhov MV, Blackledge M, Svergun DI (2007) Structural characterization of flexible proteins using small-angle X-ray scattering. J Am Chem Soc 129:5656–5664

Bernstein NK, Hammel M, Mani RS, Weinfeld M, Pelikan M, Tainer JA, Glover JN (2009) Mechanism of DNA substrate recognition by the mammalian DNA repair enzyme, polynucleotide kinase. Nucleic Acids Res 37:6161–6173

Blad H, Reiter NJ, Abildgaard F, Markley JL, Butcher SE (2005) Dynamics and metal ion binding in the U6 RNA intramolecular stem-loop as analyzed by NMR. J Mol Biol 353:540–555

Brierley I (1995) Ribosomal frameshifting on viral RNAs. J Gen Virol 76:1885–1892

Buck J, Noeske J, Wöhnert J, Schwalbe H (2010) Dissecting the influence of Mg2+ on 3D architecture and ligand-binding of the guanine-sensing riboswitch aptamer domain. Nucleic Acids Res 38:4143–4153

Butcher SE, Pyle AM (2011) The molecular interactions that stabilize RNA tertiary structure: RNA motifs, patterns, and networks. Acc Chem Res 44(12):1302–1311

Carter AP, Clemons WM, Brodersen DE, Morgan-Warren RJ, Wimberly BT, Ramakrishnan V (2000) Functional insights from the structure of the 30S ribosomal subunit and its interactions with antibiotics. Nature 407:340–348

Casiano-Negroni A, Sun X, Al-Hashimi HM (2007) Probing Na+-induced changes in the HIV-1 TAR conformational dynamics using NMR residual dipolar couplings: new insights into the role of counterions and electrostatic interactions in adaptive recognition. Biochemistry 46:6525–6535

Chen SJ (2008) RNA folding: conformational statistics, folding kinetics, and ion electrostatics. Annu Rev Biophys 37:197–214

Clos LJ 2nd, Butcher SE, Wang YX (2011) NMR spectroscopy for investigating larger nucleic acids. In Advances in Biomedical Spectroscopy: Biomolecular NMR spectroscopy, ed. Dingley, AJ, Pascal SM. IOS Press, Amsterdam, Netherlands, Vol. 3, pp 229–248

Cornish P, Giedroc D, Hennig M (2006) Dissecting non-canonical interactions in frameshift-stimulating mRNA pseudoknots. J Biomol NMR 35:209–223

Cullen BR (1986) Trans-activation of human immunodeficiency virus occurs via a bimodal mechanism. Cell 46:973–982

D'Souza V, Dey A, Habib D, Summers MF (2004) NMR structure of the 101-nucleotide core encapsidation signal of the Moloney murine leukemia virus. J Mol Biol 337:427–442

Das R, Kwok LW, Millett IS, Bai Y, Mills TT, Jacob J, Maskel GS, Seifert S, Mochrie SG, Thiyagarajan P et al (2003) The fastest global events in RNA folding: electrostatic relaxation and tertiary collapse of the tetrahymena ribozyme. J Mol Biol 332:311–319

Davis JH, Tonelli M, Scott LG, Jaeger L, Williamson JR, Butcher SE (2005) RNA helical packing in solution: NMR structure of a 30 kDa GAAA tetraloop-receptor complex. J Mol Biol 351:371–382

Davis JH, Foster TR, Tonelli M, Butcher SE (2007) Role of metal ions in the tetraloop-receptor complex as analyzed by NMR. RNA 13:76–86

Dayie KT, Brodsky AS, Williamson JR (2002) Base flexibility in HIV-2 TAR RNA mapped by solution 15N, 13C NMR relaxation. J Mol Biol 317:263–278

Debye P (1915) Zerstreuung von Röntgenstrahlen. Ann Phys 351:809–823

Dethoff EA, Hansen AL, Musselman C, Watt ED, Andricioaei I, Al-Hashimi HM (2008) Characterizing complex dynamics in the transactivation response element apical loop and motional correlations with the bulge by NMR, molecular dynamics, and mutagenesis. Biophys J 95:3906–3915

Dethoff EA, Hansen AL, Zhang Q, Al-Hashimi HM (2009) Variable helix elongation as a tool to modulate RNA alignment and motional couplings. J Magn Reson 202:117–121

Dingley AJ, Grzesiek S (1998) Direct observation of hydrogen bonds in nucleic acid base pairs by internucleotide (2)J(NN) couplings. J Am Chem Soc 120:8293–8297

Duchardt E, Schwalbe H (2005) Residue specific ribose and nucleobase dynamics of the cUUCGg RNA tetraloop motif by MNMR 13C relaxation. J Biomol NMR 32:295–308

Duchardt-Ferner E, Ferner J, Wöhnert J (2011) Rapid Identification of noncanonical RNA structure elements by direct detection of OH···O=P, NH···O=P, and NH2···O=P hydrogen bonds in solution NMR spectroscopy. Angew Chem Int Ed 50:7927–7930

Dulude D, Baril M, Brakier-Gingras L (2002) Characterization of the frameshift stimulatory signal controlling a programmed −1 ribosomal frameshift in the human immunodeficiency virus type 1. Nucleic Acids Res 30:5094–5102

Dulude D, Berchiche YA, Gendron K, Brakier-Gingras L, Heveker N (2006) Decreasing the frameshift efficiency translates into an equivalent reduction of the replication of the human immunodeficiency virus type 1. Virology 345:127–136

Edwards TE, Klein DJ, Ferré-D'Amaré AR (2007) Riboswitches: small-molecule recognition by gene regulatory RNAs. Curr Opin Struct Biol 17:273–279

ENCODE Project Consortium (2007) Identification and analysis of functional elements in 1% of the human genome by the ENCODE pilot project. Nature 447:799–816

Fang X, Littrell K, Yang XJ, Henderson SJ, Siefert S, Thiyagarajan P, Pan T, Sosnick TR (2000) Mg2+-dependent compaction and folding of yeast tRNAPhe and the catalytic domain of the B. subtilis RNase P RNA determined by small-angle X-ray scattering. Biochemistry 39:11107–11113

Farabaugh PJ (1996) Programmed translational frameshifting. Microbiol Rev 60:103–134

Fernández C, Wider G (2003) TROSY in NMR studies of the structure and function of large biological macromolecules. Curr Opin Struct Biol 13:570–580

Forster F, Webb B, Krukenberg KA, Tsuruta H, Agard DA, Sali A (2008) Integration of small-angle X-ray scattering data into structural modeling of proteins and their assemblies. J Mol Biol 382:1089–1106

Frank AT, Stelzer AC, Al-Hashimi HM, Andricioaei I (2009) Constructing RNA dynamical ensembles by combining MD and motionally decoupled NMR RDCs: new insights into RNA dynamics and adaptive ligand recognition. Nucleic Acids Res 37:3670–3679

Franke D, Svergun DI (2009) DAMMIF, a program for rapid ab-initio shape determination in small-angle scattering. J Appl Crystallogr 42:342–346

Frankel AD (1992) Activation of HIV transcription by Tat. Curr Opin Genet Dev 2:293–298

Furtig B, Buck J, Manoharan V, Bermel W, Jaschke A, Wenter P, Pitsch S, Schwalbe H (2007) Time-resolved NMR studies of RNA folding. Biopolymers 86:360–383

Gaudin C, Mazauric M-H, Traikia M, Guittet E, Yoshizawa S, Fourmy D (2005) Structure of the RNA signal essential for translational frameshifting in HIV-1. J Mol Biol 349:1024–1035

Getz M, Sun X, Casiano-Negroni A, Zhang Q, Al-Hashimi HM (2007) NMR studies of RNA dynamics and structural plasticity using NMR residual dipolar couplings. Biopolymers 86:384–402

Giedroc DP, Cornish PV (2009) Frameshifting RNA pseudoknots: structure and mechanism. Virus Res 139:193–208

Gluehmann M, Zarivach R, Bashan A, Harms J, Schluenzen F, Bartels H, Agmon I, Rosenblum G, Pioletti M, Auerbach T et al (2001) Ribosomal crystallography: from poorly diffracting microcrystals to high-resolution structures. Methods 25:292–302

Grishaev A, Ying J, Canny MD, Pardi A, Bax A (2008a) Solution structure of tRNAVal from refinement of homology model against residual dipolar coupling and SAXS data. J Biomol NMR 42:99–109

Grishaev A, Tugarinov V, Kay LE, Trewhella J, Bax A (2008b) Refined solution structure of the 82-kDa enzyme malate synthase G from joint NMR and synchrotron SAXS restraints. J Biomol NMR 40:95–106

Hall KB (2008) RNA in motion. Curr Opin Chem Biol 12:612–618

Hall KB, Tang C (1998) 13C relaxation and dynamics of the purine bases in the iron responsive element RNA hairpin. Biochemistry 37:9323–9332

Haller A, Soulière MF, Micura R (2011) The dynamic nature of RNA as key to understanding riboswitch mechanisms. Acc Chem Res 44(12):1339–1348

Hansen AL, Al-Hashimi HM (2006) Insight into the CSA tensors of nucleobase carbons in RNA polynucleotides from solution measurements of residual CSA: towards new long-range orientational constraints. J Magn Reson 179:299–307

Hansen MR, Mueller L, Pardi A (1998) Tunable alignment of macromolecules by filamentous phage yields dipolar coupling interactions. Nat Struct Mol Biol 5:1065–1074

Hart JM, Kennedy SD, Mathews DH, Turner DH (2008) NMR-assisted prediction of RNA secondary structure: identification of a probable pseudoknot in the coding region of an R2 retrotransposon. J Am Chem Soc 130:10233–10239

Henkin TM (2008) Riboswitch RNAs: using RNA to sense cellular metabolism. Genes Dev 22:3383–3390

Hennig M, Williamson JR, Brodsky AS, Battiste JL (2001) Recent advances in RNA structure determination by NMR. In: Beaucage SL et al (eds) Current protocols in nucleic acid chemistry. Wiley, New York, Chapter 7, Unit 7.7

Hermann T, Patel DJ (2000) RNA bulges as architectural and recognition motifs. Structure 8:R47–R54

Heus HA, Pardi A (1991) Novel proton NMR assignment procedure for RNA duplexes. J Am Chem Soc 113:4360–4361

Johnson JE, Hoogstraten CG (2008) Extensive backbone dynamics in the GCAA RNA tetraloop analyzed using 13C NMR spin relaxation and specific isotope labeling. J Am Chem Soc 130:16757–16769

Jonikas MA, Radmer RJ, Laederach A, Das R, Pearlman S, Herschlag D, Altman RB (2009) Coarse-grained modeling of large RNA molecules with knowledge-based potentials and structural filters. RNA 15:189–199

Juan V, Wilson C (1999) RNA secondary structure prediction based on free energy and phylogenetic analysis. J Mol Biol 289:935–947

Keene JD (2007) RNA regulons: coordination of post-transcriptional events. Nat Rev Genet 8:533–543

Kieft JS, Zhou K, Jubin R, Doudna JA (2001) Mechanism of ribosome recruitment by hepatitis C IRES RNA. RNA 7:194–206

Kierzek E, Kierzek R, Turner DH, Catrina IE (2006) Facilitating RNA structure prediction with microarrays. Biochemistry 45:581–593

Kloiber K, Spitzer R, Tollinger M, Konrat R, Kreutz C (2011) Probing RNA dynamics via longitudinal exchange and CPMG relaxation dispersion NMR spectroscopy using a sensitive 13C-methyl label. Nucleic Acids Res 39:4340–4351

Koch MHJ, Vachette P, Svergun DI (2003) Small-angle scattering: a view on the properties, structures and structural changes of biological macromolecules in solution. Q Rev Biophys 36:147–227

Konarev PV, Volkov VV, Sokolova AV, Koch MHJ, Svergun DI (2003) PRIMUS: a windows PC-based system for small-angle scattering data analysis. J Appl Crystallogr 36:1277–1282

Kulshina N, Baird NJ, Ferre-D'Amare AR (2009) Recognition of the bacterial second messenger cyclic diguanylate by its cognate riboswitch. Nat Struct Mol Biol 16:1212–1217

Latham MP, Zimmermann GR, Pardi A (2009) NMR chemical exchange as a probe for ligand-binding kinetics in a theophylline-binding RNA aptamer. J Am Chem Soc 131:5052–5053

Lee M-K, Gal M, Frydman L, Varani G (2010) Real-time multidimensional NMR follows RNA folding with second resolution. Proc Natl Acad Sci 107:9192–9197

Lemay J-F, Desnoyers G, Blouin S, Heppell B, Bastet L, St-Pierre P, Massé E, Lafontaine DA (2011) Comparative study between transcriptionally- and translationally-acting adenine riboswitches reveals key differences in riboswitch regulatory mechanisms. PLoS Genet 7:e1001278

Li PTX, Vieregg J, Tinoco I (2008) How RNA unfolds and refolds. Annu Rev Biochem 77:77–100

Lipari G, Szabo A (1982) Model-free approach to the interpretation of nuclear magnetic resonance relaxation in macromolecules. 2. Analysis of experimental results. J Am Chem Soc 104:4559–4570

Lipfert J, Sim AY, Herschlag D, Doniach S (2010) Dissecting electrostatic screening, specific ion binding, and ligand binding in an energetic model for glycine riboswitch folding. RNA 16:708–719

Liphardt J, Napthine S, Kontos H, Brierley I (1999) Evidence for an RNA pseudoknot loop-helix interaction essential for efficient −1 ribosomal frameshifting. J Mol Biol 288:321–335

Lu K, Miyazaki Y, Summers MF (2010) Isotope labeling strategies for NMR studies of RNA. J Biomol NMR 46:113–125

Lu K, Heng X, Garyu L, Monti S, Garcia EL, Kharytonchyk S, Dorjsuren B, Kulandaivel G, Jones S, Hiremath A et al (2011) NMR detection of structures in the HIV-1 5′-leader RNA that regulate genome packaging. Science 334:242–245

Lukavsky PJ, Puglisi JD (2004) Large-scale preparation and purification of polyacrylamide-free RNA oligonucleotides. RNA 10:889–893

Lukavsky PJ, Puglisi JD (2005) Structure determination of large biological RNAs. Methods Enzymol 394:399–416

Lukavsky PJ, Kim I, Otto GA, Puglisi JD (2003) Structure of HCV IRES domain II determined by NMR. Nat Struct Biol 10:1033–1038

Luo Y, Eldho NV, Sintim HO, Dayie TK (2011) RNAs synthesized using photocleavable biotinylated nucleotides have dramatically improved catalytic efficiency. Nucleic Acids Res 39(19):8559–8571

Mackerell AD Jr, Nilsson L (2008) Molecular dynamics simulations of nucleic acid-protein complexes. Curr Opin Struct Biol 18:194–199

Marcheschi RJ, Staple DW, Butcher SE (2007) Programmed ribosomal frameshifting in SIV is induced by a highly structured RNA stem-loop. J Mol Biol 373:652–663

Marcheschi RJ, Mouzakis KD, Butcher S (2009) Selection and characterization of small molecules that bind the HIV-1 frameshift site RNA. ACS Chem Biol 4(10):844–854

Mathews DH, Sabina J, Zuker M, Turner DH (1999) Expanded sequence dependence of thermodynamic parameters improves prediction of RNA secondary structure. J Mol Biol 288:911–940

Mathews DH, Disney MD, Childs JL, Schroeder SJ, Zuker M, Turner DH (2004) Incorporating chemical modification constraints into a dynamic programming algorithm for prediction of RNA secondary structure. Proc Natl Acad Sci U S A 101:7287–7292

Mathews DH, Moss WN, Turner DH (2010) Folding and finding RNA secondary structure. Cold Spring Harb Perspect Biol 2(12):A003665

Milligan JF, Uhlenbeck OC (1989) Synthesis of small RNAs using T7 RNA polymerase. Methods Enzymol 180:51–62

Miyazaki Y, Irobalieva RN, Tolbert BS, Smalls-Mantey A, Iyalla K, Loeliger K, D'Souza V, Khant H, Schmid MF, Garcia EL et al (2010) Structure of a conserved retroviral RNA packaging element by NMR spectroscopy and cryo-electron tomography. J Mol Biol 404:751–772

Muesing MA, Smith DH, Capon DJ (1987) Regulation of mRNA accumulation by a human immunodeficiency virus trans-activator protein. Cell 48:691–701

Musselman C, Zhang Q, Al-Hashimi H, Andricioaei I (2010) Referencing strategy for the direct comparison of nuclear magnetic resonance and molecular dynamics motional parameters in RNA. J Phys Chem B 114:929–939

Nikonowicz EP, Pardi A (1993) An efficient procedure for assignment of the proton, carbon and nitrogen resonances in 13C/15N labeled nucleic acids. J Mol Biol 232:1141–1156

Nikonowicz EP, Sirr A, Legault P, Jucker FM, Baer LM, Pardi A (1992) Preparation of 13C and 15N labelled RNAs for heteronuclear multi-dimensional NMR studies. Nucleic Acids Res 20:4507–4513

Noeske J, Buck J, Fürtig B, Nasiri HR, Schwalbe H, Wöhnert J (2007a) Interplay of 'induced fit' and preorganization in the ligand induced folding of the aptamer domain of the guanine binding riboswitch. Nucleic Acids Res 35:572–583

Noeske J, Schwalbe H, Wöhnert J (2007b) Metal-ion binding and metal-ion induced folding of the adenine-sensing riboswitch aptamer domain. Nucleic Acids Res 35:5262–5273

Nozinovic S, Richter C, Rinnenthal J, Fürtig B, Duchardt-Ferner E, Weigand JE, Schwalbe H (2010a) Quantitative 2D and 3D Γ-HCP experiments for the determination of the angles α and ζ in the phosphodiester backbone of oligonucleotides. J Am Chem Soc 132:10318–10329

Nozinovic S, Fürtig B, Jonker HRA, Richter C, Schwalbe H (2010b) High-resolution NMR structure of an RNA model system: the 14-mer cUUCGg tetraloop hairpin RNA. Nucleic Acids Res 38:683–694

Ottink OM, Rampersad SM, Tessari M, Zaman GJR, Heus HA, Wijmenga SS (2007) Ligand-induced folding of the guanine-sensing riboswitch is controlled by a combined predetermined-induced fit mechanism. RNA 13:2202–2212

Parisien M, Major F (2008) The MC-fold and MC-Sym pipeline infers RNA structure from sequence data. Nature 452:51–55

Paulsen RB, Seth PP, Swayze EE, Griffey RH, Skalicky JJ, Cheatham TE III, Davis DR (2010) Inhibitor-induced structural change in the HCV IRES domain IIa RNA. Proc Natl Acad Sci U S A 107:7263–7268

Pelikan M, Hura GL, Hammel M (2009) Structure and flexibility within proteins as identified through small angle X-ray scattering. Gen Physiol Biophys 28:174–189

Pennell S, Manktelow E, Flatt A, Kelly G, Smerdon SJ, Brierley I (2008) The stimulatory RNA of the Visna-Maedi retrovirus ribosomal frameshifting signal is an unusual pseudoknot with an interstem element. RNA 14:1366–1377

Pereira MJ, Behera V, Walter NG (2010) Nondenaturing purification of co-transcriptionally folded RNA avoids common folding heterogeneity. PLoS One 5:e12953

Pervushin K, Riek R, Wider G, Wuthrich K (1997) Attenuated T2 relaxation by mutual cancellation of dipole-dipole coupling and chemical shift anisotropy indicates an avenue to NMR structures of very large biological macromolecules in solution. Proc Natl Acad Sci U S A 94:12366–12371

Peterson RD, Theimer CA, Wu H, Feigon J (2004) New applications of 2D filtered/edited NOESY for assignment and structure elucidation of RNA and RNA-protein complexes. J Biomol NMR 28:59–67

Pioletti M, Schlunzen F, Harms J, Zarivach R, Gluhmann M, Avila H, Bashan A, Bartels H, Auerbach T, Jacobi C et al (2001) Crystal structures of complexes of the small ribosomal subunit with tetracycline, edeine and IF3. EMBO J 20:1829–1839

Pitt SW, Majumdar A, Serganov A, Patel DJ, Al-Hashimi HM (2004) Argininamide binding arrests global motions in HIV-1 TAR RNA: comparison with Mg2+-induced conformational stabilization. J Mol Biol 338:7–16

Pollack L (2011) Time resolved SAXS and RNA folding. Biopolymers 95:543–549

Prestegard JH, Al-Hashimi HM, Tolman JR (2000) NMR structures of biomolecules using field oriented media and residual dipolar couplings. Q Rev Biophys 33:371–424

Putnam CD, Hammel M, Hura GL, Tainer JA (2007) X-ray solution scattering (SAXS) combined with crystallography and computation: defining accurate macromolecular structures, conformations and assemblies in solution. Q Rev Biophys 40:191–285

Rambo RP, Tainer JA (2010a) Improving small-angle X-ray scattering data for structural analyses of the RNA world. RNA 16:638–646

Rambo RP, Tainer JA (2010b) Bridging the solution divide: comprehensive structural analyses of dynamic RNA, DNA, and protein assemblies by small-angle X-ray scattering. Curr Opin Struct Biol 20:128–137

Rinnenthal J, Klinkert B, Narberhaus F, Schwalbe H (2010) Direct observation of the temperature-induced melting process of the Salmonella fourU RNA thermometer at base-pair resolution. Nucleic Acids Res 38:3834–3847

Rinnenthal J, Buck J, Ferner J, Wacker A, Furtig B, Schwalbe H (2011) Mapping the landscape of RNA dynamics with NMR spectroscopy. Acc Chem Res 44(12):1292–1301

Roh JH, Guo L, Kilburn JD, Briber RM, Irving T, Woodson SA (2010) Multistage collapse of a bacterial ribozyme observed by time-resolved small-angle X-ray scattering. J Am Chem Soc 132:10148–10154

Roy R, Hohng S, Ha T (2008) A practical guide to single-molecule FRET. Nat Methods 5:507–516

Sambrook J, Russell DW (2001) Molecular cloning: a laboratory manual, 3rd edn. Cold Spring Harbor Laboratory Press, Cold Spring Harbor, NY

Schanda P, Kup e , Brutscher B (2005) SOFAST-HMQC experiments for recording two-dimensional deteronuclear correlation spectra of proteins within a few seconds. J Biomol NMR 33:199–211

Schneidman-Duhovny D, Hammel M, Sali A (2010) FoXS: a web server for rapid computation and fitting of SAXS profiles. Nucleic Acids Res 38:W540–W544

Sclavi B, Sullivan M, Chance MR, Brenowitz M, Woodson SA (1998) RNA folding at millisecond intervals by synchrotron hydroxyl radical footprinting. Science 279:1940–1943

Scott LG, Hennig M (2008) RNA structure determination by NMR. Methods Mol Biol 452:29–61

Scott LG, Tolbert TJ, Williamson JR (2000) Preparation of specifically 2H- and 13C-labeled ribonucleotides. Methods Enzymol 317:18–38

Selmer M, Dunham CM, Murphy FVT, Weixlbaumer A, Petry S, Kelley AC, Weir JR, Ramakrishnan V (2006) Structure of the 70S ribosome complexed with mRNA and tRNA. Science 313:1935–1942

Serganov A, Patel DJ (2007) Ribozymes, riboswitches and beyond: regulation of gene expression without proteins. Nat Rev Genet 8:776–790

Seth PP, Miyaji A, Jefferson EA, Sannes-Lowery KA, Osgood SA, Propp SS, Ranken R, Massire C, Sampath R, Ecker DJ et al (2005) SAR by MS: discovery of a new class of RNA-binding small molecules for the hepatitis C virus: internal ribosome entry site IIA subdomain. J Med Chem 48:7099–7102

Shajani Z, Varani G (2005) 13C NMR relaxation studies of RNA base and ribose nuclei reveal a complex pattern of motions in the RNA binding site for human U1A protein. J Mol Biol 349:699–715

Shajani Z, Varani G (2007) NMR studies of dynamics in RNA and DNA by 13C relaxation. Biopolymers 86:348–359

Shatsky IN, Dmitriev SE, Terenin IM, Andreev DE (2010) Cap- and IRES-independent scanning mechanism of translation initiation as an alternative to the concept of cellular IRESs. Mol Cells 30:285–293

Shcherbakova I, Mitra S, Beer RH, Brenowitz M (2006) Fast Fenton footprinting: a laboratory-based method for the time-resolved analysis of DNA, RNA and proteins. Nucleic Acids Res 34:e48

Shen LX, Tinoco I Jr (1995) The structure of an RNA pseudoknot that causes efficient frameshifting in mouse mammary tumor virus. J Mol Biol 247:963–978

Sklená V, Dieckmann T, Butcher SE, Feigon J (1998) Optimization of triple-resonance HCN experiments for application to larger RNA oligonucleotides. J Magn Reson 130:119–124

Spahn CM, Kieft JS, Grassucci RA, Penczek PA, Zhou K, Doudna JA, Frank J (2001) Hepatitis C virus IRES RNA-induced changes in the conformation of the 40s ribosomal subunit. Science 291:1959–1962

Staple DW, Butcher SE (2005a) Solution structure and thermodynamic investigation of the HIV-1 frameshift inducing element. J Mol Biol 349:1011–1023

Staple DW, Butcher SE (2005b) Pseudoknots: RNA structures with diverse functions. PLoS Biol 3:e213

Stoddard CD, Montange RK, Hennelly SP, Rambo RP, Sanbonmatsu KY, Batey RT (2010) Free state conformational sampling of the SAM-I riboswitch aptamer domain. Structure 18:787–797

Svergun DI (1992) Determination of the regularization parameter in indirect-transform methods using perceptual criteria. J Appl Crystallogr 25:495–503

Svergun DI (1999) Restoring low resolution structure of biological macromolecules from solution scattering using simulated annealing. Biophys J 76:2879–2886

Svergun DI, Barberato C, Koch MHJ (1995) CRYSOL—a program to evaluate X-ray solution scattering of biological macromolecules from atomic coordinates. J Appl Crystallogr 28:768–773

Talini G, Branciamore S, Gallori E (2011) Ribozymes: flexible molecular devices at work. Biochimie 93(11):1998–2005

Theis C, Reeder J, Giegerich R (2008) KnotInFrame: prediction of −1 ribosomal frameshift events. Nucleic Acids Res 36:6013–6020

Tolman JR (2001) Dipolar couplings as a probe of molecular dynamics and structure in solution. Curr Opin Struct Biol 11:532–539

Tolman JR, Ruan K (2006) NMR residual dipolar couplings as probes of biomolecular dynamics. Chem Rev 106:1720–1736

Tzakos AG, Grace CR, Lukavsky PJ, Riek R (2006) NMR techniques for very large proteins and rnas in solution. Annu Rev Biophys Biomol Struct 35:319–342

Voehler MW, Collier G, Young JK, Stone MP, Germann MW (2006) Performance of cryogenic probes as a function of ionic strength and sample tube geometry. J Magn Reson 183:102–109

Volkov VV, Svergun DI (2003) Uniqueness of ab-initio shape determination in small-angle scattering. J Appl Crystallogr 36:860–864

Walter NG (2001) Structural dynamics of catalytic RNA highlighted by fluorescence resonance energy transfer. Methods 25:19–30

Wang Y, Wills NM, Du Z, Rangan A, Atkins JF, Gesteland RF, Hoffman DW (2002) Comparative studies of frameshifting and nonframeshifting RNA pseudoknots: a mutational and NMR investigation of pseudoknots derived from the bacteriophage T2 gene 32 mRNA and the retroviral gag-pro frameshift site. RNA 8:981–996

Wang J, Zuo X, Yu P, Xu H, Starich MR, Tiede DM, Shapiro BA, Schwieters CD, Wang YX (2009) A method for helical RNA global structure determination in solution using small-angle x-ray scattering and NMR measurements. J Mol Biol 393:717–734

Wang YX, Zuo X, Wang J, Yu P, Butcher SE (2010) Rapid global structure determination of large RNA and RNA complexes using NMR and small-angle X-ray scattering. Methods 52:180–191

Weeks KM (2010) Advances in RNA structure analysis by chemical probing. Curr Opin Struct Biol 20:295–304

Weeks KM, Mauger DM (2011) Exploring RNA structural codes with SHAPE chemistry. Acc Chem Res 44:1280–1291

Woodson SA, Koculi E (2009) Analysis of RNA folding by native polyacrylamide gel electrophoresis. Methods Enzymol 469:189–208

Wyatt JR, Chastain M, Puglisi JD (1991) Synthesis and purification of large amounts of RNA oligonucleotides. Biotechniques 11:764–769

Yang S, Parisien M, Major F, Roux B (2010) RNA structure determination using SAXS data. J Phys Chem B 114:10039–10048

Yusupov MM, Yusupova GZ, Baucom A, Lieberman K, Earnest TN, Cate JHD, Noller HF (2001) Crystal structure of the ribosome at 5.5 A resolution. Science 292:883–896

Zhang Q, Al-Hashimi HM (2009) Domain-elongation NMR spectroscopy yields new insights into RNA dynamics and adaptive recognition. RNA 15:1941–1948

Zhang Q, Sun X, Watt ED, Al-Hashimi HM (2006) Resolving the motional modes that code for RNA adaptation. Science 311:653–656

Zhao P (2011) The 2009 Nobel Prize in chemistry: Thomas A. Steitz and the structure of the ribosome. Yale J Biol Med 84:125–129

Zhao R, Rueda D (2009) RNA folding dynamics by single-molecule fluorescence resonance energy transfer. Methods 49:112–117

Zuker M (2003) Mfold web server for nucleic acid folding and hybridization prediction. Nucleic Acids Res 31:3406–3415

Zuo X, Wang J, Foster TR, Schwieters CD, Tiede DM, Butcher SE, Wang YX (2008) Global molecular structure and interfaces: refining an RNA:RNA complex structure using solution X-ray scattering data. J Am Chem Soc 130:3292–3293

Zuo X, Wang J, Yu P, Eyler D, Xu H, Starich MR, Tiede DM, Simon AE, Kasprzak W, Schwieters CD et al (2010) Solution structure of the cap-independent translational enhancer and ribosome-binding element in the 3′ UTR of turnip crinkle virus. Proc Natl Acad Sci U S A 107:1385–1390

Chapter 9
Analyses of RNA–Ligand Interactions by Fluorescence Anisotropy

Aparna Kishor, Gary Brewer, and Gerald M. Wilson

9.1 Introduction

The use of fluorescence anisotropy-based techniques for the accurate, quantitative determination of biomolecular binding parameters has become increasingly common. Applications span a wide range of biological disciplines, from the fields of medical diagnostics and drug discovery to basic research. Some advantages of this technique, among others that will be discussed in Sect. 9.4, are that it does not require radioactive probes and that it is a homogenous-phase assay. Parameters including equilibrium binding constants, binding site size, and Gibb's free energy can be calculated from binding isotherms generated under various temperature and probe conditions. Additionally, monitoring fluorescence anisotropy under pre-steady-state conditions can resolve on- or off-rates of binding. For equilibrium binding experiments, data collection can often be streamlined by using plate readers configured to measure fluorescence anisotropy, several of which are commercially available. Traditional format fluorescence spectrophotometers equipped with polarizers, however, are still common; although these typically can only read one sample at a time, their sensitivity is generally superior to plate readers. Furthermore, single-cell spectrofluorometers are normally more versatile, permitting, for example, measurements under pre-steady-state conditions or across gradients of temperature.

The basic principle behind the use of anisotropy to measure macromolecular binding parameters is that the mobility of a molecule in solution will change when it

A. Kishor • G.M. Wilson (✉)
Department of Biochemistry and Molecular Biology, University of Maryland
School of Medicine, Baltimore, MD 21201, USA
e-mail: gwilson@som.umaryland.edu

G. Brewer (✉)
Department of Molecular Genetics, Microbiology, and Immunology,
UMDNJ-Robert Wood Johnson Medical School, Piscataway, NJ, USA
e-mail: brewerga@umdnj.edu

J.D. Dinman (ed.), *Biophysical Approaches to Translational Control of Gene Expression*,   173
Biophysics for the Life Sciences 1, DOI 10.1007/978-1-4614-3991-2_9,
© Springer Science+Business Media New York 2013

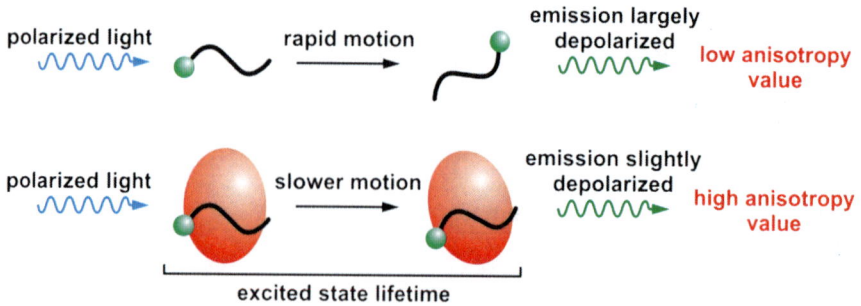

Fig. 9.1 Conceptual schematic for the use of anisotropy to measure RNA binding. The *green sphere* is the fluorescent tag. If the tagged molecule is single-stranded RNA, then both segmental and global tumbling motions will contribute to fluorophore mobility in solution. Both of these types of motion will be restricted upon binding to protein (*red* ellipsoid)

binds another molecule (it should be noted that "anisotropy" as it appears throughout this chapter is distinct from the physical chemistry term). In solutions of a given viscosity, smaller molecules will tumble faster than larger ones (Fernandes 1998). The examples described in this chapter feature interactions between small RNA substrates and the proteins that bind them. In this context where the RNA ligand is highly mobile owing to its relatively small size and large potential for segmental motion, its association with a protein or protein complex decreases its molecular motion owing in part to the increased size of the ribonucleoprotein complex.

In order to measure the protein-dependent change in the mobility of an RNA ligand, anisotropy-based assays typically use a fluorescently labeled RNA substrate (Fig. 9.1). Upon exposure of this fluorophore to plane-polarized light, only molecules that have their absorption transition dipole moment oriented appropriately will become excited (Gradinaru et al. 2010). In the absence of molecular motion, all emission from these fluorophores would be observed in the same plane as excitation. However, if the excited fluorophores are rapidly tumbling in solution during the excited state lifetime, emitted light will be depolarized. By contrast, if mobility of the excited fluorophores is limited (by association with a large macromolecular complex, for example), emitted light will be more highly polarized, since fewer excited molecules will tumble out of the plane of excitation before quantum emission (Fernandes 1998; Jameson and Ross 2010). The degree of emission depolarization is quantified by parameters termed *polarization* or *anisotropy*, both of which are defined below.

9.2 Measuring Anisotropy

9.2.1 Theory

Passing an incident light beam through a polarizer limits transmission primarily to light with an electric vector vibrating in a single plane. The direction of that vector

Fig. 9.2 System for measuring anisotropy. The z-axis is the vertical laboratory axis. The sample is placed at the origin, and readings are taken along the y-axis. Reprinted with permission from Jameson and Ross (2010). Copyright 2010 American Chemical Society

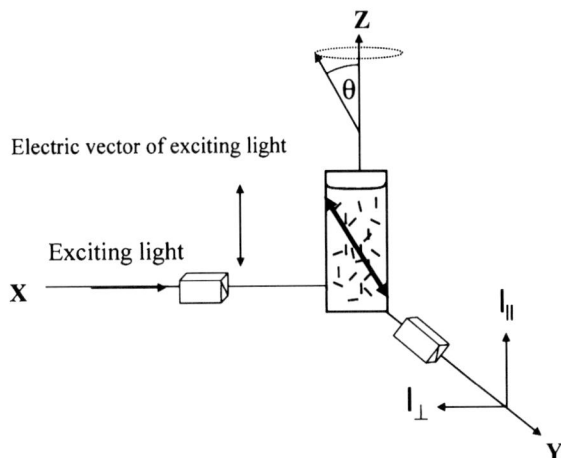

can be manipulated experimentally by rotating the polarizer. Figure 9.2 shows how these principles can be applied experimentally to measure the fluorescence anisotropy of a sample. In this case, a polarizer is inserted in a beam of light directed along the x-axis, allowing the isolation of an electric vector oriented parallel to the z-axis. When this polarized light strikes the sample at the center of the coordinate system, the fluorescently tagged molecules in solution that have an appropriately oriented absorption transition dipole moment will become excited and fluoresce. The intensity of fluorescence emission is measured 90° to the axis of excitation, along the y-axis, in the planes parallel and perpendicular to the z-axis. The relationship between these intensities gives a measure of the mobility of the fluorescently tagged probe.

In this system, anisotropy (A) is defined as the ratio of the linearly polarized component of emitted light over the total intensity or

$$A = \frac{I_{||} - I_{\perp}}{I_{||} + 2I_{\perp}} \tag{9.1}$$

Theoretically, when all emission is parallel to the z-axis ($I_{||} = 1$ and $I_{\perp} = 0$, giving $A = 1$), it means that the probe is immobile since excitation and emission are both in the same plane. Conversely, if the probe rotates at infinite speed, $I_{||}$ will be equal to $I_{\perp}$, yielding $A = 0$. Thus, A is expected to fall between 1 and 0 for most commonly used fluorophores under solution conditions, indicating complete polarization and depolarization, respectively. Under some circumstances, it is possible to observe $A < 0$, although this requires that the population of emission dipoles be heavily biased in the plane perpendicular to excitation, limiting where $I_{||} = 0$ and $I_{\perp} = 1$, giving $A = -0.5$. However, the high and low extremes of anisotropy are normally only encountered in highly structured samples such as crystals and not normally observed in solution-based biochemical assays (Jameson and Ross 2010).

Another value often seen in the literature that is based on the same principle is polarization (P). Polarization represents the fraction of light that is linearly polarized and is derived from

$$P = \frac{I_{||} - I_{\perp}}{I_{||} + I_{\perp}} \qquad (9.2)$$

Anisotropy and polarization convey similar information, but the relationship between anisotropy and the fractional concentrations of fluorescent species is more mathematically tractable and is thus preferred for data analysis. The terms can be interconverted using the function

$$A = \frac{2P}{3 - P} \qquad (9.3)$$

In practice, the maximum and minimum values of anisotropy (and polarization) depend on ensemble measurements across populations of molecules. When excited with polarized light, the probability that a given molecule will be oriented appropriately for excitation is $\cos^2 \theta$, where θ is the angle between the excitation plane and the transition dipole (Fig. 9.2). Furthermore, not all the excited molecules will be exactly parallel to the excitation beam; instead the population will be proportional to $\sin \theta$ where θ is the angle with the vertical axis. Incorporating these two additional pieces of information, the effective upper and lower limits of anisotropy for most common fluorophores are 0.4 and 0, respectively (for polarization they are 0.5 and 0) (Jameson and Ross 2010). Interestingly, the specific upper limit of polarization/anisotropy (called the limiting or intrinsic polarization) of a given fluorophore also depends on the excitation wavelength used in the experiment. This is because excitation can effect more than one electronic transition in many fluorophores, which in turn may contribute differentially to emission (Jameson and Ross 2010; Albinsson et al. 1991).

The specific polarization value of a given sample depends on the intrinsic polarization (described above) and the extent of its rotation during the excited state lifetime. The Debye rotational relaxation time is a value that is used to compare molecular rotations. It is defined as the time required for molecules of a given orientation to rotate through $\arccos(e^{-1})$ or 68.4°. This value is denoted as ρ_0 and is equal to

$$\rho_0 = \frac{3\eta V}{RT} \qquad (9.4)$$

where η is the viscosity of solution, V is the effective molar volume of the rotating unit (which is related to the specific volume of the protein and its hydration (Lakowicz 1999; Jameson and Ross 2010), R is the gas constant, and T is the absolute temperature. When considering the motion of a molecule in solution, the relationships between the determinants of the Debye rotational relaxation time seem intuitive: with increasing viscosity and increasing molecular volume, the motion of a rotating species would be retarded, increasing the relaxation time. By contrast, as temperature rises, solvents tend to become more fluid and thus permissive of motion.

Thus, temperature (in Kelvin) is inversely proportional to ρ. The rotational correlation time (θ) is a related value which is also commonly found in the literature. The two values convey the same information as $\rho = 3\theta$.

The Perrin equation relates observed anisotropy/polarization to the excited state lifetime (the time between excitation and emission) and the rotational diffusion of a fluorophore.

$$A = \frac{A_0}{1 + \left(\dfrac{3\tau}{\rho}\right)} \tag{9.5}$$

where A is observed anisotropy, A_0 is intrinsic anisotropy, and τ is the excited state lifetime. Substituting ρ into the Perrin equation, we get

$$\frac{1}{A} = \frac{1}{A_0} + \frac{\tau RT}{A_0 \eta V} \tag{9.6}$$

The intrinsic anisotropy of a system does not normally vary unless the excitation energy is transferred rather than emitted. This is commonly observed with Förster Resonance Energy Transfer (FRET), where large angles between the absorption and emission dipoles (on the FRET donor and acceptor fluorophores, respectively) can significantly decrease measured anisotropy (Jameson and Ross 2010). In addition, FRET decreases the excited state lifetime of a fluorophore, but the lifetime can also be impacted by the probe microenvironment (pH, hydrophobicity, etc.). However, if the experimental design allows T, η, τ, and A_0 to be held constant, then A will be determined solely by changes in the effective molar volume (V). The size of the rotating molecule may be altered by aggregation, degradation, or, in the cases discussed below, association/dissociation with a specific binding partner (Jameson and Ross 2010). However, there are also a few possible sources of error for A related to sample composition. For example, scattered excitation light (caused by particulate matter or large aggregates in a sample) is vertically polarized and will increase the apparent polarization if not corrected (Owicki 2000; Lakowicz 1999). Additionally, scatter from emission is possible in samples with high optical densities, which will lower observed polarization/anisotropy values (Owicki 2000; Jameson and Ross 2010).

9.2.2 Considerations Relating to the Fluorophore

Spectral properties of a fluorophore important for fluorescence applications are intrinsic polarization, lifetime, emission wavelength, and quantum yield (Yan and Marriott 2003). There are a wide variety of fluorophores that may be conjugated to RNA substrates and are compatible with anisotropy assays. However, selection of a fluorophore with a lifetime appropriate for monitoring substrate motion in the given assay conditions is essential (Jameson and Ross 2010). Many commonly used fluorophores like fluoresceins and rhodamines have fluorescence lifetimes in the low nanosecond range, a scale similar to the tumbling motions of small macromolecules in solution. Quantum yield is defined as the ratio between number of photons emitted and number of photons absorbed. As such, assays using fluorescent dyes with high quantum yields are more sensitive, permitting accurate quantitation at lower probe concentrations. For monitoring tight binding equilibria where dissociation constants (K_d) are in the low nanomolar range or below, measurement of anisotropy at sub-nanomolar substrate concentrations simplifies data analysis (described below) and is thus best served by fluorescent tags of high quantum yield. Ideally, the quantum yield of the selected fluorophore will not change as a result of interaction between the RNA substrate and its cognate binding partners; management of this issue is discussed further below.

Many proteins exhibit intrinsic fluorescence due to the presence of tryptophan, tyrosine, or phenylalanine residues. Anisotropy based on measurements of intrinsic protein fluorescence is possible and has been used in studies of protein–protein interaction or time-resolved anisotropy (Beechem and Brand 1985), but is not commonly used when studying RNA–protein binding events. The popular fluorescent proteins (GFP, etc.) are not generally useful for measurements of anisotropy in vitro as the fluorescent components are largely immobilized within the framework of these proteins and are thus relatively unresponsive to changes in molecular dynamics (Yan and Marriott 2003). In quantitative analysis of RNA–protein interactions, the RNA moiety is generally the smaller and more flexible binding partner. As a result, in fluorescence anisotropy-based analyses of these binding events, the RNA substrate is typically labeled with the fluorescent dye since this molecule normally experiences the greater net restriction of conformational motion during formation of the ribonucleoprotein complex. Most often, the fluorophore is located at one or the other end of the oligonucleotide, but internal sites may also be used (Wilson 2005). Several commercial RNA synthesis services offer a variety of options for tagging RNA oligonucleotides with these dyes. Fluorescein is popular for these applications, due to the fact that it is relatively inexpensive and that it can be readily conjugated. Challenges of using fluorescein are that it will suffer from photodegradation, particularly in environments containing oxygen, and that its spectral properties are pH-sensitive (Jameson and Ross 2010). The poor photostability of fluorescein is a particular issue in kinetics experiments, where the anisotropy of a sample will be read many times as a function of time. Other dyes that are commonly conjugated to RNA substrates include 5′-carboxytetramethylrhodamine (TAMRA) and the AlexaFluor and cyanine families of dyes.

9.2.3 Considerations Relating to Instrumentation

9.2.3.1 Conventional Spectrofluorometers

Fluorescence anisotropy is measured using a fluorescence spectrophotometer (also called a spectrofluorometer) that is equipped with polarizers on both the excitation

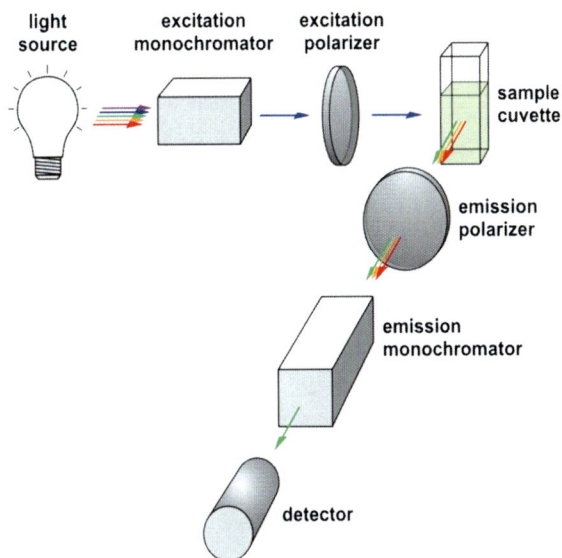

Fig. 9.3 Schematic of the L-format spectrofluorometer. This most commonly used type of fluorometer has only one emission channel. To measure anisotropy, the parallel and perpendicular components of the emission are both measured through this channel, with the emission polarizer rotating 90° between the readings

and emission channels. These devices are generally configured in one of two different formats. The most commonly used is called the L-format since it has a single emission channel oriented at a right angle to the excitation path (Fig. 9.3).

The wavelength limits for excitation and emission are set using monochromators, or in some cases, optical bandpass filters. Eliminating overlap between excitation and emission wavelengths minimizes scatter from reflected excitation light. Total fluorescence of a sample can be read directly from this configuration. However, to measure anisotropy, light from the excitation plane is first passed through a vertically oriented polarizer prior to hitting the sample cell. The anisotropy parameter is then calculated from paired measurements of emission passed through: (1) a vertical polarizer (I_{VV}; intensity from vertically polarized excitation measured through a vertical emission polarizer), and (2) a horizontal polarizer (I_{VH}). These intensity measurements are proportional to the parallel (I) and perpendicular (I) components of anisotropy, respectively (Jameson and Ross 2010). T-format spectrofluorometers are also available, so termed because they are configured with a second emission detector in the same observational plane but reading from the opposite side of the sample cuvette. These instruments thus permit both parallel and perpendicular emission to be read simultaneously, each through its own dedicated detection system. In the past, T-format instruments were favored since the readings were considered less susceptible to fluctuations in signal intensity, but that bias has been largely minimized as L-format instruments have become more sophisticated (Lakowicz 1999). However, for rapid kinetics experiments where cycling of emission polarizers is not practical (time points in low seconds range or faster), the T-format remains essential.

In L-format instruments, the emission monochromator normally has a sensitivity bias between vertically and horizontally polarized light (Jameson and Ross 2010).

For this reason, it is important to introduce a correction factor into the calculation of the anisotropy value. For L-format instruments, the correction factor (G) is given as

$$G = \frac{I_{HV}}{I_{HH}} \tag{9.7}$$

It should be noted that the horizontal excitation direction is not used for the actual measurement of the anisotropy value. However, when excitation is parallel to the direction of observation, both the vertical and horizontal measured intensities should be equal to each other and are proportional to I because both will be perpendicular to the direction of excitation (Lakowicz 1999). Hence, any differences between the two values will be due to instrument bias. The G-factor should be measured for each instrument/fluorophore combination and will vary significantly with emission wavelength. For T-format polarimeters, the correction factor is a ratio of the sensitivities of the parallel and perpendicular detectors. With the G-factor, the equations to resolve anisotropy and polarization values become, respectively

$$A = \frac{I_{VV} - GI_{VH}}{I_{VV} + 2GI_{VH}} \tag{9.8}$$

and

$$P = \frac{I_{VV} - GI_{VH}}{I_{VV} + GI_{VH}} \tag{9.9}$$

A final point about accurate measurement of fluorescence anisotropy or polarization is the importance of background subtraction. The anisotropy of a blank sample (i.e., one lacking the fluorescent ligand) cannot simply be subtracted from the anisotropy of an experimental sample. Rather, the fluorescence intensity of the blank sample taken in the parallel (I_{VV}) and perpendicular (I_{VH}) polarizer orientations must be subtracted from those of the sample containing the fluorescent ligand. From these blank-corrected intensity values, the anisotropy of the sample can then be calculated (Jameson and Ross 2010). For most instruments that measure fluorescence anisotropy, the operating software will prompt the background reading prior to addition of the fluorescent substrate and then automatically subtract those values before reporting sample anisotropy values.

9.2.3.2 Plate Readers

Plate reading technology now allows researchers to rapidly measure the fluorescence anisotropy of large numbers of samples. One popular application is to screen compound libraries for chemical inhibitors of specific macromolecular binding events.

Many of these screens are based on competition assays (Owicki 2000). By this method, a small fluorescent ligand associates with a large macromolecule resulting in a high anisotropy value owing to the large molecular volume of the complex containing the fluorophore. When added to the complex, a chemical inhibitor of this interaction would be expected to induce release of the fluorescent substrate, which is detected by a significant decrease in the measured anisotropy of the binding reaction.

For a screen that may encompass thousands or even hundreds of thousands of binding reactions, determining the effectiveness of the assay design is critical if meaningful data are to be obtained. The Z-factor was introduced to address this need (Zhang et al. 1999).

$$Z = 1 - \frac{3\,\sigma_S + 3\,\sigma_C}{|\,\mu_S - \mu_C\,|} \tag{9.10}$$

Here, σ is the standard deviation of the samples and μ is the mean. The subscripts S and C designate sample and control, respectively. When values of Z are close to 1, the means of the control and sample are well-separated with little variation between data points, signifying that the assay is designed appropriately to identify individual compounds (hits) that significantly perturb reaction binding equilibrium. Conversely, if the Z-factor is close to 0, essentially no difference can be identified between samples and negative controls.

In addition to high-throughput screening, using plate readers to measure fluorescence anisotropy is appealing because they can accommodate very small sample volumes (often down to 100 μL in 96-well mode, and even smaller in 384-well mode) (Mao et al. 2006). This feature can be particularly valuable when the components of the model system are expensive or difficult to generate. Furthermore, current generations of fluorescence anisotropy-capable plate readers are generally user friendly, require limited optimization, and often include many other detection modes. However, plate readers also suffer from some limitations with respect to measuring anisotropy. They are generally not as sensitive as cuvette-based spectrofluorometers and offer less flexibility in probe selection since excitation and emission wavelengths and bandpass limits are normally set using optical filters rather than monochromators. In addition, plate readers are not as useful for evaluation of pre-steady-state kinetics, because they are not compatible with the stopped-flow mixers necessary to initiate very rapid biochemical reactions.

9.3 Data Analysis

9.3.1 Additivity of Anisotropy

The most common applications of fluorescence anisotropy to the study of RNA–protein binding events are to quantitatively measure the affinity of these interactions.

A frequent approach is to measure the anisotropy of limiting concentrations of fluorescent substrate (in this case, RNA) across a titration of its putative binding partner. To extract equilibrium binding constants from these binding isotherms, the use of anisotropy rather than the polarization parameter presents a significant advantage. Under conditions of constant fluorescence quantum yield, the total measured anisotropy of a sample (A_t) is an additive function of the intrinsic anisotropy (A_i) and fractional concentration (f_i) of each fluorescing species.

$$A_t = \sum_i A_i f_i \qquad (9.11)$$

This conceptual framework is the basis for analytical models of RNA–protein binding equilibria discussed below. Considerations and options for cases where quantum yield is not constant throughout the binding isotherm are also described.

9.3.2 Derivation and Analysis of Binding Models

Consider a sequential association reaction model by which proteins successively bind an RNA substrate conforming to the general scheme

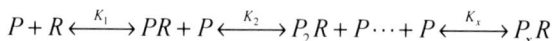

$$P + R \xrightarrow{K_1} PR + P \xrightarrow{K_2} P_2R + P \cdots + P \xrightarrow{K_x} P_xR$$

where R is the fluorescently tagged RNA molecule, P is the protein partner, and values of K_x represent the equilibrium association constants describing each binding event $(K = 1/K_d)$ (Wilson 2005). For this reaction,

$$[PR] = [R][P]K_1$$
$$[P_2R] = [R][P]^2 K_1 K_2$$
$$\downarrow$$

$$[P_xR] = [R][P]^x \prod_{i=1}^{x} K_i \qquad (9.12)$$

By conservation of mass,

$$[R]_{tot} = [R] + [PR] + \cdots + [P_xR] \qquad (9.13)$$

where $[R]_{tot}$ is the total concentration of the fluorescent RNA substrate in the binding reaction. Employing the additivity of anisotropy, we get

$$A_t = \frac{1}{[R]_{tot}}(A_R[R] + A_{PR}[PR] + \cdots + A_{PxR}[P_xR]) \qquad (9.14)$$

where A_R, A_{PR}, etc. represent the intrinsic anisotropy values of the free RNA or specific RNA:protein complexes designated by the subscripts. The general expression relating observed anisotropy to protein concentration then becomes

$$A_t = \frac{A_R + A_{PR}[P]K_1 + \cdots + A_{PxR}[P]^x \prod_{i=1}^{x} K_i}{1 + [P]K_1 + \cdots + [P]^x \prod_{i=1}^{x} K_i} \tag{9.15}$$

which incorporates the expressions describing each equilibrium association constant and sample protein concentration into the total measured anisotropy A_t. In order for this relationship to be useful, $[R]_{tot}$ must be limiting and should be at least fivefold less than any K_d ($= 1/K$) so that the free protein concentration in solution ($[P]$) approximates the total protein concentration in each reaction ($[P]_{tot}$). Explicit values of K may then be calculated by nonlinear regression of A_t vs. $[P]$, although this sequential binding function becomes increasingly complicated as the number of binding events increases and for most purposes is impractical beyond the first two discrete binding events. In some cases, however, conditions can be imposed on the system to prevent x from becoming larger than 2. One example is to limit the length of RNA so that only one or two binding events can take place (more on site size determination below). It may also be possible to generate mutant proteins that cannot form oligomeric complexes on RNA substrates (Wilson 2005).

The simplest RNA–protein binding relationships can be described by reversible one-step binding. An example of this model is given by the major inducible heat shock protein, Hsp70, which interacts with members of a discrete family of mRNA-destabilizing sequences termed AU-rich elements (AREs) (Wilson et al. 2001) found in many cytokine and proto-oncogene transcripts. Here,

$$A_t = \frac{A_R + A_{PR}K[P]}{1 + K[P]} \tag{9.16}$$

In Fig. 9.4, the upper asymptote of the isotherm represents A_{PR} and the lower asymptote resolves A_R. The concentration of probe used to generate this isotherm was 0.15 nM, appropriate for a reaction with a K_d of 25 nM, but still high enough to present a favorable signal-to-noise ratio. Although the anisotropy isotherm describing Hsp70 binding to this RNA substrate is consistent with a 1:1 binding model, a similar result would also be observed if multiple thermodynamically equivalent binding events were occurring. To discriminate between these possibilities, we have found it useful to perform electrophoretic mobility shift assays (EMSAs) in parallel with binding experiments resolved by fluorescence anisotropy. In the case of Hsp70 binding the TNFα ARE substrate, EMSAs spanning a range of protein concentrations resolved a single RNA–protein binding event, supporting a 1:1 binding equilibrium (Wilson et al. 2001). As a general rule, EMSAs are unreliable as a quantitative tool for measuring binding affinities owing to ribonucleoprotein complex dissociation during loading and fractionation through the gel; this is particularly troublesome for highly dynamic binding equilibria. However, we find that EMSAs remain

Fig. 9.4 Binding curve of His$_6$-tagged Hsp70 to a fluorescein-conjugated RNA substrate encoding the ARE from tumor necrosis factor α (TNFα) mRNA (*solid circles*). The best-fit function is consistent with a 1:1 binding interaction with $K = 4.0 \times 10^7$ M^{-1} ($K_d = 25$ nM). A residual runs test (*bottom*) shows that this binding model is consistent with the observed anisotropy data across the entire protein titration. A separate protein titration experiment shows no significant Hsp70 binding to a similarly sized fragment of β-globin mRNA (*open circles*)

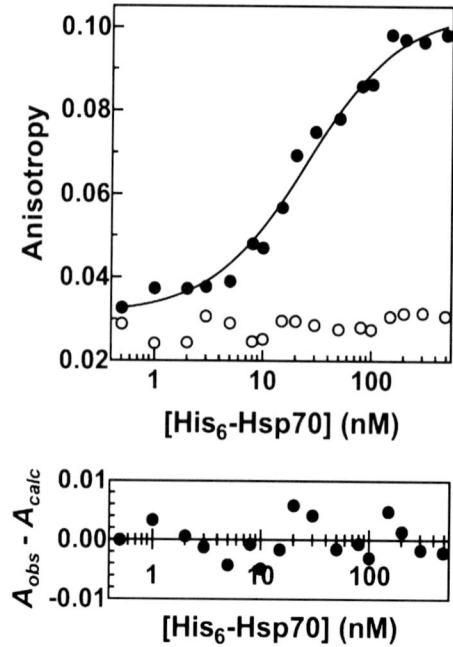

useful for identifying the minimum number of RNA-containing complexes formed during protein binding reactions, which in turn yields a valuable starting point for defining binding models that will then be quantitatively characterized by anisotropy-based assays.

In Fig. 9.4 and all other binding curves presented in this chapter, the protein concentration has been plotted on a logarithmic scale instead of a linear scale. When anisotropy isotherms are presented in the literature, however, some investigators will opt to present the x-axis on a linear scale. This method of presentation is completely accurate, but complicates facile assessments of binding mode and makes assessment of deviations between data and the model across the range of tested protein concentrations more difficult.

For reactions best described by two rounds of protein binding to a common RNA substrate, the equation becomes

$$A_t = \frac{A_R + A_{PR}K_1[P] + A_{P2R}K_1K_2[P]^2}{1 + K_1[P] + K_1K_2[P]^2} \tag{9.17}$$

Binding of the human p37^{AUF1} protein to an RNA substrate follows this two-step binding model as shown in Fig. 9.5. AUF1 is a family of four dimeric proteins generated by alternative splicing of a common pre-mRNA (Wagner et al. 1998) that has diverse functions in control of mRNA decay, translation, and telomere maintenance (Lu et al. 2006; Liao et al. 2007; Sarkar et al. 2011; Eversole and Maizels 2000).

Fig. 9.5 Binding of His$_6$-tagged p37^{AUF1} to a fluorescein-conjugated TNFα ARE substrate. In this experiment the RNA substrate concentration was 0.2 nM. The solid line shows the nonlinear regression solution to a two-step binding model (K_d values reported in the text). The *dotted line* represents the best fit to a single-site binding model, which significantly diverges from the experimentally observed anisotropy values at both low and high protein concentrations

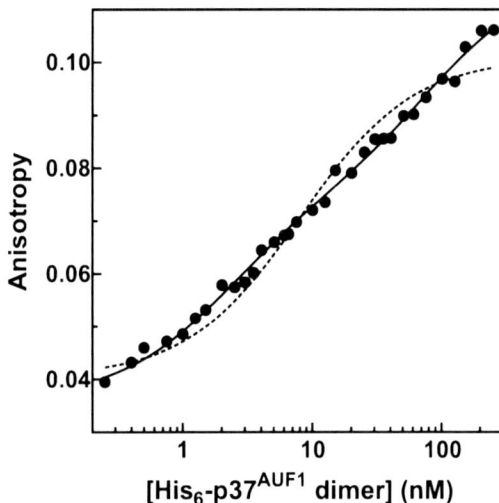

When p37^{AUF1} dimers are incubated with a fluorescein-tagged TNFα ARE substrate, a high affinity complex is formed at low protein concentrations ($K_{d1} = 1/K_1 = 1.7 \pm 0.3$ nM). However, as protein concentration increases further, a second p37^{AUF1} dimer binds with lower affinity ($K_{d2} = 1/K_2 = 74 \pm 16$ nM) to generate a tetrameric protein complex on this RNA ligand (Zucconi et al. 2010).

Using this two-step model, equilibrium binding parameters can readily be calculated as long as saturation of the second binding event can be attained and the affinity of the second binding event is at least fivefold weaker than the first. The latter condition allows regression software packages to explicitly calculate the intrinsic anisotropy of the intermediate species (A_{PR}), which can be graphically estimated from the plateau linking the high and low affinity phases of the binding curve. In cases where this criterion is not met (i.e., in reactions where K_2 approaches or exceeds K_1), it becomes very difficult to confidently resolve each binding constant using the two-step function alone. However, in some cases it may be possible to determine A_{PR} or K_2 using some other strategy. With these values in hand, the remaining constants can be determined. For example, we recently used macromolecular binding density analysis (MBDA) to resolve the intrinsic anisotropy of an intermediate RNA:protein complex, which in turn permitted resolution of explicit values for K_1 and K_2 in the cooperative assembly of an oligomeric ribonucleoprotein (Zucconi et al. 2010). This approach is discussed further below.

For multistep RNA–protein binding events where individual binding constants are not resolvable or perhaps not explicitly required, an alternative analysis strategy is to employ a variation on the Hill model, which, when converted to a form which relates total measured anisotropy (A_t) to protein concentration becomes

$$A_t = A_R + (A_{PxR} - A_R) \times \left[\frac{([P]/[P]_{1/2})^h}{1 + ([P]/[P]_{1/2})^h} \right] \qquad (9.18)$$

Fig. 9.6 Binding of HuR to the fluorescein-conjugated TNFα ARE (0.2 nM RNA substrate). In this experiment $[P]_{1/2}$ was calculated to be 2.2 nM and the Hill coefficient was 1.49. The *solid line* indicates the cooperative fit, and the *dotted line* indicates the best fit possible using a single-site binding model

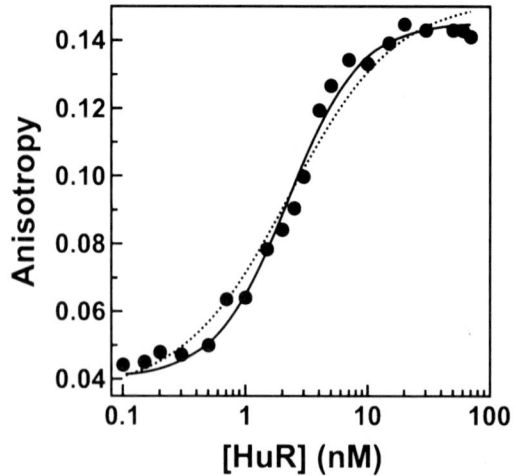

where A_R and A_{PxR} are the intrinsic anisotropy values of free RNA and the saturated RNA:protein complex, respectively, $[P]_{1/2}$ is the concentration of protein at which the reaction achieves half-maximal binding, and h is the Hill coefficient (Wilson 2005). For this scheme, the standard assay requirements of limiting RNA concentration and constant quantum yield still apply. However, for regression solutions resolving $h = 1$, this function simplifies to a single-site binding algorithm, with $[P]_{1/2} = K_d$.

The association of the mRNA-stabilizing protein HuR with the TNFα ARE substrate, shown in Fig. 9.6, illustrates an application of this modified Hill function. The solution of $h > 1$ indicates that HuR proteins assemble cooperatively into oligomeric complexes on this RNA substrate. As indicated on the plot, the one-step binding model does not resolve these data well, an assertion validated by nonrandom distributions of residuals for the single-site solution and by pairwise statistical comparisons of cooperative and single-site fits using the F test (Fialcowitz-White et al. 2007).

When the affinity of an RNA–protein binding event is very high (i.e., $K_d < 1$ nM), it may not be possible to accurately measure the anisotropy of RNA substrates at concentrations low enough for the free protein concentration $[P]$ to approximate the total protein added to each binding reaction. In these situations, binding algorithms that incorporate ligand depletion must be employed, which are normally expressed in terms of total protein ($[P]_{tot}$) and RNA substrate ($[R]_{tot}$) concentrations. Shown below is the ligand depletion model for reversible one-step binding (Wilson 2005).

$$A_t = A_R + (A_{PR} - A_R) \times$$

$$\left[\frac{1 + K[R]_{tot} + K[P]_{tot} - \sqrt{(1 + K[R]_{tot} + K[P]_{tot})^2 - 4[R]_{tot}[P]_{tot}K^2}}{2K[R]_{tot}} \right] \quad (9.19)$$

All data analysis strategies discussed to this point rely on constant fluorescence quantum yield across the anisotropy isotherm, ensuring that both bound and unbound fractions of the fluorescent RNA substrate make equivalent contributions to total measured anisotropy. While this simplifies data analysis, situations do arise where the quantum yield of the fluorescent RNA substrate changes as a function of protein concentration; this is detected as a protein-dependent change in total (i.e., non-polarized) fluorescence intensity. Protein binding can influence the quantum yield of labeled RNA substrates in several different ways, including direct contact between the protein and the fluorophore, disruption of its solvent accessibility, alteration of microenvironmental pH or ionic strength, or changes in local RNA structure (Wilson 2005). Whatever the cause, the most straightforward solution is to modify the identity or location of the fluorescent probe on the RNA substrate. For instance, using TAMRA instead of fluorescein may abrogate variations in probe quantum yield. Alternatively, conjugating the fluorophore to the 5′-end of the RNA rather than the 3′-end or vice versa is an option. Another possibility is to increase the distance between the fluorophore and the protein binding site on the RNA substrate by adding a few irrelevant nucleotides. However, in the event that protein-dependent changes in probe quantum yield cannot be avoided, it is often possible to extract binding parameters by calculating the contributions of fluorescence emission from bound and unbound RNA to the overall measured anisotropy of the sample. One example of this approach is given by the following equation

$$x = \frac{A_{\mathrm{t}} - A_{R}}{A_{PR} - A_{R} + (f - 1)(A_{PR} - A_{\mathrm{t}})} \tag{9.20}$$

where x is the fraction of bound ligand, A_{PR} is anisotropy of the protein-bound RNA species, A_{R} is the anisotropy of the free RNA, and f is the relative change in quantum yield resulting from protein binding, calculated as the ratio of total fluorescence emission from the bound RNA ligand (measured under saturating protein concentrations) relative to the free RNA (measured in the absence of binding proteins) (Jameson and Ross 2010). Using this relationship, x can be calculated for every point on a titration curve. Any determination of binding constants will be calculated from the resulting function.

It is important to provide justification when applying specific analytical models to RNA–protein binding events. It is often difficult to predict a priori how a protein will interact with an RNA substrate in vitro. As described above, an EMSA approach can often provide some insight into appropriate binding model selection by defining a lower limit on the number of RNA:protein complexes formed (Anderson et al. 2008). Other tests for the suitability of specific binding models can usually be performed using data analysis software packages. We have found PRISM (GraphPad) a particularly versatile and user-friendly platform for analysis of fluorescence anisotropy-based binding isotherms. A high coefficient of determination (R^2 approaching 1) is one metric that suggests that a binding function fits experimental data well. However, another validation approach that we always include is a residual runs test, which returns a P value reflecting the likelihood that deviations of data from the regression solution are nonrandom. If $P < 0.05$, then the model is likely biased for subsets of the plotted data, and thus does not reflect the protein binding mechanism across the entire range of concentrations tested. When comparing two different binding models, the F test can determine whether the improvement in sum-of-squares deviation obtained by fitting to a more complex binding function justifies the decrease in degrees of freedom associated with that model. This comparison is reported by a P value that reflects the probability that the improved fit to the more complex model is the result of random data scatter. This is particularly useful for comparing variants of a binding scheme, such as one- vs. two-site binding, or cooperative vs. noncooperative interactions. In the example shown in Fig. 9.6, cooperative formation of HuR oligomers on the TNFα ARE substrate was favored over the single-site binding model with $P < 0.0001$ (Fialcowitz-White et al. 2007). Finally, it is important to remember that the behavior of biomolecules in vitro does not necessarily predict their behavior in the cellular milieu. As such, cell-based assay systems should be used to define the biological relevance of binding analyses wherever possible.

9.4 Advantages and Disadvantages of Anisotropy-Based Assay Systems

Compared to other methods for detecting and quantifying RNA–protein interactions, fluorescence anisotropy offers several distinct advantages. First, fluorescence-based approaches do not require the radioactive compounds commonly used for EMSAs or filter-binding assays, which simplifies the logistics of conducting these experiments (Jameson and Ross 2010). Second, fluorescence anisotropy measurements can be taken under both equilibrium and pre-steady-state conditions, permitting separate analysis of thermodynamic and kinetic binding parameters. Third, many anisotropy-based assays can be easily automated, making this an ideal platform for high-throughput screening (HTS) in drug discovery applications.

A fourth and perhaps most important advantage of fluorescence anisotropy-based binding strategies is that they are homogenous phase assays, meaning that measurements are taken while all binding partners and resultant complexes are present together in solution (Jameson and Sawyer 1995; Wilson 2005). Neither filter-binding assays, in which the RNA:protein complex is retained on the filter and unbound species are washed away, nor EMSAs, where the various reaction products are separated by passage through a native acrylamide gel, provide an environment where dynamic equilibria can be maintained or monitored in real time. EMSAs are particularly impacted by this complication, since dynamic ribonucleoprotein complexes can dissociate during electrophoresis resulting in an underestimation of binding affinity. Surface plasmon resonance (SPR) is another method by which RNA–protein binding events have been monitored, but require tethering one partner to the assay tube which limits its rotational and translational motion. Heterogeneity in ligand binding to this surface can result in the appearance of distinct subpopulations of binding activities (Sadana 2001). Furthermore, SPR approaches customarily resolve binding constants from the ratio between binding on- and off-rates. For the induced-fit binding mechanisms normally associated with biomolecular interactions, extraction of multiple on- and off-rates (which include both intramolecular as well as intermolecular events) from SPR data provides a significant challenge. By contrast, the mathematical relationships relating fluorescence anisotropy to the fractional concentrations of fluorophore-tagged reactants and products are well established and lend themselves to quantitative characterization using several different models of complex formation (Wilson 2005).

Several potential drawbacks to the application of anisotropy-based assays to RNA–protein binding studies also exist. First, the RNA substrate must be fluorescent, which requires covalent modification to incorporate the fluorescent dye. This introduces the risk that the dye itself may contribute to or inhibit protein binding. This possibility can be tested by comparing the K_d value resolved using the fluorescent RNA substrate with the K_i value calculated from a competition experiment with the unlabeled RNA ligand (Anderson et al. 2008). Second, for relatively weak binding events (i.e., $K_d > 1$ μM), generation of binding isotherms requires large amounts of protein to approach saturation. Besides the logistical issues of producing and purifying large quantities of some RNA-binding proteins, at high protein concentrations sample viscosity is no longer constant and must be accounted for during data analysis. One approach that can remedy this complication is to run a parallel binding isotherm but using the dye alone as the substrate; provided the dye does not bind the protein itself, protein-dependent changes in sample viscosity will be detected by increases in probe anisotropy. Finally, synthetic RNA substrates containing fluorescent labels remain fairly expensive (Mao et al. 2006). This cost can be compounded in cases where several rounds of optimization are required, such as situations where several probes and/or linkage positions must be screened to circumvent adverse interactions between dyes and binding proteins, or protein-dependent changes in fluorescence quantum yield.

9.5   Applications of Fluorescence Anisotropy

In recent years, many groups have used fluorescence anisotropy to investigate various biological pathways in which RNA and protein must interact. Below are a few examples, along with some useful extensions of the technique.

9.5.1   Examples of RNA:Protein Complexes Analyzed by Fluorescence Anisotropy

Targeted modification of an RNA substrate is a common strategy to identify sequence and/or structural preferences of specific binding proteins and can also provide insights into binding mechanisms. One study shows that Pumilio (Pum), an RNA-binding protein that recognizes Nanos Response Elements (NREs) in the

3'-untranslated region (UTR) of *hunchback* mRNA and suppresses its translation in early *Drosophila* embryos, binds to this RNA target sequence with 2:1 protein:RNA stoichiometry (Gupta et al. 2009). This determination was made by measuring binding between recombinant Pum protein and the wild-type NRE substrate, then comparing these data to binding isotherms involving NRE substrates where specific subdomains defined as Box A (GUUGU) and Box B (AUUGUA) were selectively mutated. Protein–RNA binding events were monitored using both fluorescence anisotropy and EMSA approaches, and interestingly, this study demonstrates discrepancies that may arise between data collected using these two techniques. Anisotropy isotherms showed that Pum can bind NRE sequences containing either Box A or Box B, although with a marked preference for substrates containing only the Box B sequence. The EMSAs, however, showed no convincing Pum binding to RNA substrates lacking the Box B sequence, although they did demonstrate the formation of two distinct protein:RNA complexes on the WT sequence. It is likely that the lower affinity complex containing Pum and the Box A substrate could not be retained during gel loading and electrophoretic separation.

A distinct family of RNA-binding proteins studied using fluorescence anisotropy are the ARE-binding proteins (ARE-BPs), which target the ARE sequences responsible for regulating the cytoplasmic decay of many labile transcripts in mammalian cells (described above). In Fig. 9.4, Hsp70 was shown to bind the ARE from TNFα mRNA in a 1:1 complex. However, removing the adenosine residues from this substrate to generate a uridylate homopolymer allowed multiple copies of Hsp70 to bind in a highly cooperative fashion, with a Hill coefficient approaching 1.7 (Wilson et al. 2001). The functional significance of this change in binding mechanism is currently unknown, but highlights the utility of fluorescence anisotropy in detecting even subtle distinctions in the mechanics of protein association with RNA substrates.

Another common approach that exploits the rigorous discrimination of RNA–protein binding affinities offered by anisotropy-based assays involves screening protein mutants to identify subdomains contributing to the stability of ribonucleoprotein complexes. For example, the mRNA-stabilizing protein HuR binds cooperatively to ARE-like sequences of 18 nucleotides or longer (Fig. 9.6). The HuR protein consists of three RNA recognition motifs (RRMs); two positioned in tandem at the N-terminus and the third at the C-terminus, linked to the other two by a flexible hinge region. HuR mutants lacking the C-terminal RRM retain high affinity RNA-binding activity, but lose the ability to cooperatively form oligomeric complexes on RNA substrates. By contrast, HuR proteins lacking both the C-terminal RRM and hinge domain exhibit very poor RNA-binding activity (Fialcowitz-White et al. 2007). These experiments have helped assign specific functions to HuR subdomains in RNA recognition and higher-order ribonucleoprotein complex assembly. However, beyond screening mutants of RNA-binding proteins, fluorescence anisotropy-based approaches have also identified alterations in RNA–protein interactions resulting from covalent protein modifications like phosphorylation (Wilson et al. 2003) or non-covalent events like co-factor binding (Bernstein et al. 2008).

A different study used truncation and site-directed mutants of RNA helicase A (RHA) to localize subdomains responsible for recognition of structured RNA domains called post-transcriptional control elements (PCEs). PCEs are located in the 5′ UTRs of some retroviral and cellular mRNAs. Association of RHA with these transcripts is required for ribosome loading during translational initiation. Fluorescence anisotropy-based binding assays using extended (96- and 98-nt) RNA substrates containing PCE sequences showed that the protein domain responsible for specific recognition of these structured RNA substrates was localized to the N-terminus of RHA (Ranji et al. 2011). Interestingly, an RG-rich domain at the RHA C-terminus also exhibited RNA-binding activity, but was not specific for the PCE substrates, indicating that it may contribute nonspecific binding energy in the formation of RHA:PCE complexes.

Some final examples highlight cases where binding reactions may contain multiple components, not solely the RNA substrate and cognate protein. These applications are important because the functional relevance of some binding interactions may require or be regulated by the presence of ancillary factors. However, even in these cases RNA-binding events can sometimes be analyzed using one-step reversible models, such as situations where ancillary factors are in vast excess or if the protein components form stable complexes in the absence of the RNA probe. An example of the latter case is the association of the 43S translation preinitiation complex with RNA substrates. Previous anisotropy studies tracking protein–protein interactions involved in 43S assembly showed that its substituent components assemble into stable complexes (Maag and Lorsch 2003). As such, if complex members are preincubated prior to addition of fluorescent RNA substrates, changes in anisotropy readings will reflect RNA targeting by the preassembled complex. Using this approach, the role of the mRNA 7-methylguanosine 5′ cap structure in 43S recruitment was quantitatively characterized by measuring the association kinetics of fluorescent RNA substrates. Although eIF4E is the canonical cytoplasmic mRNA cap-binding protein (Marcotrigiano et al. 1997), optimal recognition of capped RNA substrates only occurred for 43S particles containing eIF3, eIF4A, eIF4B, and the eIF4E:eIF4G complex (Mitchell et al. 2010). By contrast, withdrawing any of these ancillary factors dramatically decreased the rate of 43S binding to capped RNA substrates. Curiously, uncapped RNA ligands were recruited to 43S substrates lacking various eIF partners, prompting a model whereby the 5′ cap structure specifically selects 43S complexes with their full complement of co-factors, which are thus optimally primed for initiation of translation. Conversely, uncapped mRNAs may be sequestered by a variety of incomplete 43S complexes, which may prevent them from assembling into translationally active polysomes. These experiments provide an elegant example where quantitative, anisotropy-based in vitro studies, starting with binary equilibria but then developing through addition of ancillary factors, ultimately defined a global model of translational preinitiation complex assembly and targeted mRNA recruitment.

9.5.2 Using Fluorescence Anisotropy to Determine Binding Site Size and Thermodynamic Constants

Resolving the length of RNA required to bind a specific protein partner is very useful when defining features of the association mechanism and provides a biochemical basis for discriminating novel binding sites. For simple binary binding events (i.e., one protein binding to one RNA substrate), the site size can often be estimated by measuring complex affinity across a series of RNA substrate truncation mutants. However, for RNA-binding proteins that form multiple complexes on RNA substrates, particularly if any allosteric events are involved, it is very difficult to determine the size of individual protein binding sites by this approach alone.

When using fluorescence anisotropy-based assays, one useful strategy for estimating the binding site size of specific RNA-binding proteins is the MBDA approach first described by Wlodek Bujalowski (Lohman and Bujalowski 1991; Bujalowski and Jezewska 2000). Calculation of the binding site size using the MBDA approach begins by generating anisotropy binding isotherms across titrations of protein and a given RNA substrate. Several independent isotherms are required, each using a different concentration of RNA selected so that they span values below and above the K_d describing formation of that specific RNA:protein complex (or at least one K_d value for multistep binding mechanisms). An example of such a dataset is shown in Fig. 9.7a. MBDA holds that each possible RNA:protein complex confers a distinct change in the intrinsic anisotropy of the fluorescent RNA ligand ($\Delta A_i = A_i - A_R$ where A_R is the intrinsic anisotropy of the RNA ligand in the absence of protein) that contributes to the population weighted average (ΔA_{obs}). A common value of ΔA_{obs} in multiple samples containing different total RNA concentrations ($[R]_{T1}$, $[R]_{T2}$, ..., $[R]_{Tx}$) indicates that each system possesses similar distributions of complexes, yielding the same average binding density (Σv, defined as the average number of protein molecules bound per RNA molecule) and free protein concentration $[P]_F$ at the corresponding total protein concentrations ($[P]_{T1}$, $[P]_{T2}$, ..., $[P]_{Tx}$). If paired ($[P]_T$, $[R]_T$) concentrations obtained at common values of ΔA_{obs} across all isotherms are then plotted, the slope of the linear regression solution yields Σv. Applying this approach to many values of ΔA_{obs} distributed across the dynamic range of the isotherms thus reveals the relationship between ΔA_{obs} and Σv, such as that displayed in Fig. 9.7b. In practice, it is often difficult to reliably resolve values of Σv at ΔA_{obs} values greater than 70–80 % of the asymptotic maximum (ΔA_{sat}) for each anisotropy isotherm owing to the contributions of nonspecific binding events at very high protein concentrations. As such, determination of the protein binding density at saturation (Σv_{sat}) generally requires extrapolation of the resulting plot to the anisotropy value observed at saturation (ΔA_{sat}), which is defined by the upper asymptote of the original binding isotherms. The binding site size (n) is then calculated from the ratio of RNA nucleotide length (N) to saturation binding density as $n = N/\Sigma v_{sat}$. In the case of HuR binding to the ARE substrate in Fig. 9.7, linear extrapolation of Σv to the maximal ΔA_{obs} yielded Σv_{sat} of 4.2±0.3 HuR molecules bound per fluorescent ARE substrate (Fialcowitz-White et al. 2007). Given that this RNA substrate was 38 nucleotides in length, the HuR site size was thus resolved to 9.1±0.3 nucleotides at

Fig. 9.7 Binding site size determination for HuR to the TNFα ARE using MBDA. (**a**) HuR binding isotherms using reactions containing 0.2 nM (*solid black circles*), 1 nM (*open green circles*), or 5 nM (*red triangles*) fluorescein-conjugated RNA substrate. (**b**) The relationship of Σv to ΔA_{obs} resolved from the dataset shown in (**a**). The *red dashed line* shows a linear extrapolation to the ΔA_{sat} value of 0.12 permitting solution of the HuR binding site size at saturation

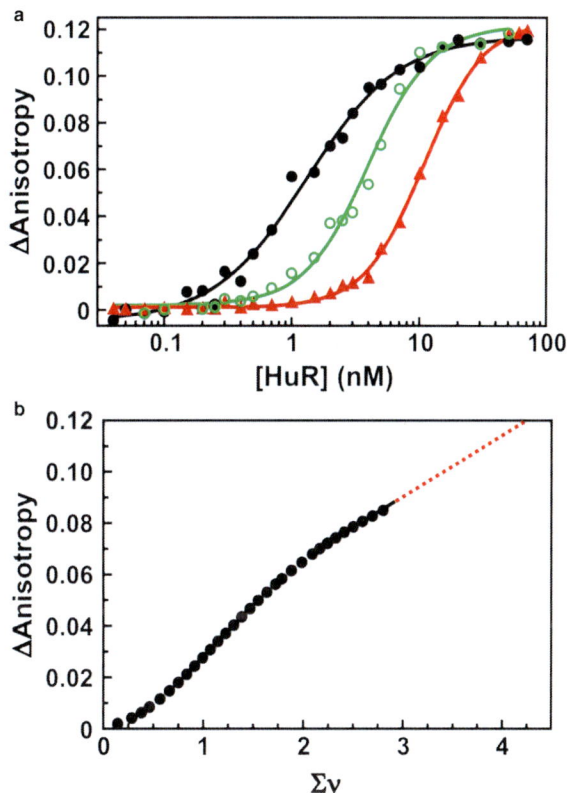

binding saturation. These findings were supported by subsequent observations that two HuR molecules cooperatively associate with an 18-nt ARE substrate, but that only a single HuR binding event is possible on substrates shorter than 18-nt (Fialcowitz-White et al. 2007).

Plots of anisotropy vs. Σv that do not approximate linearity across the complete range of Σv (as in Fig. 9.7b) indicate the likelihood of multiple, nonequivalent binding modes. A benefit of this phenomenon is that the amplitude of ΔA_{obs} associated with individual binding modes may be estimated. In the two-step protein binding model described above, a significant handicap was that explicit values for K_1 and K_2 could not be resolved if K_2 approaches or exceeds K_1, owing to redundancy in the impact of A_{PR} and K_2 on the regression solution. However, using MBDA the average change in anisotropy contributed by the first binding step (ΔF_1) can be estimated by the slope of the low density binding regime ($\Delta F_1 = \partial \Delta A_{obs} / \partial \Sigma v = A_{PR} - A_R$ in the two-step binding model). The A_{PR} value resolved by this approach can then be set constant in the two-step binding algorithm and permits confident solutions of K_1 and K_2. Using this strategy, the affinities of both steps of p42[AUF1] binding to an ARE substrate were resolved, which quantitatively defined a cooperative relationship between these RNA-binding events (Zucconi et al. 2010).

Beyond resolving RNA site size, equilibrium binding constants resolved by fluorescence anisotropy-based assays can also be used to determine the enthalpic

(ΔH) and entropic (ΔS) contributions to RNA–protein binding events using van't Hoff analysis. These parameters are often useful for discriminating molecular events that drive ribonucleoprotein assembly. For example, enthalpically driven binding events normally feature extensive ionic or hydrogen bond formation, while favorable entropic changes may result from liberation of ions or water molecules driven by hydrophobic surface burial in the RNA:protein complex (Beaudette and Langerman 1980; Draper 1995). The van't Hoff approach requires that protein binding isotherms be generated across a range of reaction temperatures. From these data sets, association equilibrium constants (K) for the RNA:protein complex are calculated at each temperature, and then plotted as $\ln K$ vs. $1/T$, where T is the absolute temperature (in Kelvin). Given that

$$\Delta G = -RT \ln K = \Delta H - T\Delta S \qquad (9.21)$$

where R is the gas constant, a linear plot of $\ln K$ vs. $1/T$ will yield a slope equal to $-\Delta H/R$, and a vertical intercept equal to $\Delta S/R$. Linear solutions of the van't Hoff plot indicate that ΔH is essentially constant across the temperature range tested. Numerous examples in the literature are well described by this model, including the binding of translation initiation factor complexes to a pseudoknot structure from an internal ribosome entry site within the tobacco etch virus mRNA (Yumak et al. 2010) and the association of the *trp* RNA-binding attenuation protein from *Bacillus subtilis* to its cognate binding site (Baumann et al. 1996).

In other cases, however, the relationship between $\ln K$ and $1/T$ is clearly nonlinear. Figure 9.8 shows the van't Hoff plot describing Hsp70 binding to the TNFα

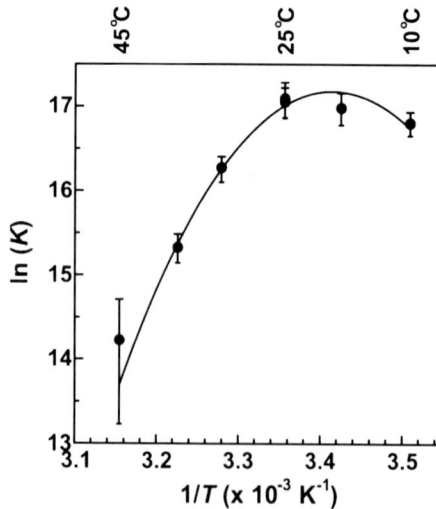

Fig. 9.8 A van't Hoff plot for Hsp70 binding to the fluorescein-conjugated TNFα ARE substrate. The nonlinear relationship indicates that the enthalpic contributions to binding energy are not constant over this temperature range. The fit permits calculation of the change in the specific heat capacity associated with Hsp70:ARE ribonucleoprotein formation and separate temperatures at which ΔH and ΔS are zero, from which the enthalpic and entropic contributions to binding can be calculated for any given temperature

ARE substrate. The substantial downward curvature of the plot indicates that ΔH is not constant across the range of temperatures tested. However, these data may be resolved by considering a change in the specific heat capacity of this system associated with RNA–protein binding using

$$\ln K = (\Delta C_{\mathrm{P,obs}} / R)[(T_{\mathrm{H}} / T) - \ln(T_{\mathrm{S}} / T) - 1] \qquad (9.22)$$

where T_{H} and T_{S} are the temperatures at which enthalpy and entropy contribute no energy to the system, respectively, and $\Delta C_{\mathrm{P,obs}}$ is the change in molar heat capacity. In this example, $\Delta C_{\mathrm{P,obs}}$ resolves to -2.3 ± 0.9 kcal mol^{-1} K^{-1}, while $T_{\mathrm{H}} = 293 \pm 2$ K and $T_{\mathrm{S}} = 297 \pm 2$ K (all values are quoted ± 95 % confidence interval) (Wilson et al. 2001). The large negative change in heat capacity is not uncommon with protein–nucleic acid interactions and is thought to indicate conformational transitions associated with induced-fit binding mechanisms (Ha et al. 1989). With these values in hand, the contributions of enthalpy and entropy to RNA–protein binding at any given temperature can be calculated using

$$\Delta H = \Delta C_{\mathrm{P,obs}}(T - T_{\mathrm{H}}) \qquad (9.23)$$

and

$$T\Delta S = T \cdot \Delta C_{\mathrm{P,obs}} \cdot \ln(T / T_{\mathrm{S}}) \qquad (9.24)$$

9.5.3 High-Throughput Competition Screens

Competition screens are a valuable tool for identifying novel compounds that disrupt macromolecular interactions and for measuring the effectiveness of these agents. In general, concentrations of a fluorescent ligand (e.g., fluorophore-conjugated RNA) and binding protein are selected for which the bound substrate fraction is between 0.5 and 0.8 and ligand depletion effects are avoided. In order for the screen to be maximally sensitive to a broad range of potential competitors, theoretical models recommend selecting a fluorescent ligand that binds the target protein very tightly (Huang 2003). Under these conditions, a compound that binds with lower affinity than the fluorescent ligand will display an IC$_{50}$ value that is independent of the ligand binding parameters and linearly proportional to the K_{d} value of the competitor ($= K_{\mathrm{i}}$). Conversely, inhibitory molecules that bind target proteins much better than the fluorescent ligand will exhibit IC$_{50}$ values that are independent of K_{i}, but rather will be proportional to the affinity of the fluorescent ligand. As such, the dynamic range of complex inhibition is greater when using fluorescent ligands with low K_{d} values. This principle holds for competition assays using standard format spectrofluorometers in addition to high density plate readers.

Recent studies using HTS approaches have provided several additional insights into the design of fluorescence anisotropy-based binding assays for plate readers

(Mao et al. 2006). First, binding of long fluorescent RNA substrates (95 nt or more, MW > 30 kDa) can often be well resolved for even relatively small proteins (70 kDa), although larger differences between protein and ligand size are still preferred, since the local segmental motions within a long RNA strand will be better restricted by a more massive binding partner. Second, RNA–protein binding reactions can often be designed such that they are minimally impacted by modest concentrations of solvents commonly used for drug libraries (e.g., DMSO, methanol). Finally, improvements in the sensitivity and resolution of microplate readers now permit reproducible fluorescence anisotropy readings from samples as small as 5 μL containing as little as 1 nM of a fluorescent ligand. In fact, smaller sample volumes can improve the precision of anisotropy measurements because they minimize the optical path through the sample, which reduces the contributions of library compounds to optical density (Owicki 2000). The constantly evolving nature of microplate reader technology, including the development of more sophisticated optics (Owicki 2000), is continually improving their utility.

Often, when screening a large number of compounds for inhibition of protein–RNA interactions, the mechanism by which the compound disrupts the complex is initially less important to the investigator than the knowledge that it has an effect at all. However, downstream validation and drug characterization studies can reveal novel and unexpected mechanisms for inhibiting RNA–protein association. For example, a fluorescence anisotropy-based HTS survey for drugs that could inhibit ARE binding by the mRNA-stabilizing factor HuR identified several potent lead compounds. Subsequent mechanistic studies applying a wide range of potential binding models revealed that some compounds blocked HuR binding by preventing protein dimerization that may be required for RNA binding, while another bound HuR directly (Meisner et al. 2007). It remains to be seen whether these compounds or their derivatives will ultimately be useful for suppressing expression of ARE-containing mRNAs in vivo or have therapeutic utility in patients.

9.6   Conclusions

The quantitative characterization of protein–RNA interactions in vitro can provide critical validation for the functional significance of putative RNA-binding *trans*-regulatory proteins identified using cellular or genetic assays. Reciprocally, resolving binding mechanisms and RNA substrate requirements using in vitro approaches can help direct downstream cellular studies, including screens for novel RNA targets, drug design, and the interface between RNA recognition mechanisms and cellular signaling pathways. Fluorescence anisotropy-based binding assays have proven to be a sensitive and flexible platform for rigorous quantitative biochemical analyses of ribonucleoprotein complex formation and are revealing features of RNA–protein interactions that are not resolvable by conventional biochemical approaches.

Acknowledgements This work was supported by NIH grants CA052443 (to G.B.) and CA102428 (to G.M.W.).

References

Albinsson B, Ericksson S, Lyng R, Kubista M (1991) The electronically excited states of 2-phenylindole. Chem Phys 151:157

Anderson BJ, Larkin C, Guja K, Schildbach JF (2008) Using fluorophore-labeled oligonucleotides to measure affinities of protein-DNA interactions. Methods Enzymol 450:253–272

Baumann C, Otridge J, Gollnick P (1996) Kinetic and thermodynamic analysis of the interaction between TRAP (*trp* RNA-binding attenuation protein) of *Bacillus subtilis* and *trp* leader RNA. J Biol Chem 271:12269–12274

Beaudette NV, Langerman N (1980) The thermodynamics of nucleotide binding to proteins. CRC Crit Rev Biochem 9:145–169

Beechem JM, Brand L (1985) Time-resolved fluorescence of proteins. Annu Rev Biochem 54:43–71

Bernstein J, Patterson DN, Wilson GM, Toth EA (2008) Characterization of the essential activities of *Saccharomyces cerevisiae* Mtr4p, a 3′ to 5′ helicase partner of the nuclear exosome. J Biol Chem 283:4930–4942

Bujalowski W, Jezewska MJ (2000) Quantitative determination of equilibrium binding isotherms for multiple ligand-macromolecule interactions using spectroscopic methods. In: Gore MG (ed) Spectrophotometry and spectrofluorimetry. Oxford University Press, Oxford, UK, pp 141–165

Draper DE (1995) Protein-RNA recognition. Annu Rev Biochem 64:593–620

Eversole A, Maizels N (2000) *In vitro* properties of the conserved mammalian protein hnRNPD suggest a role in telomere maintenance. Mol Cell Biol 20:5425–5432

Fernandes PB (1998) Technological advances in high-throughput screening. Curr Opin Chem Biol 2:597–603

Fialcowitz-White EJ, Brewer BY, Ballin JD, Willis CD, Toth EA, Wilson GM (2007) Specific protein domains mediate cooperative assembly of HuR oligomers on AU-rich mRNA-destabilizing sequences. J Biol Chem 282:20948–20959

Gradinaru CC, Marushchak DO, Samim M, Krull UJ (2010) Fluorescence anisotropy: from single molecules to live cells. Analyst 135:452–459

Gupta YK, Lee TH, Edwards TA, Escalante CR, Kadyrova LY, Wharton RP, Aggarwal AK (2009) Co-occupancy of two Pumilio molecules on a single hunchback NRE. RNA 15:1029–1035

Ha JH, Spolar RS, Record MT Jr (1989) Role of the hydrophobic effect in stability of site-specific protein-DNA complexes. J Mol Biol 209:801–816

Huang X (2003) Fluorescence polarization competition assay: the range of resolvable inhibitor potency is limited by the affinity of the fluorescent ligand. J Biomol Screen 8:34–38

Jameson DM, Ross JA (2010) Fluorescence polarization/anisotropy in diagnostics and imaging. Chem Rev 110:2685–2708

Jameson DM, Sawyer WH (1995) Fluorescence anisotropy applied to biomolecular interactions. Methods Enzymol 246:283–300

Lakowicz JR (1999) Fluorescence anisotropy. In: Principles of fluorescence spectroscopy. Kluwer, New York, pp 291–319

Liao B, Hu Y, Brewer G (2007) Competitive binding of AUF1 and TIAR to MYC mRNA controls its translation. Nat Struct Mol Biol 14:511–518

Lohman TM, Bujalowski W (1991) Thermodynamic methods for model-independent determination of equilibrium binding isotherms for protein-DNA interactions: spectroscopic approaches to monitor binding. Methods Enzymol 208:258–290

Lu JY, Sadri N, Schneider RJ (2006) Endotoxic shock in AUF1 knockout mice mediated by failure to degrade proinflammatory cytokine mRNAs. Genes Dev 20:3174–3184

Maag D, Lorsch JR (2003) Communication between eukaryotic translation initiation factors 1 and 1A on the yeast small ribosomal subunit. J Mol Biol 330:917–924

Mao C, Flavin KG, Wang S, Dodson R, Ross J, Shapiro DJ (2006) Analysis of RNA-protein interactions by a microplate-based fluorescence anisotropy assay. Anal Biochem 350:222–232

Marcotrigiano J, Gingras AC, Sonenberg N, Burley SK (1997) Cocrystal structure of the messenger RNA 5′ cap-binding protein (eIF4E) bound to 7-methyl-GDP. Cell 89:951–961

Meisner NC, Hintersteiner M, Mueller K, Bauer R, Seifert JM, Naegeli HU, Ottl J, Oberer L, Guenat C, Moss S, Harrer N, Woisetschlaeger M, Buehler C, Uhl V, Auer M (2007) Identification and mechanistic characterization of low-molecular-weight inhibitors for HuR. Nat Chem Biol 3:508–515

Mitchell SF, Walker SE, Algire MA, Park EH, Hinnebusch AG, Lorsch JR (2010) The 5′-7-methylguanosine cap on eukaryotic mRNAs serves both to stimulate canonical translation initiation and to block an alternative pathway. Mol Cell 39:950–962

Owicki JC (2000) Fluorescence polarization and anisotropy in high throughput screening: perspectives and primer. J Biomol Screen 5:297–306

Ranji A, Shkriabai N, Kvaratskhelia M, Musier-Forsyth K, Boris-Lawrie K (2011) Features of double-stranded RNA-binding domains of RNA helicase A are necessary for selective recognitions and translation of complex mRNAs. J Biol Chem 286:5328–5337

Sadana A (2001) A kinetic study of analyte-receptor binding and dissociation, and dissociation alone, for biosensor applications: a fractal analysis. Anal Biochem 291:34–47

Sarkar S, Han J, Sinsimer KS, Liao B, Foster RL, Brewer G, Pestka S (2011) RNA-binding protein AUF1 regulates lipopolysaccharide-induced IL-10 expression by activating IκB kinase complex in monocytes. Mol Cell Biol 31:602–615

Wagner BJ, DeMaria CT, Sun Y, Wilson GM, Brewer G (1998) Structure and genomic organization of the human AUF1 gene: alternative pre-mRNA splicing generates four protein isoforms. Genomics 48:195–202

Wilson GM (2005) RNA folding and RNA-protein binding analyzed by fluorescence anisotropy and resonance energy transfer. In: Geddes CD, Lakowicz JR (eds) Reviews in fluorescence, vol 2. Springer, New York, pp 223–243

Wilson GM, Sutphen K, Bolikal S, Chuang K, Brewer G (2001) Thermodynamics and kinetics of Hsp70 association with A+U-rich mRNA-destabilizing sequences. J Biol Chem 276: 44450–44456

Wilson GM, Lu J, Sutphen K, Suarez Y, Sinha S, Brewer B, Villanueva-Feliciano EC, Ylsa RM, Charles S, Brewer G (2003) Phosphorylation of p40[AUF1] regulates binding to A+U-rich mRNA-destabilizing elements and protein-induced changes in ribonucleoprotein structure. J Biol Chem 278:33039–33048

Yan Y, Marriott G (2003) Analysis of protein interactions using fluorescence technologies. Curr Opin Chem Biol 7:635–640

Yumak H, Khan MA, Goss DJ (2010) Poly(A) tail affects equilibrium and thermodynamic behavior of tobacco etch virus mRNA with translation initiation factors eIF4F, eIF4B and PABP. Biochim Biophys Acta 1799:653–658

Zhang JH, Chung T, Oldenburg K (1999) A simple statistical parameter for use in evaluation and validation of high throughput screening assays. J Biomol Screen 4:67–73

Zucconi BE, Ballin JD, Brewer BY, Ross CR, Huang J, Toth EA, Wilson GM (2010) Alternatively expressed domains of AU-rich element RNA-binding protein 1 (AUF1) regulate RNA-binding affinity, RNA-induced protein oligomerization, and the local conformation of bound RNA ligands. J Biol Chem 285:39127–39139

Chapter 10
Approaches for the Identification and Characterization of RNA–Protein Interactions

Saiprasad Palusa and Jeffrey Wilusz

10.1 Stabilizing Cellular RNA–Protein Interactions for Improved Analysis

10.1.1 Introduction

While some RNA–protein interactions can be analyzed directly, it is advisable (particularly at the discovery phase) to increase the stability of these interaction to permit their ready identification and characterization. This is generally approached by using one of two methods: Nonreversible UV cross-linking (with or without the inclusion of a photactivatable modified nucleoside) or reversible cross-linking using formaldehyde. While both methods are presented below, please note that they are independent approaches to the same goal of stabilizing RNA–protein interactions to allow for more effective downstream characterization.

10.1.2 Specialized Materials

1. Cultured cells of interest.
2. UV light source (e.g., Stratalinker 2400; Stratagene).
3. 4-Thio uridine (Sigma-Aldrich).
4. 37% Formaldehyde (Sigma-Aldrich).
5. Phosphate Buffered Saline (PBS) (Cellgro, Mediatech, Inc.).
6. Glycine (Sigma-Aldrich).

S. Palusa • J. Wilusz (✉)
Department of Microbiology, Immunology and Pathology, Colorado State University,
Fort Collins, CO 80523, USA
e-mail: jeffrey.wilusz@colostate.edu

J.D. Dinman (ed.), *Biophysical Approaches to Translational Control of Gene Expression*, 199
Biophysics for the Life Sciences 1, DOI 10.1007/978-1-4614-3991-2_10,
© Springer Science+Business Media New York 2013

10.1.3   Methods

10.1.3.1   Stabilization of RNA–Protein Complexes Using UV Cross-Linking

1. Grow the cells in the appropriate culture medium on a 150 mm plate to approximately 80% confluence. If desired (e.g., see Note 1 of Sect. 10.1.4), add 100 μM 4-thio uridine to the media and let the cells grow overnight.
2. Wash the cells with cold PBS.
3. Remove the PBS, leaving the cells attached to the plate.
4. Irradiate the cells with 0.15 J/cm² of 254 nm UV light using a Stratalinker 2400 system or equivalent. If the cells were grown in the presence of 4-thio uridine, use 0.15 J/cm² of a longer wavelength (365 nm) UV light.

10.1.3.2   Stabilization of RNA–Protein Complexes Using Formaldehyde

1. Remove cells from the culture plate, wash with PBS, and resuspend in 10 mL of cold PBS. Add formaldehyde to a final concentration of 1%.
2. Incubate at room temperature with slow mixing for 10 min.
3. To quench the reaction, add glycine to a final concentration of 0.25 M and incubate at room temperature for 5 min.
4. Harvest the cells by centrifugation at $250 \times g$ for 4 min at 4 °C.
5. Wash cell pellet twice in ice-cold PBS.
6. Cell pellet can be used immediately or stored at −80 °C for future use.

10.1.4   Notes

1. UV cross-linking of unmodified RNA to proteins is highly specific but notoriously inefficient due to the short-lived nature of the free radical intermediate formed by irradiation with short wavelength UV light. The inclusion of 4-thio uridine can dramatically increase the efficiency of cross-linking (Hafner et al. 2010).
2. Formaldehyde concentration and incubation time can be adjusted to achieve maximum cross-linking while minimizing noise due to over-cross-linking. In our experience, formaldehyde concentrations and the time of incubation are the two variables that should be optimized empirically. Formaldehyde concentrations are generally varied from 0.3 to 1.0%. The range of fixation times is usually between 5 min and 1 h.
3. Additional approaches using bifunctional cross-linking reagents and chemical inducers of dimerization (CIDs) (Harvey et al. 2002) exist to stabilize RNA–protein interactions but are not commonly used. Many bifunctional cross-linking reagents available favor protein–protein cross-linking and thus are not suitable for

RNA–protein applications. EDC (1-ethyl-3(3-dimethylaminopropyl) carbodiimide) has been used successfully to stabilize ribosomal protein–RNA interactions (Chiaruttini et al. 1989).

4. Analysis of the material generated by these approaches can be found in the methods presented below.

10.2 Immunoprecipitation of RNA–Protein Complexes and Analysis of Interactions with Select Targets

10.2.1 Introduction

A common first step in assessing whether a specific RNA binding protein is involved in the regulation of an mRNA of interest is to determine whether or not the protein interacts with the transcript in living cells. Assuming that antibody reagents are available, this question is effectively approached by determining whether the alleged mRNA target can be co-immunoprecipitated with the protein. An example of the successful use of the technique can be found in Sokoloski et al. (2010) and Niranjanakumari et al. (2002).

10.2.2 Specialized Materials

1. Cultured cells of interest.
2. Antibody binding matrix (Protein A or Protein G Sepharose beads).
3. Antibody specific for protein of interest and a nonspecific control antibody.
4. Sonicator (Fisher Scientific, Model 100).
5. RNase OUT (Fermentas).
6. DNase 1—RNase-free (New England Biolabs).
7. Glycogen (20 mg/mL) (Fermentas).
8. Random hexamers (IDT).
9. Reverse transcriptase (Im Prom-II™).
10. Protease inhibitor cocktail (Roche).

Buffers:

1. *RIPA buffer*: 50 mM Tris-Cl pH 7.5, 1% Nonidet P-40 (NP-40), 0.5% sodium deoxycholate, 0.05% sodium dodecyl sulfate (SDS), 1 mM ethylenediaminetetraacetic acid (EDTA), 150 mM NaCl. Protease inhibitors (e.g., 0.2 mM phenylmethylsulfonyl fluoride [PMSF]) can also be added to this solution if desired.
2. *High stringency RIPA buffer*: 50 mM Tris-Cl pH 7.5, 1% NP-40, 1% sodium deoxycholate, 0.1% SDS, 1 mM EDTA, 1 M NaCl, 1 M urea, and 0.2 mM PMSF.

3. *TEDS buffer:* 50 mM Tris-Cl pH 7.0, 5 mM EDTA, 10 mM dithiothreitol (DTT), and 1% SDS.
4. *NT2 buffer:* 50 mM Tris-Cl, pH 7.4, 150 mM NaCl, 1 mM $MgCl_2$, 0.05% NP-40.

10.2.3   Methods

10.2.3.1   Collecting Cells

1. Culture cells on 150 mm plates to more than 80% confluence; wash with cold PBS, then harvest cells by scraping or using trypsin.
2. Centrifuge the cell suspension at $250 \times g$ for 5 min to pellet the cells.
3. Wash the cell pellet twice with PBS.
4. Resuspend the pellet in 10 mL PBS.
5. Check cell viability using a dye exclusion assay (e.g., trypan blue staining) and count the cells using a hemacytometer (or equivalent).
6. Stabilize RNA–protein complexes using formaldehyde as described above in Sect. 10.1.3.2.

10.2.3.2   Cell Lysis

1. Resuspend the fixed cell pellet in 1 mL RIPA buffer.
2. Divide the suspension into two tubes (500 μL each).
3. Lyse the cells by five rounds of sonication using a probe sonicator (~3 s each at an amplitude of 8–9 W). Tubes should be placed on ice for 1 min between each round of sonication.
4. Check for cell lysis under microscope to assess the efficiency of step 3.
5. Centrifuge at $16,000 \times g$ for 10 min at 4 °C to remove insoluble material. Transfer supernatant to a new tube and discard pellet. Use 500 μL of supernatant (lysate) and store the rest at −80 °C.

10.2.3.3   Preparation of Sepharose Beads

1. Swell protein A or protein G Sepharose beads in RIPA buffer for at least 30 min.
2. Wash five times with 500 μL RIPA buffer. Add RIPA buffer to create a 50% (W/V) slurry. Note that the slurry can be stored overnight at 4 °C.
3. Bind specific and control antibodies to individual aliquots of the swollen beads. This is done by incubation of the antibodies and beads in RIPA buffer at 4 °C for 2 h. Wash the antibody-bound beads twice with RIPA buffer.

10.2.3.4   Preclearing of the Cell Lysate

1. Mix 500 μL of lysate with ~50 μL of packed protein A or protein G Sepharose bead slurry (either without pre-bound antibodies or bound with a control IgG) and tRNA (100 μg/mL).
2. Rotate the mixture for 1 h at 4 °C.
3. Centrifuge at $1,300 \times g$ for 5 min at 4 °C.
4. Transfer the now precleared supernatant to a clean tube. Transfer 25 μL of the precleared supernatant into another tube for RNA extraction (this 5% aliquot represents the input sample for quantification). Keep this sample on ice until it is processed.

10.2.3.5   Immunoprecipitation of Cross-Linked mRNP Complexes

1. Evenly divide the lysate by mixing 225 μL of precleared cell lysate with 225 μL RIPA buffer in two tubes.
2. Add the Sepharose beads containing either specific or nonspecific antibody to the diluted precleared cell lysates.
3. Rotate at 4 °C for 90 min.
4. Centrifuge at $16,000 \times g$ for 1 min at 4 °C (if analysis of the "unbound" fraction is desired, save the ~450 μL of supernatant for RNA extraction).
5. Wash the beads five times with 1 mL of high stringency RIPA buffer.
6. Centrifuge at $16,000 \times g$ for 1 min. Discard supernatant and collect the beads in 100 μL of TEDS buffer. Incubate at 70 °C for 45 min. Extract RNA from each immunoprecipitated sample as well as the precleared input sample (25 μL) as outlined below.

10.2.3.6   RNA Isolation

1. Add 125 μL NT2 buffer to the input sample. Adjust all samples to 0.1% SDS. Add 20 μg of proteinase K. Incubate for 15 min at 37 °C.
2. Add 150 μL TRIzol to samples, at this point you can freeze samples at −80 °C or directly proceed to next step.
3. Add 80 μL chloroform to the samples and mix thoroughly by vortexing.
4. Centrifuge at $13,000 \times g$ for 10 min at 4 °C.
5. Transfer the upper aqueous phase to a new tube.
6. Add 1 μL glycogen and 150 μL isopropanol.
7. Centrifuge at $13,400 \times g$ for 20 min at 4 °C.
8. Discard supernatant carefully, wash the pellet with 500 μL 80% ethanol, centrifuge again at full speed for 1 min in a microcentrifuge. Discard the supernatant and dry the pellet.
9. Resuspend the pellet in 89 μL RNase-free water. Add 10 μL of 10× DNAse I buffer and 1 μL of DNase 1 (RNase-free); incubate at 37 °C for 30 min.

10. Add 100 μL phenol:chloroform:isoamyl alcohol (25:24:1).
11. Mix thoroughly by vortexing and centrifuge for 1 min in a microcentrifuge.
12. Transfer the top aqueous phase to a new tube.
13. Adjust samples to 2 N ammonium acetate and 250 μL 100% ethanol.
14. Place at −80 °C for 20 min or longer.
15. Centrifuge for 10 min at full speed in a microcentrifuge.
16. Discard supernatant carefully.
17. Wash the pellet with 100 μL of 80% ethanol.
18. Air dry the pellet and resuspend in 4 μL of RNase-free water.

10.2.3.7   Analysis of Immunoprecipitated RNA: cDNA Preparation and PCR

1. Use 1 μL of RNA (one-quarter of the sample) for reverse-transcription (RT) to make a cDNA.
2. Add 1 μL of either random hexamers (500 ng/μL) or oligo (dT) (500 ng/μL) or a gene-specific primer (10 μM) and 3 μL water in a 5 μL final reaction volume.
3. To anneal the primers, incubate at 70 °C for 5 min then keep on ice for another 5 min.
4. For the RT reaction, mix the 5 μL annealing reaction with 4 μL of 5× (RT) buffer, 1 μL dNTPs (10 mM stock), 2.4 μL $MgCl_2$ (25 mM stock), 1 μL RNase inhibitor and 1 μL reverse transcriptase, and 5.6 μL water for a final reaction volume of 20 μL.
5. Incubate at 25 °C for 5 min, 42 °C for 1 h, and finally 70 °C for 10 min.
6. The cDNA is now ready for use in PCR. Specific PCR conditions will be determined by the nature of the primers for your mRNA of interest. Analyze PCR reaction products by agarose gel electrophoresis, visualizing the bands by staining with ethidium bromide or SYBR green. Alternatively, qPCR can be performed to provide more accurate quantification.

10.2.4   Notes

1. In our experience, the use of bath sonicators results in inefficient and inconsistent cell lysis and thus is discouraged.
2. Keeping the cells on ice during each sonication round is important since heat can reverse RNA–protein cross-links.
3. If necessary, antibodies (particularly if they are not purified) can be treated with RNAse inhibitor (e.g., RNasin) prior to loading onto Sepharose beads. A 10 min incubation will significantly reduce background RNA degradation due to common RNase contaminants.
4. In a useful extension of this technique, note that UV cross-linking can also be applied to determine the RNA binding site on a protein of interest using mass spectrometry (Urlaub et al. 2008).

10.3 Immunoprecipitation of RNA–Protein Complexes and Global Analyses of Target Interactions (e.g., RIP-CHIP)

10.3.1 Introduction

Oftentimes the need arises to take a more global look at mRNAs that are associated with a particular protein factor. This can be approached by a variety of techniques including RIP-CHIP and CLIP analyses (e.g., HITS-CLIP, PAR-CLIP, etc.). All of these approaches rely upon the specific immunoprecipitation of RNA–protein complexes. RIP-CHIP involves the precipitation of intact mRNAs bound to proteins followed by analysis of bound RNAs by either microarray or deep sequencing (Tenenbaum et al. 2002; Jain et al. 2011; Galgano and Gerber 2011; Lee et al. 2010). HITS-CLIP (Darnell 2010; Jensen and Darnell 2008) and PAR-CLIP (Hafner et al. 2010) rely on the immunoprecipitation of RNAse-treated UV cross-linked RNA–protein complexes and analysis of the small RNA fragments by deep sequencing. An attractive feature of PAR-CLIP is that RNAs are modified using 4-thio uridine in cells prior to UV cross-linking. In addition to increasing the efficiency of cross-linking, cross-linked U residues get converted to C residues during the RT step, allowing one to pinpoint the exact site of RNA–protein interaction. A universally applicable RIP-CHIP protocol is presented below. Please note that a number of recent excellent reviews have focused on the "CLIP" strategies (e.g., Kishore et al. 2011).

10.3.2 Specialized Materials

1. Tissues/cells of interest.
2. Antibody specific for protein of interest and a nonspecific antibody.
3. Antibody binding matrix (Protein G or protein A Sepharose beads)
4. Glycogen (20 mg/mL) (Fermentas).
5. RNAse OUT (Fermantas).
6. Dnase 1—RNase-free (Fermantas).
7. Micro-spin columns (Pierce).
8. Random hexamers (IDT).
9. Reverse transcriptase (Im Prom-II™).
10. Protease inhibitor cocktail (Roche).

Buffers:

1. *RNP lysis buffer (RLB)*: 100 mM KCl, 5 mM $MgCl_2$, 10 mM HEPES, pH 7.0, 0.5% NP-40; *To be added at the time of use*; 1 mM DTT, 100 units/mL RNase OUT, 0.2% vanadyl ribonucleoside complexes, 0.2 mM PMSF.
2. *NT2 buffer*: 50 mM Tris-Cl, pH 7.4, 150 mM NaCl, 1 mM $MgCl_2$, 0.05% Nonidet P-40.

10.3.3 Methods

10.3.3.1 Collecting Cells and Preparation of Cell Lysate

1. Culture cells on 150 mm plates. When cells are more than 80% confluent, collect cells as described in Sect. 10.2.3.1.
2. Resuspend cell pellet in RLB in a volume equal to cell pellet volume.
3. Snap freeze. The cell lysate is ready for the immunoprecipitation step or may be stored at −80 °C until ready for use.

10.3.3.2 Preparation of Protein A or Protein G Sepharose Beads

1. Swell and wash protein A or protein G Sepharose beads as described in Sect. 10.2.3.3 using NT2 buffer instead of RIPA buffer.

10.3.3.3 Immunoprecipitation Reaction

1. If necessary, thaw cell lysates on ice.
2. Clear the lysate by centrifuging at full speed in a bench top microcentrifuge for 10 min at 4 °C.
3. Transfer supernatant to new tube and repeat step 2 to ensure that all insoluble material has been removed.
4. Remove 10 μL of the lysate and place at 4 °C until processed. This is your "Input" sample that will be used as a standard to compare to immunoprecipitated samples.
5. Add 100 μL cleared lysate to either the protein-specific antibody of interest or 5–7 μg of normal IgG/pre-bleed serum (as negative control). Incubate on ice for 1 h.
6. Spin at full speed in a microcentrifuge for 10 min at 4 °C to remove any material that has become insoluble during the incubation/handling.
7. Add 20 μL of washed protein A or protein G Sepharose bead slurry and incubate for 1 h at 4 °C.
8. Pellet antibody–antigen complexes on beads using a 30 s spin in a microcentrifuge; remove the supernatant (save for later analysis as the "unbound fraction").
9. Wash the pellet three times with 250 μL of ice-cold NT2 buffer. Transfer beads to a micro-spin column and wash four more times with 200 μL of ice-cold NT2 buffer.
10. Isolate RNA from the immunoprecipitated samples as described in Sect. 10.2.3.6.

10.3.3.4   Preparation of RNA Samples and Microarray Hybridization/ Sequencing

1. Use 50–100 ng RNA to generate labeled cDNA fragments for hybridization to Affimetrix gene Chip arrays (e.g., Human Tiling 1.0R Array Set (P/N 900774) or GeneChip Mouse Tiling 1.1R Array Set (P/N 900853)) following the manufacturer's protocol. Alternatively, samples may be analyzed by next-generation deep sequencing.

10.3.4 Notes

1. Harvesting cells when they are $\geq 80\%$ confluency ($1-3 \times 10^6$ cells) is important to get sufficiently concentrated total protein (20–50 mg) in the cell lysate.
2. Do not overdry the RNA as it may be difficult to get the nucleic acid back into solution. If you have trouble resuspending the RNA pellet, incubate at 65 °C for 5 min.
3. We prefer to use random hexamers to prepare cDNA rather than oligo(dT) priming for several reasons. One key reason is that oligo (dT) will only prime mRNAs, so rRNA cannot be used as a reference gene. In addition to this, oligo (dT) priming introduces a 3′ end bias to samples which could influence the readout.
4. Deep sequencing can be used as an alternative to microarrays for RNA analysis and is likely to become the method of choice in the near future.

10.4   In Vitro Analyses of RNA–Protein Interactions

In this section we describe three approaches to complement the immunoprecipitation-based approaches outlined above in the analysis of RNA–protein interactions. These approaches are often useful in validating and extending results obtained from RNA–protein analyses in tissue culture cells. Importantly, they can also be used as a discovery tool to identify novel, functionally important RNA–protein interactions.

10.4.1 Electrophoretic Mobility Shift Assays

When purified or recombinant RNA binding proteins are available, RNA Electrophoretic Mobility Shift Assays (EMSA) or "gel shift" assays are particularly useful to both confirm the results of co-immunoprecipitation assays as well as to study biochemical aspects of the RNA–protein interaction — including measuring binding affinity (by calculating the dissociation constant K_d) and mapping binding sites (by surveying a series of variant RNA substrates containing deletions/ insertions).

10.4.1.1 Specialized Materials

1. Labeled (radioactive or fluorescent) in vitro transcribed or chemical synthesized RNA.
2. Purified protein of interest.
3. Spermidine (100 mM) (Sigma-Aldrich).
4. RNAse OUT (Fermentas).
5. Heparin Sulfate (40 mg/mL) (Sigma-Aldrich).

Buffers:

1. 5× *Gel shift buffer*: 70 mM HEPES pH 7.9, 450 mM KCl, 11 mM $MgCl_2$, 28% glycerol.
2. *Lysis buffer*: 50 mM HEPES pH 7.9, 150 mM KCl, 1 mM $MgCl_2$, 1% Triton X − 100, 10% glycerol.
3. 6× *Loading buffer*: 30% glycerol, 0.3% bromophenol blue, 0.3% xylene cyanol.
4. 10× *TBE (1 L)*: 108 g tris base, 55 g boric acid, 9.3 g EDTA.

10.4.1.2 Methods

1. Prepare the following reaction in a microcentrifuge tube (final reaction volume is 14 μL): 1.5 μL of 1 mM spermidine (0.1 mM final concentration), 3 μL of 5× gel shift buffer, 0.5 μL of RNase Inhibitor, X μL of Lysis buffer (as determined by protein volume), 1 μL of radiolabeled RNA (we generally use ~5 fmoles; 50–100 K cpm/μL), and recombinant protein (100 nM).
2. Incubate at 30 °C for 5 min.
3. Add 1 μL of Heparin Sulfate (40 mg/mL) and transfer the reactions to ice for 5 min. This will reduce nonspecific interactions.
4. Add 3 μL 6× loading buffer. Load the reaction products on pre-run 5% non-denaturing gel (native gel) at 200 V for 2–3 h to achieve good separation between the input RNA and protein–RNA complexes.
5. Transfer the gel to Whatman filter paper, cover with plastic wrap, and dry with a gel-drying apparatus.
6. The dried gel should be exposed to the phosphor screen and analyzed by phosphorimaging.

10.4.1.3 Notes

1. The concentration of proteins and RNA must be known to calculate binding constants. For a detailed description, see Ryder et al. (2008).
2. RNA structure can strongly influence the migration of the transcript in native gels. The addition of spermidine can help minimize effects of RNA structural isoforms on migration of the transcript in the gel.

3. If the RNA–protein complex is sticking in the wells, the addition of bovine serum albumin may help to minimize this effect.

10.4.2 UV Cross-Linking of RNA to Proteins from Cellular Extracts or Recombinant Sources

In addition to its use outlined above in stabilizing RNA–protein complexes, UV cross-linking can also be used to identify RNA–protein interactions. Complexes are allowed to form between proteins and radiolabeled RNAs. Irradiation with short wavelength UV light forms covalent bonds between proteins and closely associated RNA molecules. Reactions are treated with RNAse and proteins with covalently associated short radioactive RNA tags are analyzed by conventional SDS-PAGE and visualized by phosphorimaging. In this fashion, the approximate molecular weight of candidate RNA binding proteins can be quickly determined and used to predict candidate proteins from available databases. When coupled with extract-based functional assays (e.g., splicing, polyadenylation, or RNA decay) and competition analysis, one can rapidly associate the functional relevance of RNA–protein interactions and prioritize candidates for further study.

10.4.2.1 Specialized Materials

1. Cellular extracts (prepared as described, e.g., Sokoloski et al. 2008)/recombinant protein.
2. Stratalinker 2400 (Stratagene).
3. RNase OUT.
4. RNase ONE (Promega).

10.4.2.2 Method

1. Produce radiolabeled RNA through in vitro transcription using bacteriophage polymerases (e.g., T7, SP6) and $^{32}P-UTP$.
2. Incubate radiolabeled RNAs (~5 fmoles; 100–200 K cpm) with the extract or recombinant protein for 5–10 min at 30 °C.
3. Transfer the sample to a well of a microtiter dish on ice.
4. Cross-link using 254 nm UV light for 1–5 min or 1,800 μJ in a Stratalinker 2400. The light should be 1–2 cm from the dish.
5. Pipette irradiated samples back to 1.5 mL tubes containing 0.5 μL RNase ONE (a broad spectrum ribonuclease). Incubate at 37 °C for 15 min to degrade the input RNA.

6. Add an equal volume of 2× SDS loading buffer, mix the samples by vortexing, heat to 100 °C for 2 min, and after brief centrifugation (to remove any precipitated proteins) electrophorese on a 10% SDS-polyacrylamide gel.
7. Run the gel until sufficient resolution has been obtained.
8. Transfer the gel to Whatman filter paper, cover with plastic wrap, and dry with a gel-drying apparatus.
9. The dried gel should be exposed to the phosphor screen and analyzed by phosphorimaging.

10.4.2.3   Notes

1. We typically use 10–15 μL total reaction volumes in the protocol. While amounts of individual components need to be determined empirically, 50–100 μg of extract (or 100 ng to 1 μg of recombinant protein) and 50–100 K cpm of ^{32}P-labeled RNA are good starting points.
2. Reaction mixtures can also be immunoprecipitated following irradiation and RNAse treatment to confirm the identity of the cross-linked protein. The protocol used for this is similar to that found in Sect. 10.2.3.5, steps 1–4. Following washing of the beads, simply resuspend the sample in 2× SDS loading buffer and follow steps 6–9 above.

10.4.3   In Vitro Affinity Purification of RNA Binding Proteins

A common strategy to uncover the identity of an RNA binding protein is to purify it by taking advantage of its ability to specifically interact with an RNA substrate. Following purification, the protein is identified using mass spectrometry. The method described below focuses on a small scale purification using short pieces of RNA that will likely interact with a single or small number of proteins. It is important to appreciate that a similar approach can also be applied to larger RNAs and global proteomic analysis using Ribo-Trap technology (Beach and Keene 2010).

10.4.3.1   Specialized Materials

1. Biotinylated specific and nonspecific RNA target oligomers (IDT).
2. Cellular extracts (prepared as described for example in Sokoloski et al. 2008).
3. Streptavidin agarose resin (Thermo Scientific).

Buffers:

1. *Buffer D*: 20 mM HEPES-KOH, 20% glycerol, 100 mM KCl, 0.2 mM EDTA, 1 mM DTT.
2. *HSCB*: 400 mM NaCl, 25 mM Tris-Cl, pH 7.6, 0.1% SDS.

10.4.3.2   Methods

1. In separate tubes, add 5 μg of biotinylated specific or nonspecific RNA oligomers to ~3 mg of HeLa cytoplasmic extract (Sokoloski et al. 2008). Adjust reaction mixtures to 2.5 mM EDTA to help stabilize the RNA oligomers by blocking exonuclease activities commonly found in extracts.
2. Incubate for 1 h at 4 °C with rotation. Centrifuge at $16,000 \times g$ at 4 °C to remove proteins that have precipitated during handling and incubations. Transfer the supernatants to new tubes.
3. Add 20 μL streptavidin agarose beads (that have been equilibrated in buffer D) to each tube; incubate 15 min at 4 °C with rotation.
4. Centrifuge at $4,000 \times g$ for 5 min at 4 °C to pellet streptavidin beads bound with RNA oligomers. Remove supernatant and discard.
5. Wash beads five times with buffer D at 4 °C.
6. Add 300 μL of HSCB solution to the beads and incubate for 5 min at room temperature to release proteins from the bound RNA. If necessary, 1 mM DTT can be added and the mixture heated to 100 °C to release tightly associated proteins. Remove the supernatant to a fresh microcentrifuge tube.
7. Add 100 μL water, 400 μL of methanol, and 100 μL of chloroform; vortex.
8. Centrifuge at $16,000 \times g$ for 2 min.
9. Remove the upper aqueous layer and add another 400 μL of methanol.
10. Centrifuge at $16,000 \times g$ for 2 min.
11. Remove all the supernatant carefully, dry the pellet.
12. Add 2× protein loading dye to the pellet, heat the sample at 65 °C for 5 min, and spin at maximum speed in a microcentrifuge for 15 min.
13. Resolve proteins on a 10% SDS-PAGE gel and visualize bands by silver staining.
14. Excise protein bands of interest and analyze via tandem mass spectrometry following trypsin digestion.

10.4.3.3   Notes

1. For maximal efficiency, titrate the specific RNA oligo to identify the amount needed to provide a sufficient number of binding sites to saturate the available protein of interest. This could be done, for example, by competition analysis using a radiolabeled RNA substrate in a UV cross-linking assay as described in Sect. 10.4.2 above.
2. By altering salt conditions, etc., one may create selective conditions to maximize the amount of specific RNA–protein interactions while minimizing the amount of nonspecific interactions. The stability of the RNA–protein interaction of interest to a variety of monovalent and divalent cations, etc., can readily be assayed by UV cross-linking as described in Sect. 10.4.2 above.

Acknowledgements We wish to thank Ashley Neff, Stephanie Moon, Jerome Lee, John Anderson, and other members of the Wilusz laboratories for helpful comments and critical suggestions. Related RNA–protein work in the laboratory is supported by NIH grant GM072481 to J.W.

References

Beach DL, Keene JD (2010) Ribotrap:targeted purification of RNA-specific RNPs from cell lysates through immunoaffinity precipitation to identify regulatory proteins and RNAs. Methods Mol Biol 419:68–91

Chiaruttini C, Milet M, Hayes DH, Expert-Bezancon A (1989) Crosslinking of ribosomal proteins S4, S5, S7, S8, S11, S12 and S18 to domains 1 and 2 of 16S rRNA in the Escherichia coli 30S particle. Biochimie 71:839–852

Darnell RB (2010) HITS-CLIP: panoramic views of protein–RNA regulation in living cells. Wiley Interdiscip Rev RNA 1:266–286

Galgano A, Gerber AP (2011) RNA-binding protein immunopurification-microarray (RIP-chip) analysis to profile localized RNAs. Methods Mol Biol 714:369–385

Hafner M, Landthaler M, Burger L, Khorshid M, Hausser J, Berninger P, Rothballer A, Ascano M Jr, Jungkamp AC, Munschauer M, Ulrich A, Wardle GS, Dewell S, Zavolan M, Tuschl T (2010) Transcriptome-wide identification of RNA-binding protein and microRNA target sites by PAR-CLIP. Cell 141:129–411

Harvey I, Garneau P, Pelletier J (2002) Forced engagement of a RNA/protein complex by a chemical inducer of dimerization to modulate gene expression. Proc Natl Acad Sci U S A 99:1882–1887

Jain R, Devine T, George AD, Chittur SV, Baroni TE, Penalva LO, Tenenbaum SA (2011) RIP-chip analysis: RNA-binding protein immunoprecipitation-microarray (chip) profiling. Methods Mol Biol 703:247–263

Jensen KB, Darnell RB (2008) CLIP: crosslinking and immunoprecipitation of in vivo RNA targets of RNA-binding proteins. Methods Mol Biol 488:85–98

Kishore S, Jaskiewicz L, Burger L, Hausser J, Khorshid M, Zavolan M (2011) A quantitative analysis of CLIP methods for identifying binding sites of RNA-binding proteins. Nat Methods. doi:10.1038/nmeth.1608.

Lee JE, Lee JY, Wilusz J, Tian B, Wilusz CJ (2010) Systematic analysis of cis-elements in unstable mRNAs demonstrates that CUGBP1 is a key regulator of mRNA decay in muscle cells. PLoS One 5:e11201

Niranjanakumari S, Lasda E, Brazas R, Garcia-Blanco MA (2002) Reversible cross-linking combined with immunoprecipitation to study RNA-protein interactions in vivo. Methods 26:182–190

Ryder SP, Recht MI, Williamson JR (2008) Quantitative analysis of protein-RNA interactions by gel mobility shift. Methods Mol Biol 488:99–115

Sokoloski KJ, Wilusz J, Wilusz CJ (2008) The preparation and applications of cytoplasmic extracts from mammalian cells for studying aspects of mRNA decay. Methods Enzymol 448:139–163

Sokoloski KJ, Dickson AM, Chaskey EL, Garneau NL, Wilusz CJ, Wilusz J (2010) Sindbis virus usurps the cellular HuR protein to stabilize its transcripts and promote productive infections in mammalian and mosquito cells. Cell Host Microbe 8:196–207

Tenenbaum SA, Lager PJ, Carson CC, Keene JD (2002) Ribonomics: identifying mRNA subsets in mRNP complexes using antibodies to RNA-binding proteins and genomic arrays. Methods 26:191–198

Urlaub H, Kühn-Hölsken E, Lührmann R (2008) Analyzing RNA-protein crosslinking sites in unlabeled ribonucleoprotein complexes by mass spectrometry. Methods Mol Biol 488:221–245

Chapter 11
A Multidisciplinary Approach to RNA Localisation

Russell S. Hamilton, Graeme Ball, and Ilan Davis

11.1 Introduction

11.1.1 mRNA Localisation

mRNA localisation and localised translation work hand in hand as post-transcriptional modes of targeting protein production to its site of function. These mechanisms of regulation of gene expression were initially thought to only operate in large cells such as oocytes, but are now recognised as relevant to all model organisms including small cells such as yeast (Rosbash and Singer 1993) and bacteria (Nevo-Dinur et al. 2011). The field was initially founded as a consequence of discoveries in cell and developmental biology using genetics and histochemical staining methods in conjunction with in situ hybridisation and antibody staining (Levsky and Singer 2003). However, with time, the frontiers of the field of mRNA localisation and localised translation have moved to include interdisciplinary approaches, including advanced microscopy methods, and structural and computational biology. In recent years, these have become central in the field, especially for the characterisation of the molecular mechanisms of localisation and translational regulation with an emphasis on the *cis*-acting RNA sequences required and the *trans*-acting proteins involved in their interpretation. Some classical examples of RNA localisation targeting proteins to their site of function in a wide range of organisms are: *Ash1* mRNA localisation to the bud tip of the daughter cells of dividing *Saccharomyces cerevisiae* cells (Bobola et al. 1996) (Fig. 11.1a). *Vg1*, and *VegT*, localising to the vegetal poles of *Xenopus* oocytes to set up the body axis (Mowry and Melton 1992) (Fig. 11.1b). *β-actin* localising to the leading edge in motile chicken fibroblasts and

R.S. Hamilton (✉) • G. Ball • I. Davis
Department of Biochemistry, University of Oxford, South Parks Road, Oxford, OX1 3QU, UK
e-mail: Russell.Hamilton@bioch.ox.ac.uk; Ilan.Davis@bioch.ox.ac.uk

J.D. Dinman (ed.), *Biophysical Approaches to Translational Control of Gene Expression*, 213
Biophysics for the Life Sciences 1, DOI 10.1007/978-1-4614-3991-2_11,
© Springer Science+Business Media New York 2013

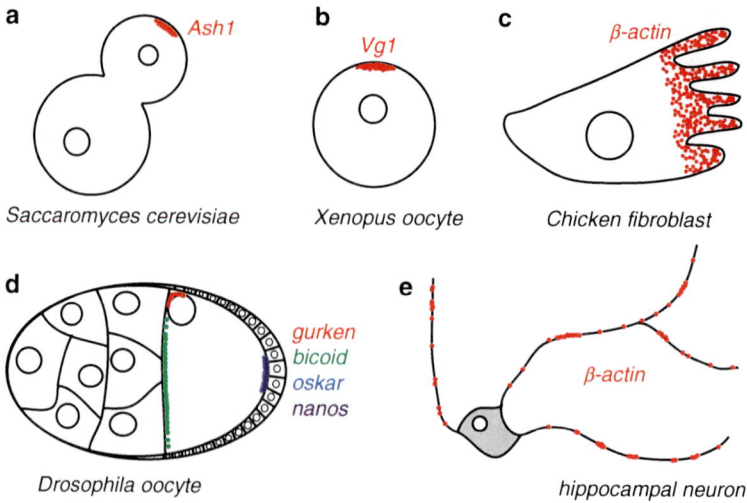

Fig. 11.1 Examples of RNA localisation. (**a**) *Ash1* mRNA localises in the budding tip of dividing *Saccharomyces cerevisiae*. (**b**) *Vg1* mRNA localisation in the *Xenopus* oocyte. (**c**) β-Actin is localised in chicken fibroblasts. (**d**) In the *Drosophila* oocyte *gurken*, *bicoid*, *oskar* and *nanos* localise to set up the axis of the developing cell. *Nanos* is localised at a later stage in the development of the oocyte. (**e**) β-actin mRNA localised in hippocampal neurons

has roles in establishing polarity and directing cell migration (Condeelis and Singer 2005) (Fig. 11.1c). The axes of the developing *Drosophila* oocyte are specified by *bicoid* (*bcd*), *oskar* (*osk*), *gurken* (*grk*) and in later stages by *nanos* (*nos*) (St Johnston 2005; Tekotte and Davis 2002) (Fig. 11.1d). In the nervous system, *β-actin* mRNA has been shown to localise in developing neurites during growth (Bassell et al. 1998) (Fig. 11.1e). RNA localisation and translational control in the nervous system allows rapid translation of proteins as required, the extended nature of nerve cells precludes the diffusion of mRNAs or the protein products. RNA localisation in neurons has been linked to memory and learning.

11.1.2   Visualisation of mRNA Localisation

The first visualisation of the asymmetrical distribution of mRNAs was achieved through in situ hybridisation using radio-labelled probes (Levsky et al. 2002; Jeffery and Wilson 1983). Fluorescent in situ hybridisation (FISH) was pioneered for improved visualisation of mRNAs and to study the colocalisation with the *trans-*acting factors required for localisation (Bauman et al. 1980). The technique was further developed for use in a large-scale screen of 3,000 transcripts in the *Drosophila* blastoderm embryo where 71 % of the transcripts showed a non-uniform distribution (Lecuyer et al. 2007). Electron microscopy (EM) offers a high-resolution approach for the detection of RNAs coupled to gold particles in vivo and has the advantage of visualising cellular components at the ultra-structural level. Refinement of the technique has seen the use of antisense probes to specific RNAs, labelled with antibodies conjugated to gold. Different size gold particles allow several RNAs and/ or proteins to be visualised simultaneously (Herpers et al. 2010; Delanoue et al. 2007; Herpers and Rabouille 2004). The visualisation of mRNA in fixed material continues to be an essential technique for investigating the localisation of mRNAs and colocalisation with protein factors and cell structures. However, in order to reveal the underlying mechanisms underlying RNA localisation and to observe the RNA on its journey rather than just at its destination the RNA must be visualised in real time using live cell imaging techniques.

For large cells, often the simplest method for visualising mRNA with live cell imaging is via the micro-injection of fluorescently labelled RNA. The RNA is synthesised in vitro and can be conjugated to a fluorophore of choice. Careful selection of fluorophores can allow several RNAs to be visualised simultaneously through multi-colour imaging. This technique has been used to image *Vg1* mRNA localisation in the *Xenopus* oocyte (Yisraeli et al. 1990), *oskar* localisation in the *Drosophila* oocyte and to show that *wingless* and the *pair-rule* transcripts localise along microtubules and require the molecular motor Dynein (Wilkie and Davis 2001). Micro-injection has also been used to map *cis*-acting signals in *gurken* and *I* factor mRNAs in *Drosophila* oocytes and embryos (Van De Bor et al. 2005) and identify protein transacting factors, e.g. Egalitarian, required for RNA localisation (Bullock and Ish-Horowicz 2001). The main disadvantage of introducing in vitro transcribed fluorescently labelled RNA into a cell is that the endogenous RNA is not labelled. The MS2 system for visualising native mRNAs in vivo was originally developed to label *Ash1* mRNA in *S. cerevisiae* (Bertrand et al. 1998). mRNAs are genetically engineered to contain multiple copies of the MS2 loop, which then bind an MS2-coat protein fused to green fluorescent protein (MCP-GFP), also genetically engineered to be expressed. The MS2-coat protein specifically binds the MS2 loops in the RNA with high affinity. Typically 6, 12 or 18 copies of the MS2 loops in the RNA ensure a bright fluorescent signal from each RNA copy for live cell imaging. The MS2 system has been successfully employed for the visualisation of *nanos*, *gurken*, *bicoid* and *oskar* mRNAs in *Drosophila* (Forrest and Gavis 2003; Jaramillo et al. 2008; Weil et al. 2006; Zimyanin et al. 2008).

Molecular motor-based RNA localisation is reliant on the cells underlying cytoskeleton. As well as localisation occurring via the microtubule network, actin filaments are also of importance. For example actin mRNA is transported on the actin network, and *oskar* mRNA anchoring is actin dependent (St Johnston 2005). In the *Drosophila* oocyte, the microtubule network can be visualised with the EB1 protein labelling the growing plus end of microtubules. By imaging and tracking these microtubule tips and the RNA cargoes it has been possible to unravel the mechanisms of cell polarisation (Zimyanin et al. 2008). Visualisation of the microtubule network appears random, highly dynamic and undirected. However, through the use of particle tracking and analysis tools the polarity of the microtubule network can be unravelled (Parton et al. 2011; Hamilton et al. 2010).

11.2   Identification, Characterisation and Searching for Localising RNAs

11.2.1   Computational Identification of Localising RNAs

There has been much interest in the development of computational approaches for the investigation of localising RNAs, ranging from the prediction of secondary structures to the identification and searching for localisation motifs (Mathews et al. 2010; Hamilton and Davis 2007, 2011). These methods have been extensively reviewed elsewhere, so this chapter will highlight some of the more recent developments in RNA motif searches. A recent method, RNAcontext, uses a learning-based approach to identify the binding preferences of RNA binding proteins (Kazan et al. 2010). The method takes into account the RNA sequence, binding affinities and RNA secondary structure preferences and was applied successfully to a set of known examples and was able to predict new binding preferences for a second set of proteins. RNADetect searches for the small non-canonical motifs; G-bulge loops, kink-turns, C-loops and tandem-GAs (Cruz and Westhof 2011). The method is most powerful when used in conjunction with secondary structure methods and new non-canonical motifs can be generated with RNABuild. RNApromo utilises Stochastic Context-Free Grammars (SCFGs) to search for local RNA structure motifs, the secondary structures can be derived from minimum free energy (MFE) predictions or experimentally determined (Rabani et al. 2008). RNApromo was used to study RNA localisation by being applied to an in situ database of RNAs found to localise in *Drosophila* embryos structural motifs were found in 9 out of the 94 classes in the database (Lecuyer et al. 2007).

11.2.2   Mapped RNA Cis-Acting Signals

For several known localising mRNAs in the *Drosophila* oocyte, the signal within the RNAs necessary and sufficient for localisation has been mapped (Fig. 11.2). The *gurken* and *I* factor retrotransposon RNA signals have been mapped to hairpins of 64 and 58 nucleotides (nt) respectively (Van De Bor et al. 2005). These hairpins were then used as search templates to find other localising retrotransposon signals, resulting in the discovery of the *G2* and *Jockey* signals (Hamilton et al. 2009). In the case of bicoid, the localisation signal(s) have been mapped to a region of ~600 nt (Macdonald and Kerr 1998). Within this region there are five stable hairpins and, although none are absolutely required for localisation, stems IV and V are each sufficient for localisation. fs(1)K10 mRNA localises to the anterior of the *Drosophila* oocyte and has been mapped to a 44 nt stem-loop (Serano and Cohen 1995). The K10 stem-loop was the first localisation signal to have its tertiary structure determined (see below). A similarly structured localisation signal has also been characterised in *orb* mRNA (Cohen et al. 2005). The *hairy* pair-rule mRNA localisation

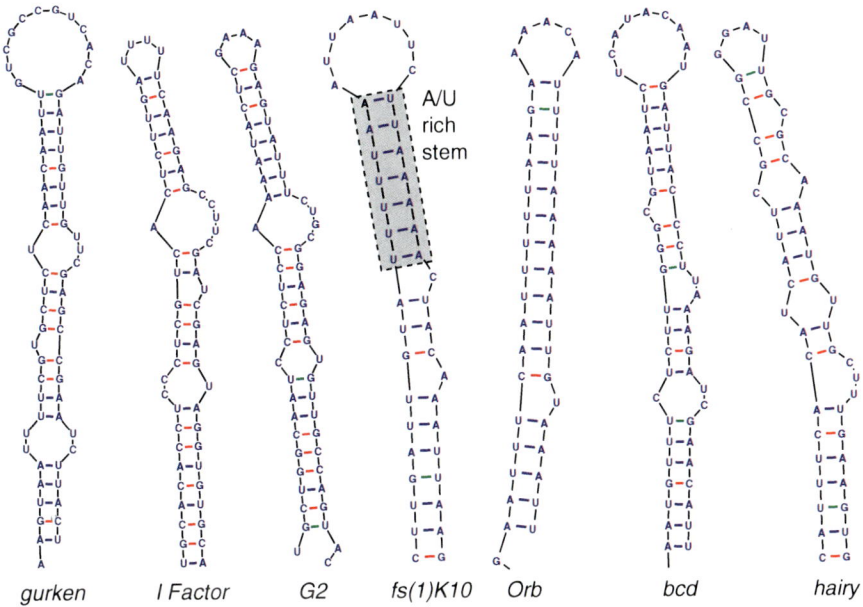

Fig. 11.2 Secondary structure predictions for known *cis*-acting signals in *Drosophila* oocytes/ embryos. Each signal is necessary and sufficient to localise the RNA. The *dashed box* around *fs(1) K10* indicates a purine (A/U) rich helical stem, allowing specific recognition by transport particle components. The *bicoid* signal is part of a larger sequence mapped that includes several putative *cis*-acting signals

element has been mapped to a 121 nt region in the 3' UTR containing two important stem-loop secondary structures (Bullock et al. 2003). In yeast, the localisation signal of β-*actin* mRNA has been mapped to 54 nt and contains a tandem repeat of ACACCC recognised by Zbp1 (Kislauskis et al. 1994; Ross et al. 1997). *Ash1* mRNA has been mapped to find four localisation signals in the coding region and 3' UTR, each of which localises *Ash1* to the bud tip of dividing yeast cells (Chartrand et al. 1999; Gonzalez et al. 1999). A spatial arrangement of two cytosine nucleotides within the signals was shown to be required for localisation, and further RNAs containing the motif were found computationally using MC-Search (Olivier et al. 2005).

11.2.3   RNA Cis-Acting Signals

RNA folds to form secondary structures, which then assemble into tertiary structures. RNA secondary structure is built on canonical Watson–Crick (A:U, U:A, G:C, C:G) and non-canonical (e.g. G:U, G:A, see Fig. 11.3) pairings between bases. The RNA can either be double stranded, when there are complimentary base pairings across the strands, or single stranded, when there are no base pairings made. The base pairings allow a wide range of structures, including helices, loops, bulges, internal loops

a

Interacting Edges

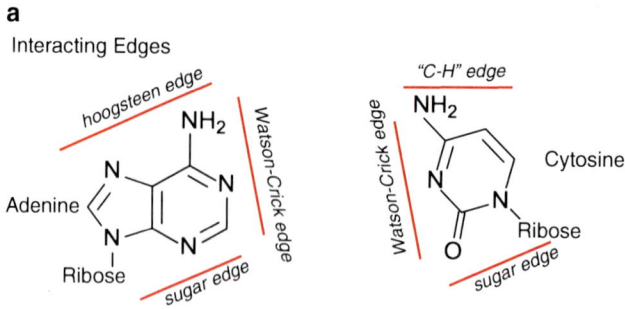

b

Canonical / Watson-Crick Base Pairings

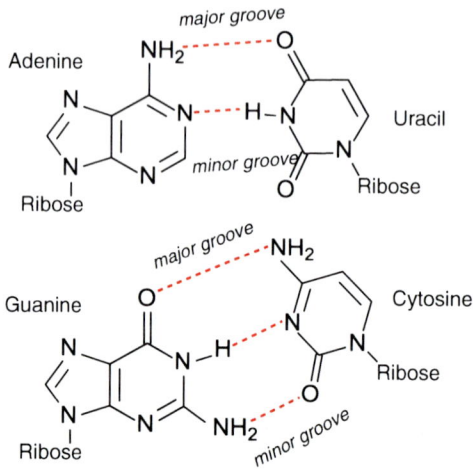

c

Non canonical A:G Base Pairing through Watson-Crick edge

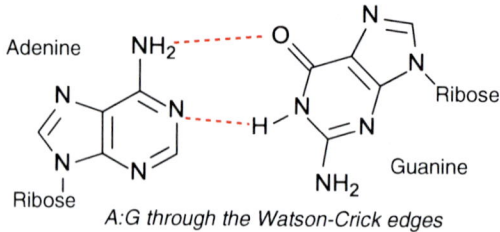

A:G through the Watson-Crick edges

d

Classic Kink turn motif

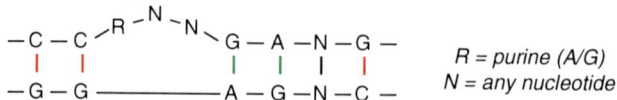

R = purine (A/G)
N = any nucleotide

Fig. 11.3 Base pair conformations. (**a**) Nucleotide bases can form hydrogen bonds to pair with other bases through one of their three edges; Watson–Crick, Hoogsteen/C–H and sugar. (**b**) The base pairings are most commonly canonical/Watson–Crick (A:U, G:C, U:A, C:G) via the Watson–Crick edges. (**c**) Non-canonical base pairings are less commonly observed between bases other than the ones considered canonical (e.g. A:G) and/or hydrogen bonds are formed though the Hoogsteen/C–H and sugar edges. (**d**) A classic kink-turn motif characterised by an internal loop of three bases linked to two G:A non-canonical base pairings flanked by canonical base pairings forming a stem

and pseudoknots to be formed. As RNA structure is important to protein–RNA recognition, bases important for maintaining RNA secondary structure are under evolutionary pressure and are usually well conserved between homologous sequences. Compensatory mutations in the RNA sequence allow the maintenance of specific base pairings and of the secondary structure of the RNA. The RNA tertiary structures are organised from the secondary structure elements, with inter-helical contacts, contact between loops, docking of adenine-platforms and other interactions (Lunde et al. 2007; Hamilton and Davis 2007, 2011). RNA binding proteins recognise tertiary and secondary structural elements, but the most common recognition site for RNA binding proteins is in the single stranded RNA regions joining structured areas.

A well-characterised *cis*-acting spatial code, defined by non-canonical base pairings that forms a kink-turn (KT) motif, has been identified in several mRNAs localising in the nervous system (Tiedge 2006) (Fig. 11.3d). The kink-turn motif is characterised by a pair of non-canonical G:A base pairings adjacent to an internal loop of 3 nt and flanked by canonical, often G:C rich helical stems. In low Ca^{2+} or Mg^{2+} concentrations the kink-turn motif is in an extended conformation with a minimal kink. The kink is induced by high concentrations of Ca^{2+}, found during innervations decreasing in concentration away from the synapse towards the soma, and Mg^{2+} found during depolymerisation events. The KT motif, known as a dendritic targeting element (DTE), specifically binds to heteronuclear ribonucleoprotein A2 (hnRNP A2), also required for correct localisation (Tiedge 2006). Myelin basic protein (MBP) mRNA contains a KT motif in the 3′ UTR in a 21 nt region previously identified as an hnRNP A2 response element (A2RE). BC1, a member of the non-coding small brain-specific cytoplasmic class of RNAs, contains a KT motif in its 5′ UTR and its transport has been shown to be microtubule dependent (Cristofanilli et al. 2006). Activity-regulated cytoskeleton-associated (Arc) mRNA contains two KT motifs in its 3′ UTR responsible for targeting (Kobayashi et al. 2005). Protein kinase M δ (PKMδ) mRNA contains a KT motif in a 44 nt stem-loop in the 3′ UTR. CamKII also has a predicted KT motif in a region of the 3′ UTR previously shown to be responsible for targeting (Blichenberg et al. 2001). A reverse KT motif has recently been described where the kink opens up the major groove rather than the minor groove found and differs by just one nucleotide from the classical KT motif (Antonioli et al. 2010). There is also a database of sequences containing KT motifs from experimental and theoretical structures (Schroeder et al. 2010).

There are several neurodegenerative diseases linked to defective RNA localisation including Fragile X-associated tremor/ataxia syndrome (FXTAS), characterised by the expansion of CGG repeats in the fragile X mental retardation 1 (FMR1) mRNA (Jacquemont et al. 2007). The CGG repeats form stable stem-loop structures containing non-canonical G:G base pairings and have been shown to interact with hnRNP A2 (Muslimov et al. 2011). The CGG repeats were shown to compete with hnRNP A2 binding to the KT motifs of BC1 and PKMδ, and in the case of the expanded CGG repeats (>55) this leads to reduced targeting of mRNAs containing kink-turn motifs.

11.2.4 RNA Secondary Structure Prediction

Traditionally, RNA secondary structure prediction algorithms, such as Mfold and RNAfold, have only predicted canonical (plus G:U) base pairings (Zuker 1994; Hofacker 2003). However, it is becoming clear that RNA non-canonical base pairings such as those in the kink-turn motif can be crucial to their function. There are several computational approaches capable of predicting non-canonical interactions. MC-Fold uses "nucleotide cyclic motifs", which are small structural fragments from experimentally derived RNA structures, used to predict lowest free energy secondary structures (Parisien and Major 2008). The CONTRAfold approach is also able to predict non-canonical base pairings through the implementation of probabilistic models (Do et al. 2006). Conditional log-linear models (CLLMs) are a generalised form of SCFGs, and include parameters from MFE models. This enhancement to pure probabilistic models enables CONTRAfold to outperform pure SCFG methods. A more recent method, RNAwolf, also predicts extended RNA secondary structures including non-canonical base pairings (zu Siederdissen et al. 2011). The method is based on a statistical analysis of the frequencies of the 12 basic types of base pairings of Leontis and Westhof in experimentally derived structures (Leontis and Westhof 2001, 2002). A classification of the pseudo-pairs between nucleotide bases and amino acids in 446 nucleotide–protein complexes (including DNA as well as RNA) was undertaken to quantify preferences in the interactions (Kondo and Westhof 2011). It was found that the majority of bases interact through canonical interactions to guanine and asparagine residues. The Hoogsteen edge was mainly presented by adenine and guanine bases and interactions with the sugar edge were rarely seen (Fig. 11.3a).

11.3 Identification and Characterisation of RNA Binding Trans-Acting Factors

Identification of RNA binding *trans*-acting factors can be achieved in three different ways: First, biochemical pull down approaches using a known localising RNA as bait typically attached to beads in a column, to fish out the proteins binding the RNA from cell extracts. Proteins directly binding the RNA, as well as those indirectly binding, are eluted from the column. The proteins eluted can then be identified using mass spectrometry. One example of such a technique is GRNA chromatography (Czaplinski et al. 2005). Second, carrying out a genetic screen relying on an RNA (mis)localisation phenotype. In the case of the developing *Drosophila* oocyte for example, many of the *trans*-acting factors implicated in the localisation of the key axis determining mRNAs *gurken, oskar, bicoid* and *nanos* were found to belong to a group of genes earlier identified as "maternal-effect" female sterile mutations involved in the same process (Schupbach and Wieschaus 1989; Luschnig et al. 2004). A similar screen was also performed in zebrafish for mutations affecting

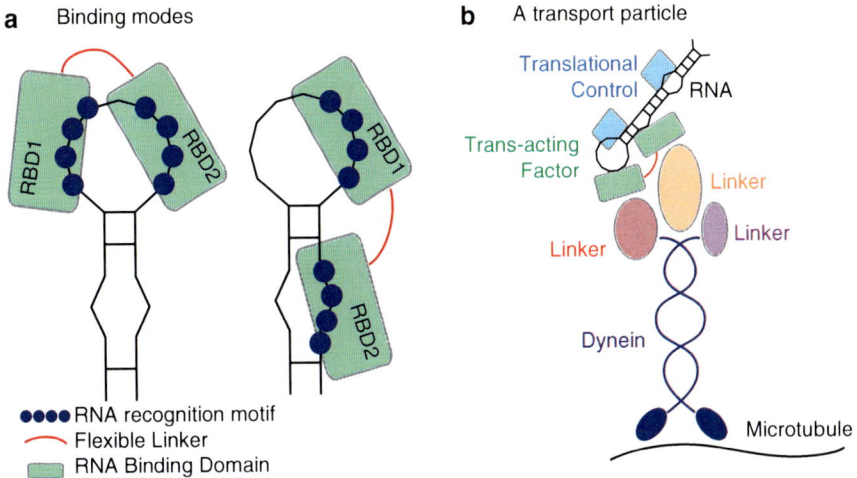

Fig. 11.4 Binding modes of RNA binding proteins and the composition of a transport particle. (**a**) RNA binding proteins bind their RNA targets through two or more RNA binding domains. Each domain typically interacts with 4–8 nucleotides in a predominantly single stranded region. Flexible domain linkers allow different RNAs to be specifically bound. (**b**) A schematic of the composition of a transport particle. An RNA (e.g. *gurken*) is linked to a molecular motor (e.g. Dynein) via protein *trans*-acting factors (e.g. Squid) and linkers (e.g. Egalitarian and Bicaudal D). The translation of the RNA is suppressed during transport via bound translational regulators

development (Geisler et al. 2007). Third, a direct visual screen of RNA/protein colocalisation by fluorescence light microscopy or cryo-immuno EM.

11.3.1 How RNA Binding Proteins Bind Their RNA Targets

RNA binding proteins recognising single stranded regions usually contain multiple RNA binding domains (RBDs). Single domain recognition of unstructured (single stranded) RNA is normally low affinity and cooperative binding of multiple domains is necessary to enhance the specificity and affinity of the interactions (Lunde et al. 2007). In this binding mode, each domain interacts with short regions of the RNA and the inter-domain linkers influence the specificity and the affinity of the binding (Fig. 11.4a). The linkers may also directly participate in the interaction with the RNA, further increasing the binding affinity or simply tether different RNA binding units, with short linkers leading to higher overall affinity than long linkers. In contrast to RNA, DNA forms more conservative secondary structures with DNA binding proteins binding to single contiguous *cis*-acting target sequences. There is a reasonably complete picture of protein–RNA interactions for only a small number of systems. Probably the best example is the Nova (Neuro-oncological ventral antigen) RNA binding protein, which acts as a neuronal splicing factor and contains three KH (K-homology) RBDs. This protein has not been directly linked to RNA localisation, but contains the KH domains found in many *trans*-acting factors involved in RNA transport. A key question in the recognition of an RNA by a protein is how specific recognition is achieved. RNA binding proteins interact with the RNA through two or more of their RBDs, each binding 4–8 nt, increasing the affinity and specificity of the binding (Fig. 11.4a).

11.3.2 Experimental Characterisation of RNA: Protein Binding Preferences

There are several techniques for the determination of the binding preferences of specific RNA binding proteins. These vary from in vitro-based techniques such as Systematic Evolution of Ligands by Exponential Enrichment (SELEX) and NMR to in vivo characterisation using the cross-linking and immunoprecipitation (CLIP)-based methods. The Zbp1 protein binds β-actin mRNA through a "zipcode" in the 3' UTR and is required for localisation. An electrophoretic mobility shift assay (EMSA) with the third and fourth KH domains revealed that Zbp1 binds just the first 28 nt of the zipcode (Chao et al. 2010). This study also involved X-ray structure determination discussed below. The binding preferences of five *Drosophila* hnRNP A/B proteins (hrp36, hrp38, hrp40/Sqd, hrp48) were determined using a SELEX combined with RIP-Chip (RNA binding protein immunoprecipitation coupled to a microarray chip) approach (Blanchette et al. 2009). SELEX revealed the nucleotide binding preferences of the purified proteins, identifying sequence profiles for each protein. However, SELEX is not an ideal method for the characterisation of multi-RBD RNA binding proteins as the recognition is through two or more short 4–8 nt target sequences which, independently, are too low affinity and specificity to attain reliable results. An alternative to address this limitation, Scaffold-independent analysis (SIA) is an NMR technique to determine the binding preference for ssRBDs (Beuth et al. 2007). A pool of randomised RNA sequences together with the protein is investigated through changes in the NMR chemical shifts. The interactions with individual amino acids in the protein for each of the nucleotides are quantified to give the degree of specificity. SIA has been used to characterise the individual binding preferences of different KH RBDs of the KSRP protein (Garcia-Mayoral et al. 2008). Based on this and other structural information, a general model for KSRP-RNA binding has been proposed. The KH3 domain was also characterised using CLIP (Ule et al. 2005a) and the sequence specificity was confirmed using Scaffold Independent Analysis (SIA) (Beuth et al. 2007). In addition, microarray analysis with brain tissue and the Nova (1 and 2) proteins revealed 41 in vivo RNA targets (Ule et al. 2005b).

Biochemical methods such as CLIP have been used to identify RNAs binding a protein of interest and also map the binding sites on a transcriptome scale. The RNA binding protein is cross-linked to a target RNA by the formation of a covalent bond when exposed to UV light. The cross-linking is performed in vivo and so represents endogenous RNA binding interactions. Purification of the RNA:protein complexes is then followed by sequencing of the RNAs. A digestion step also allows the exact RNA fragment binding the protein to be identified (Konig et al. 2010). One limitation, however, is that the cross-linking only works for single stranded RNA targets, while many RNA binding targets contain structured elements. There are also biases for uracil tracts in the RNA (Ule et al. 2005b), certain amino acids in the protein (Williams and Konigsberg 1991) and requirement of a favourable topology of the interaction to make the covalent bond (Maquat and Kiledjian 2008). Photoactivatable ribonucleoside-enhanced CLIP (PAR-CLIP) method uses modified photoactivatable nucleotides with increased cross-linking efficiency to overcome some of these limitations (Hafner et al. 2010). A recent analysis of the CLIP and PAR-CLIP methods found both to be able to resolve binding sites on a single nucleotide level, but also found biases introduced by the digestion steps (Kishore et al. 2011).

11.3.3 Computational Characterisation of RNA: Protein Binding Preferences

There are several computational methods for the prediction of the amino acids in RNA binding proteins that directly interact with the RNA: RNABindR (Terribilini et al. 2006), NAPS (Carson et al. 2010) and PiRaNhA (Murakami et al. 2010). These methods are trained on databases of experimentally determined structures of RNA:protein complexes and are based on amino acid properties such as hydrophobicity, accessibility and substitution scores. However none of these methods directly take into account the RNA itself, so the RNA sequence and secondary structure information are largely ignored. The inclusion of the secondary structure prediction of the RNA targets has been shown to greatly enhance the prediction of the binding sites of RNA binding proteins (Li et al. 2010), as the nucleotides in the target RNA must be accessible to bind to the individual RBDs. The inclusion of binding affinity data further improves the determination of binding preferences for RNA binding proteins. However this is at the level of single RBDs and their individual RNA target motifs.

11.3.4 Structure Determination of RNA Localisation Transport Particle Components

The first structure to be determined for a transport particle component was that of Staufen (Ramos et al. 2000; Bycroft et al. 1995). Staufen contains five dsRBDs and was also the first protein component to be identified being required for RNA localisation (St Johnston 1995). The NMR structure of the third dsRDB domain from *Drosophila* Staufen revealed the critical residues for RNA binding and the mutation of these residues abolished localisation with the full length protein. However the

RNA used in the structure determination was optimised for NMR rather than genuine in vivo targets such as *oskar*, *prospero* and *bicoid*. The dsRDB domain was thought to be a non-sequence specific interaction as the nucleotides are not accessible; however, two recent studies suggest this may not be the case. The first study is the NMR structure of the 44 nt fs(1)K10 mRNA localisation signal and is also the first such RNA signal to have its structure determined (Bullock et al. 2010). K10 localises to the anterior in *Drosophila* oocytes helping set up the dorsoventral axis. The structure of the stem-loop was consistent with the secondary structure predictions and unexpectedly revealed that the stem forms A' RNA. Typically double stranded RNA is A form; however, the runs of stacked purines (As and Gs) in the stem lead to a widened major groove allowing access to the bases and hence specific recognition by RNA binding proteins (Fig. 11.2). Another study has shown that adenosine deaminase (ADAR2) is able to specifically recognise double stranded RNA in a stem-loop (Stefl et al. 2010). Although ADAR2 is not involved in RNA localisation, it shows that RNA binding proteins are able to specifically recognise double stranded A-form RNA.

Finally, zipcode binding protein 1 (Zbp1) has been shown to bind and be required for the localisation of β-actin mRNA (Ross et al. 1997). Zbp1 contains two RNA recognition motif (RRM) and four KH domains and specifically recognises a 28 nt region of the β-actin 3' UTR. The structure of the human ortholog of ZBP1, IMP1 KH34, reveals that the RNA loops around the third and forth KH domains (Chao et al. 2010). The looping is required as the RNA binding faces of the KH domains are on opposite sides of the protein, and is thought to aid in the assembly of additional proteins involved in post-transcriptional regulation.

11.3.5 RNA Tertiary Structure Prediction

The experimental determination of RNA structure by X-ray crystallography and NMR spectroscopy is both costly and time consuming, resulting in an ever-widening gap between the sequence and structure databases. This has led to the development of computational methods to bridge this gap. RNA folds in a hierarchical manner, first into secondary structures then building up into tertiary structures (Brion and Westhof 1997). As with the RNA secondary structure methods, the non-helical regions provide the greatest challenges. In the last 5 years there have been several methods published for the prediction of RNA tertiary structures, with the most promising methods outlined below. MC-Sym is part of a pipeline and generates RNA tertiary structures from the secondary structures predicted by MC-Fold (Parisien and Major 2008). Databases of experimentally derived structures are used to generate small fragments, which are built up to generate first a set of secondary structures, then a full tertiary structure. A recent comparison between an NMR structure and an MC-Sym prediction of a 4×4 internal loop highlights the promise of the computational prediction of tertiary structure (Lerman et al. 2011). However a close match to the NMR structure was only selected with the inclusion of low-resolution NMR restraints in the prediction. Fragment assembly of RNA (FARNA) uses a Rosetta-based method from the protein structure prediction field (Das and Baker 2007). The method utilises a "knowledge-based" energy function for nucleotide bases and backbone conformations. The resulting structures are considered to be de novo as no phylogenetic or experimental restraints are used in the predictions. However the inclusion of secondary structure improves the accuracy of the predictions. Coarse-grained molecular modelling of RNAs has been implemented in several methods (e.g. NAST (Jonikas et al. 2009b)); however, they lack atomic detail. The "coarse to atomic" (C2A) method rebuilds coarse grain models to include full atomic detail derived from high resolution X-ray structures (Jonikas et al. 2009a). For the assessment of RNA tertiary structure prediction methods, it is necessary to compare the models to experimentally determined structures; this is usually achieved through root mean squared deviation (RMSD). If the structures being compared are of different sizes, then the RMSD must be normalised accordingly (Masquida et al. 2010). A statistical method for RMSD significance indicates that 25 Å for a de novo model and 14 Å for a secondary structure augmented model represent "good" models (Hajdin et al. 2010). There are several shortcomings of the methods mainly due to their computer intensive nature limiting the size of the RNAs that can be modelled and the relatively small number of experimentally derived RNA structures available in the databases. However, these methods show great promise for the future prediction of accurate RNA tertiary structures.

11.4 Towards the Composition/Modelling of an RNA Transport Particle

So far there has been great progress in the identification of localising RNAs and the mapping of their *cis*-acting signals. Furthermore the identification and characterisation of the binding preference of RNA binding proteins has led to a good understanding of RNA transport. However it has yet to be explained how RNA binding to a *trans*-acting factor which is in turn bound, possibly via intermediate protein factors, to a molecular motor determines the precise destination for the RNAs. The components of transport particles are thought to be highly dynamic with different factors contributing at different stages of the transport (Dreyfuss et al. 2002). The *trans*-acting factors can have roles in direct linkage to motor components, influencing motor activity, translational control, anchoring and defining destination.

So far a wealth of *trans*-acting factors have been identified through genetic screens and biochemical pull-downs. One of the best studied RNA transport particles is that of *gurken* mRNA which is localised to the dorso-anterior corner in *Drosophila* oocytes forming a cap around the nucleus (Fig. 11.4b). Squid (Sqd/ Hrp40) has been identified as a component of *gurken* transport particles, shown to bind *gurken* directly and be required for anchoring (Delanoue et al. 2007). Sqd has also been shown to repress translation of *gurken* during transport (Caceres and Nilson 2005), and in combination with PABP, Encore and Cup (Clouse et al. 2008).

Sqd also interacts with Bruno and Cup (Norvell et al. 1999; Nakamura et al. 2004). The K10 protein has also been linked to interactions with *gurken* mRNA and Sqd and is thought to be required for its localisation and anchoring (Norvell et al. 1999). Hrb27C (Hrb48) binds *gurken* mRNA 3′ UTR and interacts with Sqd and Ovarian tumour (Otu). Mutants and over-expression of these proteins suggest roles in localisation (Goodrich et al. 2004). Imp interacts with the Sqd/Hrb27C complex and binds the *gurken* UTRs, to contribute to translational repression and localisation. UAP56, a nuclear export factor, has also been shown to be required for *gurken* localisation (Meignin and Davis 2008). Dynein light chain has also been shown to be able to bind *gurken* directly (Rom et al. 2007) and recruits the Dynein associated proteins Egalitarian (Egl) and Bicaudal D (BicD) in the *Drosophila* oocyte nurse cells (Clark et al. 2007; Dienstbier et al. 2009). A key problem to address is the composition of transport particles at particular times and locations to understand how changes affect the destination and dynamics.

Although many components have been identified for microtubule dependent localisation of RNAs, such as *gurken*, *bicoid* and *oskar*, it is likely there are as yet undiscovered protein components involved in specifying destinations, translational control and anchoring. One intriguing possibility is that ratios of the components in a transport particle determine localisation destination, termed the "combinatorial theory" (Bullock 2007). For example the ratio of the plus-end directed motors, Dynein, to the minus-end directed motors, Kinesin, may dictate the final destination of the particle. As the *trans*-acting protein factors bind the *cis*-acting RNA signals through two or more RBDs the exact binding stoichiometry is unknown. However, with the experimentally or computationally determined structures of the RNA and *trans*-acting factor(s), and the sequence/structure requirements of the binding interface, there are a wealth of restraints to guide molecular modelling. This will be especially important in determining whether the RBD linkers provide the flexibility to simultaneously bind at least two RNA target sequences. Coarse-grained molecular modelling approaches have great potential to elucidate the binding modes of the RBDs to different RNA targets (Fig. 11.4a). A recent development in the coarse-grained modelling approach is the adaption of collision detection algorithms (Katsimitsoulia and Taylor 2010; Taylor and Katsimitsoulia 2010b), and has been successfully applied to the motor Myosin V moving along Actin filaments (Taylor and Katsimitsoulia 2010a).

The development of computational approaches to modelling macromolecular assemblies has the potential to finally uncover the determinants of localisation destination. A recent computational method successfully combined spatial restraint data from diverse sources to model large macromolecular assemblies (Alber et al. 2007a). This was applied to the 50 MDa nuclear pore complex (NPC), which contains 456 individual proteins (Alber et al. 2007b). Structural, proteomic, biochemical and electron microscopy data all provide restraints that together are sufficient to model the large complex. The dynamic nature of macromolecules is not captured by methods such as X-ray crystallography, electron microscopy, or even molecular modelling; however, they can be investigated using computer simulations that are now possible on millisecond timescales (Russel et al. 2009).

11.5 Concluding Remarks

The field of mRNA localisation and localised translation has continued to extend into interdisciplinary areas of research in addition to including the more traditional genetic and cell biology approaches. For example, the use of microscopy has undergone a transformation to include new and revolutionary the so-called super-resolution and super-precision methods. Furthermore, the development of less noisy camera technologies that operate at significantly improved time resolution allows the observation of the dynamics of RNA motility, in some cases at the single molecule level. Computational methods are also becoming increasingly important for the discovery and characterisation of new localising RNAs. RNA search tools are routinely applied on genome-wide scales, made possible by the continued development of algorithms coupled to advancements in computer power. As the use of next generation sequencing, particularly RNA-Seq, is becoming more widespread, this necessitates the development of new bioinformatics tools to analyse the wealth of data being produced by this powerful technique. The structure determination of the components of transport particles promises to provide crucial insights into the sorting of RNAs to their sites of function in the cell. Techniques such as the NMR-based SIA and iCLIP can now be used to determine the binding preferences of individual RNA binding proteins. Finally, to fully understand the dynamics and specificity of transport particle behaviour, it will also be necessary to develop new approaches for computer simulations and molecular modelling.

Acknowledgements This work was supported by a Wellcome Trust Senior Research Fellowship (Grant number 081858) to ID for RSH and ID. GB is supported through a Wellcome Trust Strategic Award (091911).

References

Alber F, Dokudovskaya S, Veenhoff LM, Zhang W, Kipper J, Devos D, Suprapto A, Karni-Schmidt O, Williams R, Chait BT, Rout MP, Sali A (2007a) Determining the architectures of macromolecular assemblies. Nature 450(7170):683–694. doi:nature06404 [pii] 10.1038/nature06404

Alber F, Dokudovskaya S, Veenhoff LM, Zhang W, Kipper J, Devos D, Suprapto A, Karni-Schmidt O, Williams R, Chait BT, Sali A, Rout MP (2007b) The molecular architecture of the nuclear pore complex. Nature 450(7170):695–701. doi:nature06405 [pii] 10.1038/nature06405

Antonioli AH, Cochrane JC, Lipchock SV, Strobel SA (2010) Plasticity of the RNA kink turn structural motif. RNA 16(4):762–768. doi:rna.1883810 [pii] 10.1261/rna.1883810

Bassell GJ, Zhang H, Byrd AL, Femino AM, Singer RH, Taneja KL, Lifshitz LM, Herman IM, Kosik KS (1998) Sorting of beta-actin mRNA and protein to neurites and growth cones in culture. J Neurosci 18(1):251–265

Bauman JG, Wiegant J, Borst P, van Duijn P (1980) A new method for fluorescence microscopical localization of specific DNA sequences by in situ hybridization of fluorochromelabelled RNA. Exp Cell Res 128(2):485–490

Bertrand E, Chartrand P, Schaefer M, Shenoy SM, Singer RH, Long RM (1998) Localization of ASH1 mRNA particles in living yeast. Mol Cell 2(4):437–445

Beuth B, Garcia-Mayoral MF, Taylor IA, Ramos A (2007) Scaffold-independent analysis of RNA-protein interactions: the Nova-1 KH3-RNA complex. J Am Chem Soc 129(33):10205–10210. doi:10.1021/ja072365q

Blanchette M, Green RE, MacArthur S, Brooks AN, Brenner SE, Eisen MB, Rio DC (2009) Genome-wide analysis of alternative pre-mRNA splicing and RNA-binding specificities of the Drosophila hnRNP A/B family members. Mol Cell 33(4):438–449. doi:S1097-2765(09)00066-5 [pii] 10.1016/j.molcel.2009.01.022

Blichenberg A, Rehbein M, Muller R, Garner CC, Richter D, Kindler S (2001) Identification of a cis-acting dendritic targeting element in the mRNA encoding the alpha subunit of Ca2+/calmodulin-dependent protein kinase II. Eur J Neurosci 13(10):1881–1888. doi:ejn1565 [pii]

Bobola N, Jansen RP, Shin TH, Nasmyth K (1996) Asymmetric accumulation of Ash1p in postanaphase nuclei depends on a myosin and restricts yeast mating-type switching to mother cells. Cell 84(5):699–709. doi:S0092-8674(00)81048-X [pii]

Brion P, Westhof E (1997) Hierarchy and dynamics of RNA folding. Annu Rev Biophys Biomol Struct 26:113–137. doi:10.1146/annurev.biophys.26.1.113

Bullock SL (2007) Translocation of mRNAs by molecular motors: think complex? Semin Cell Dev Biol 18(2):194–201. doi:S1084-9521(07)00026-2 [pii] 10.1016/j.semcdb.2007.01.004

Bullock SL, Ish-Horowicz D (2001) Conserved signals and machinery for RNA transport in Drosophila oogenesis and embryogenesis. Nature 414(6864):611–616. doi:10.1038/414611a

Bullock SL, Zicha D, Ish-Horowicz D (2003) The Drosophila hairy RNA localization signal modulates the kinetics of cytoplasmic mRNA transport. EMBO J 22(10):2484–2494. doi:10.1093/emboj/cdg230

Bullock SL, Ringel I, Ish-Horowicz D, Lukavsky PJ (2010) A'-form RNA helices are required for cytoplasmic mRNA transport in Drosophila. Nat Struct Mol Biol 17(6):703–709. doi:nsmb.1813 [pii] 10.1038/nsmb.1813

Bycroft M, Grunert S, Murzin AG, Proctor M, St Johnston D (1995) NMR solution structure of a dsRNA binding domain from Drosophila staufen protein reveals homology to the N-terminal domain of ribosomal protein S5. EMBO J 14(14):3563–3571

Caceres L, Nilson LA (2005) Production of gurken in the nurse cells is sufficient for axis determination in the Drosophila oocyte. Development 132(10):2345–2353

Carson MB, Langlois R, Lu H (2010) NAPS: a residue-level nucleic acid-binding prediction server. Nucleic Acids Res 38(suppl):W431–W435. doi:gkq361 [pii] 10.1093/nar/gkq361

Chao JA, Patskovsky Y, Patel V, Levy M, Almo SC, Singer RH (2010) ZBP1 recognition of beta-actin zipcode induces RNA looping. Genes Dev 24(2):148–158. doi:24/2/148 [pii] 10.1101/gad.1862910

Chartrand P, Meng XH, Singer RH, Long RM (1999) Structural elements required for the localization of ASH1 mRNA and of a green fluorescent protein reporter particle in vivo. Curr Biol 9(6):333–336

Clark A, Meignin C, Davis I (2007) A dynein-dependent shortcut rapidly delivers axis determination transcripts into the Drosophila oocyte. Development 134(10):1955–1965

Clouse KN, Ferguson SB, Schupbach T (2008) Squid, Cup, and PABP55B function together to regulate gurken translation in Drosophila. Dev Biol 313(2):713–724. doi:S0012-1606(07)01537-0 [pii] 10.1016/j.ydbio.2007.11.008

Cohen RS, Zhang S, Dollar GL (2005) The positional, structural, and sequence requirements of the Drosophila TLS RNA localization element. RNA 11(7):1017–1029

Condeelis J, Singer RH (2005) How and why does beta-actin mRNA target? Biol Cell 97(1):97–110. doi:BC20040063 [pii] 10.1042/BC20040063

Cristofanilli M, Iacoangeli A, Muslimov IA, Tiedge H (2006) Neuronal BC1 RNA: microtubule-dependent dendritic delivery. J Mol Biol 356(5):1118–1123. doi:S0022-2836(05)01539-1 [pii] 10.1016/j.jmb.2005.11.090

Cruz JA, Westhof E (2011) Sequence-based identification of 3D structural modules in RNA with RMDetect. Nat Methods 8(6):513–521. doi:nmeth.1603 [pii] 10.1038/nmeth.1603

Czaplinski K, Kocher T, Schelder M, Segref A, Wilm M, Mattaj IW (2005) Identification of 40LoVe, a Xenopus hnRNP D family protein involved in localizing a TGF-beta-related mRNA during oogenesis. Dev Cell 8(4):505–515. doi:S1534-5807(05)00016-X [pii] 10.1016/j. devcel.2005.01.012

Das R, Baker D (2007) Automated de novo prediction of native-like RNA tertiary structures. Proc Natl Acad Sci U S A 104(37):14664–14669. doi:0703836104 [pii] 10.1073/pnas.0703836104

Delanoue R, Herpers B, Soetaert J, Davis I, Rabouille C (2007) Drosophila Squid/hnRNP helps dynein switch from a gurken mRNA transport motor to an ultrastructural static anchor in sponge bodies. Dev Cell 13(4):523–538. doi:S1534-5807(07)00344-9 [pii] 10.1016/j. devcel.2007.08.022

Dienstbier M, Boehl F, Li X, Bullock SL (2009) Egalitarian is a selective RNA-binding protein linking mRNA localization signals to the dynein motor. Genes Dev 23(13):1546–1558. doi:gad.531009 [pii] 10.1101/gad.531009

Do CB, Woods DA, Batzoglou S (2006) CONTRAfold: RNA secondary structure prediction without physics-based models. Bioinformatics 22(14):e90–e98. doi:22/14/e90 [pii] 10.1093/bioinformatics/btl246

Dreyfuss G, Kim VN, Kataoka N (2002) Messenger-RNA-binding proteins and the messages they carry. Nat Rev Mol Cell Biol 3(3):195–205. doi:10.1038/nrm760 nrm760 [pii]

Forrest KM, Gavis ER (2003) Live imaging of endogenous RNA reveals a diffusion and entrapment mechanism for nanos mRNA localization in Drosophila. Curr Biol 13(14):1159–1168. doi:S0960982203004512 [pii]

Garcia-Mayoral MF, Diaz-Moreno I, Hollingworth D, Ramos A (2008) The sequence selectivity of KSRP explains its flexibility in the recognition of the RNA targets. Nucleic Acids Res 36(16):5290–5296. doi:gkn509 [pii] 10.1093/nar/gkn509

Geisler R, Rauch GJ, Geiger-Rudolph S, Albrecht A, van Bebber F, Berger A, Busch-Nentwich E, Dahm R, Dekens MP, Dooley C, Elli AF, Gehring I, Geiger H, Geisler M, Glaser S, Holley S, Huber M, Kerr A, Kirn A, Knirsch M, Konantz M, Kuchler AM, Maderspacher F, Neuhauss SC, Nicolson T, Ober EA, Praeg E, Ray R, Rentzsch B, Rick JM, Rief E, Schauerte HE, Schepp CP, Schonberger U, Schonthaler HB, Seiler C, Sidi S, Sollner C, Wehner A, Weiler C, Nusslein-Volhard C (2007) Large-scale mapping of mutations affecting zebrafish development. BMC Genomics 8:11. doi:1471-2164-8-11 [pii] 10.1186/1471-2164-8-11

Gonzalez I, Buonomo SB, Nasmyth K, von Ahsen U (1999) ASH1 mRNA localization in yeast involves multiple secondary structural elements and Ash1 protein translation. Curr Biol 9(6):337–340

Goodrich JS, Clouse KN, Schupbach T (2004) Hrb27C, Sqd and Otu cooperatively regulate gurken RNA localization and mediate nurse cell chromosome dispersion in Drosophila oogenesis. Development 131(9):1949–1958. doi:10.1242/dev.01078 dev.01078 [pii]

Hafner M, Landthaler M, Ascano M, Khorshid M, Burger L, Zavolan M, Tuschl T, Hausser J, Berninger P, Rothballer A, Jungkamp A-C, Munschauer M, Ulrich A, Wardle GS, Dewell S (2010) PAR-CliP—a method to identify transcriptome-wide the binding sites of RNA binding proteins. J Vis Exp 41:e2034

Hajdin CE, Ding F, Dokholyan NV, Weeks KM (2010) On the significance of an RNA tertiary structure prediction. RNA 16(7):1340–1349. doi:rna.1837410 [pii] 10.1261/rna.1837410

Hamilton RS, Davis I (2007) RNA localization signals: deciphering the message with bioinformatics. Semin Cell Dev Biol 18(2):178–185

Hamilton RS, Davis I (2011) Identifying and searching for conserved RNA localisation signals. Methods Mol Biol 714:447–466. doi:10.1007/978-1-61779-005-8_27

Hamilton RS, Hartswood E, Vendra G, Jones C, Van De Bor V, Finnegan D, Davis I (2009) A bioinformatics search pipeline, RNA2DSearch, identifies RNA localization elements in Drosophila retrotransposons. RNA 15(2):200–207. doi:15/2/200 [pii] 10.1261/rna.1264109

Hamilton RS, Parton RM, Oliveira RA, Vendra G, Ball G, Nasmyth K, Davis I (2010) ParticleStats: open source software for the analysis of particle motility and cytoskeletal polarity. Nucleic Acids Res 38(Web Server issue):W641–646. doi:gkq542 [pii]10.1093/nar/gkq542

Herpers B, Rabouille C (2004) mRNA localization and ER-based protein sorting mechanisms dictate the use of transitional endoplasmic reticulum-golgi units involved in gurken transport in Drosophila oocytes. Mol Biol Cell 15(12):5306–5317. doi:10.1091/mbc.E04-05-0398 E04-05-0398 [pii]

Herpers B, Xanthakis D, Rabouille C (2010) ISH-IEM: a sensitive method to detect endogenous mRNAs at the ultrastructural level. Nat Protoc 5(4):678–687. doi:nprot.2010.12 [pii] 10.1038/nprot.2010.12

Hofacker IL (2003) Vienna RNA secondary structure server. Nucleic Acids Res 31(13):3429–3431

Jacquemont S, Hagerman RJ, Hagerman PJ, Leehey MA (2007) Fragile-X syndrome and fragile X-associated tremor/ataxia syndrome: two faces of FMR1. Lancet Neurol 6(1):45–55. doi:S1474-4422(06)70676-7 [pii] 10.1016/S1474-4422(06)70676-7

Jaramillo AM, Weil TT, Goodhouse J, Gavis ER, Schupbach T (2008) The dynamics of fluorescently labeled endogenous gurken mRNA in Drosophila. J Cell Sci 121(Pt 6):887–894. doi:jcs.019091 [pii] 10.1242/jcs.019091

Jeffery WR, Wilson LJ (1983) Localization of messenger RNA in the cortex of Chaetopterus eggs and early embryos. J Embryol Exp Morphol 75:225–239

Jonikas MA, Radmer RJ, Altman RB (2009a) Knowledge-based instantiation of full atomic detail into coarse-grain RNA 3D structural models. Bioinformatics 25(24):3259–3266. doi:btp576 [pii] 10.1093/bioinformatics/btp576

Jonikas MA, Radmer RJ, Laederach A, Das R, Pearlman S, Herschlag D, Altman RB (2009b) Coarse-grained modeling of large RNA molecules with knowledge-based potentials and structural filters. RNA 15(2):189–199. doi:15/2/189 [pii] 10.1261/rna.1270809

Katsimitsoulia Z, Taylor WR (2010) A hierarchic collision detection algorithm for simple Brownian dynamics. Comput Biol Chem 34(2):71–79. doi:S1476-9271(10)00002-2 [pii] 10.1016/j.compbiolchem.2010.01.001

Kazan H, Ray D, Chan ET, Hughes TR, Morris Q (2010) RNAcontext: a new method for learning the sequence and structure binding preferences of RNA-binding proteins. PLoS Comput Biol 6:e1000832. doi:10.1371/journal.pcbi.1000832

Kishore S, Jaskiewicz L, Burger L, Hausser J, Khorshid M, Zavolan M (2011) A quantitative analysis of CLIP methods for identifying binding sites of RNA-binding proteins. Nat Methods 8(7):559–564. doi:nmeth.1608 [pii] 10.1038/nmeth.1608

Kislauskis EH, Zhu X, Singer RH (1994) Sequences responsible for intracellular localization of beta-actin messenger RNA also affect cell phenotype. J Cell Biol 127(2):441–451

Kobayashi H, Yamamoto S, Maruo T, Murakami F (2005) Identification of a cis-acting element required for dendritic targeting of activity-regulated cytoskeleton-associated protein mRNA. Eur J Neurosci 22(12):2977–2984. doi:EJN4508 [pii] 10.1111/j.1460-9568.2005.04508.x

Kondo J, Westhof E (2011) Classification of pseudo pairs between nucleotide bases and amino acids by analysis of nucleotide-protein complexes. Nucleic Acids Res 39:8628–8637. doi:gkr452 [pii] 10.1093/nar/gkr452

Konig J, Zarnack K, Rot G, Curk T, Kayikci M, Zupan B, Turner DJ, Luscombe NM, Ule J (2010) iCLIP reveals the function of hnRNP particles in splicing at individual nucleotide resolution. Nat Struct Mol Biol 17(7):909–915. doi:nsmb.1838 [pii] 10.1038/nsmb.1838

Lecuyer E, Yoshida H, Parthasarathy N, Alm C, Babak T, Cerovina T, Hughes TR, Tomancak P, Krause HM (2007) Global analysis of mRNA localization reveals a prominent role in organizing cellular architecture and function. Cell 131(1):174–187. doi:S0092-8674(07)01022-7 [pii] 10.1016/j.cell.2007.08.003

Leontis NB, Westhof E (2001) Geometric nomenclature and classification of RNA base pairs. RNA 7(4):499–512

Leontis NB, Westhof E (2002) The annotation of RNA motifs. Comp Funct Genomics 3(6):518–524. doi:10.1002/cfg.213

Lerman YV, Kennedy SD, Shankar N, Parisien M, Major F, Turner DH (2011) NMR structure of a 4 x 4 nucleotide RNA internal loop from an R2 retrotransposon: identification of a three purine-purine sheared pair motif and comparison to MC-SYM predictions. RNA 17:1664–1677. doi:rna.2641911 [pii] 10.1261/rna.2641911

Levsky JM, Singer RH (2003) Fluorescence in situ hybridization: past, present and future. J Cell Sci 116(pt 14):2833–2838. doi:10.1242/jcs.00633 116/14/2833 [pii]

Levsky JM, Shenoy SM, Pezo RC, Singer RH (2002) Single-cell gene expression profiling. Science 297(5582):836–840. doi:10.1126/science.1072241 297/5582/836 [pii]

Li X, Quon G, Lipshitz HD, Morris Q (2010) Predicting in vivo binding sites of RNA-binding proteins using mRNA secondary structure. RNA 16(6):1096–1107. doi:rna.2017210 [pii] 10.1261/rna.2017210

Lunde BM, Moore C, Varani G (2007) RNA-binding proteins: modular design for efficient function. Nat Rev Mol Cell Biol 8(6):479–490. doi:nrm2178 [pii] 10.1038/nrm2178

Luschnig S, Moussian B, Krauss J, Desjeux I, Perkovic J, Nusslein-Volhard C (2004) An F1 genetic screen for maternal-effect mutations affecting embryonic pattern formation in Drosophila melanogaster. Genetics 167(1):325–342. doi:167/1/325 [pii]

Macdonald PM, Kerr K (1998) Mutational analysis of an RNA recognition element that mediates localization of bicoid mRNA. Mol Cell Biol 18(7):3788–3795

Maquat LE, Kiledjian M (2008) RNA turnover in eukaryotes: analysis of specialized and quality control RNA decay pathways. Preface. Methods Enzymol 449:xvii–xviii. doi:S0076-6879(08)02422-1[pii]10.1016/S0076-6879(08)02422-1

Masquida B, Beckert B, Jossinet F (2010) Exploring RNA structure by integrative molecular modelling. N Biotechnol 27(3):170–183. doi:S1871-6784(10)00394-8 [pii] 10.1016/j.nbt.2010.02.022

Mathews DH, Moss WN, Turner DH (2010) Folding and finding RNA secondary structure. Cold Spring Harb Perspect Biol 2(12):a003665. doi:cshperspect.a003665 [pii] 10.1101/cshperspect. a003665

Meignin C, Davis I (2008) UAP56 RNA helicase is required for axis specification and cytoplasmic mRNA localization in Drosophila. Dev Biol 315(1):89–98. doi:S0012-1606(07)01574-6 [pii] 10.1016/j.ydbio.2007.12.004

Mowry KL, Melton DA (1992) Vegetal messenger RNA localization directed by a 340-nt RNA sequence element in Xenopus oocytes. Science 255(5047):991–994

Murakami Y, Spriggs RV, Nakamura H, Jones S (2010) PiRaNhA: a server for the computational prediction of RNA-binding residues in protein sequences. Nucleic Acids Res 38(suppl):W412–W416. doi:gkq474 [pii] 10.1093/nar/gkq474

Muslimov IA, Patel MV, Rose A, Tiedge H (2011) Spatial code recognition in neuronal RNA targeting: role of RNA-hnRNP A2 interactions. J Cell Biol 194:441–457. doi:jcb.201010027 [pii] 10.1083/jcb.201010027

Nakamura A, Sato K, Hanyu-Nakamura K (2004) Drosophila cup is an eIF4E binding protein that associates with Bruno and regulates oskar mRNA translation in oogenesis. Dev Cell 6(1):69–78. doi:S1534580703004003 [pii]

Nevo-Dinur K, Nussbaum-Shochat A, Ben-Yehuda S, Amster-Choder O (2011) Translation-independent localization of mRNA in E. coli. Science 331(6020):1081–1084. doi:331/6020/1081[pii]10.1126/science.1195691

Norvell A, Kelley RL, Wehr K, Schupbach T (1999) Specific isoforms of squid, a Drosophila hnRNP, perform distinct roles in Gurken localization during oogenesis. Genes Dev 13(7):864–876

Olivier C, Poirier G, Gendron P, Boisgontier A, Major F, Chartrand P (2005) Identification of a conserved RNA motif essential for She2p recognition and mRNA localization to the yeast bud. Mol Cell Biol 25(11):4752–4766

Parisien M, Major F (2008) The MC-fold and MC-Sym pipeline infers RNA structure from sequence data. Nature 452(7183):51–55. doi:nature06684 [pii] 10.1038/nature06684

Parton RM, Hamilton RS, Ball G, Yang L, Cullen F, Lu W, Ohkura H et al (2011) A PAR-1-dependent orientation gradient of dynamic microtubules directs posterior cargo transport in the Drosophila oocyte. J Cell Biol, 194(1):121–135. doi:10.1083/jcb.201103160

Rabani M, Kertesz M, Segal E (2008) Computational prediction of RNA structural motifs involved in posttranscriptional regulatory processes. Proc Natl Acad Sci U S A 105(39):14885–14890. doi:0803169105 [pii] 10.1073/pnas.0803169105

Ramos A, Grunert S, Adams J, Micklem DR, Proctor MR, Freund S, Bycroft M, St Johnston D, Varani G (2000) RNA recognition by a Staufen double-stranded RNA-binding domain. EMBO J 19(5):997–1009. doi:10.1093/emboj/19.5.997

Rom I, Faicevici A, Almog O, Neuman-Silberberg FS (2007) Drosophila Dynein light chain (DDLC1) binds to gurken mRNA and is required for its localization. Biochim Biophys Acta 1773(10):1526–1533. doi:S0167-4889(07)00119-X [pii] 10.1016/j.bbamcr.2007.05.005

Rosbash M, Singer RH (1993) RNA travel: tracks from DNA to cytoplasm. Cell 75(3):399–401. doi:0092-8674(93)90373-X [pii]

Ross AF, Oleynikov Y, Kislauskis EH, Taneja KL, Singer RH (1997) Characterization of a beta-actin mRNA zipcode-binding protein. Mol Cell Biol 17(4):2158–2165

Russel D, Lasker K, Phillips J, Schneidman-Duhovny D, Velazquez-Muriel JA, Sali A (2009) The structural dynamics of macromolecular processes. Curr Opin Cell Biol 21(1):97–108. doi:S0955-0674(09)00031-3 [pii] 10.1016/j.ceb.2009.01.022

Schroeder KT, McPhee SA, Ouellet J, Lilley DM (2010) A structural database for k-turn motifs in RNA. RNA 16(8):1463–1468. doi:rna.2207910 [pii] 10.1261/rna.2207910

Schupbach T, Wieschaus E (1989) Female sterile mutations on the second chromosome of Drosophila melanogaster. I. Maternal effect mutations. Genetics 121(1):101–117

Serano TL, Cohen RS (1995) A small predicted stem-loop structure mediates oocyte localization of Drosophila K10 mRNA. Development 121(11):3809–3818

St Johnston D (1995) The intracellular localization of messenger RNAs. Cell 81(2):161–170. doi:0092-8674(95)90324-0 [pii]

St Johnston D (2005) Moving messages: the intracellular localization of mRNAs. Nat Rev Mol Cell Biol 6(5):363–375. doi:10.1038/nrm1643

Stefl R, Oberstrass FC, Hood JL, Jourdan M, Zimmermann M, Skrisovska L, Maris C, Peng L, Hofr C, Emeson RB, Allain FH (2010) The solution structure of the ADAR2 dsRBM-RNA complex reveals a sequence-specific readout of the minor groove. Cell 143(2):225–237. doi:S0092-8674(10)01074-3 [pii] 10.1016/j.cell.2010.09.026

Taylor WR, Katsimitsoulia Z (2010a) A coarse-grained molecular model for actin-myosin simulation. J Mol Graph Model 29(2):266–279. doi:S1093-3263(10)00084-7 [pii] 10.1016/j.jmgm.2010.06.004

Taylor WR, Katsimitsoulia Z (2010b) A soft collision detection algorithm for simple Brownian dynamics. Comput Biol Chem 34(1):1–10. doi:S1476-9271(09)00136-4 [pii] 10.1016/j.compbiolchem.2009.11.003

Tekotte H, Davis I (2002) Intracellular mRNA localization: motors move messages. Trends Genet 18(12):636–642. doi:10.1016/S0168-9525(02)02819-6

Terribilini M, Lee JH, Yan C, Jernigan RL, Honavar V, Dobbs D (2006) Prediction of RNA binding sites in proteins from amino acid sequence. RNA 12(8):1450–1462. doi:rna.2197306 [pii] 10.1261/rna.2197306

Tiedge H (2006) K-turn motifs in spatial RNA coding. RNA Biol 3(4):133–139. doi:3249 [pii]

Ule J, Jensen K, Mele A, Darnell RB (2005a) CLIP: a method for identifying protein-RNA interaction sites in living cells. Methods 37(4):376–386. doi:S1046-2023(05)00178-7 [pii] 10.1016/j.ymeth.2005.07.018

Ule J, Ule A, Spencer J, Williams A, Hu JS, Cline M, Wang H, Clark T, Fraser C, Ruggiu M, Zeeberg BR, Kane D, Weinstein JN, Blume J, Darnell RB (2005b) Nova regulates brain-specific splicing to shape the synapse. Nat Genet 37(8):844–852. doi:ng1610 [pii] 10.1038/ng1610

Van De Bor V, Hartswood E, Jones C, Finnegan D, Davis I (2005) gurken and the I factor retrotransposon RNAs share common localization signals and machinery. Dev Cell 9(1):51–62

Weil TT, Forrest KM, Gavis ER (2006) Localization of bicoid mRNA in late oocytes is maintained by continual active transport. Dev Cell 11(2):251–262. doi:S1534-5807(06)00262-0 [pii] 10.1016/j.devcel.2006.06.006

Wilkie GS, Davis I (2001) Drosophila wingless and pair-rule transcripts localize apically by dynein-mediated transport of RNA particles. Cell 105(2):209–219. doi:10.1016/S0092-8674(01)00312-9

Williams KR, Konigsberg WH (1991) Identification of amino acid residues at interface of protein-nucleic acid complexes by photochemical cross-linking. Methods Enzymol 208:516–539

Yisraeli JK, Sokol S, Melton DA (1990) A two-step model for the localization of maternal mRNA in Xenopus oocytes: involvement of microtubules and microfilaments in the translocation and anchoring of Vg1 mRNA. Development 108(2):289–298

Zimyanin VL, Belaya K, Pecreaux J, Gilchrist MJ, Clark A, Davis I, St Johnston D (2008) In vivo imaging of oskar mRNA transport reveals the mechanism of posterior localization. Cell 134(5):843–853. doi:S0092-8674(08)00841-6 [pii] 10.1016/j.cell.2008.06.053

zu Siederdissen CH, Bernhart SH, Stadler PF, Hofacker IL (2011) A folding algorithm for extended RNA secondary structures. Bioinformatics 27(13):i129–i136. doi:btr220[pii]10.1093/bioinformatics/btr220

Zuker M (1994) Prediction of RNA secondary structure by energy minimization. Methods Mol Biol 25:267–294

Chapter 12
Virtual Screening for RNA-Interacting Small Molecules

Hyun-Ju Park and So-Jung Park

12.1 Virtual Screening and Docking Tools

Computational virtual screening strategy is useful in the very early stage of the drug discovery pipeline and provides a powerful tool for rapid discovery of small biologically active molecules. Such strategy can decrease the number of candidate compounds providing a good starting point for chemical synthesis and biological screening. Therefore, in many cases, virtual screening can be used prior to expensive experimental highthroughput screening. For this reason, since the terms of computational virtual screening came out in the late 1990s, it has been considered as a novel and essential technology in drug discovery. Ligand- and structure-based virtual screenings have been successfully applied to drug discovery programs in various disease areas. After prospective results for various protein targets were obtained, the utility of virtual screening to identify compounds for nucleic acid-based receptors has been the focus of much attention.

Many computational docking programs that can automatically dock small molecules into a binding site of a target receptor with minimum input from an operator have been developed and their applicability has been proven (McInnes 2007). In addition, the development of various free or commercially available databases allows for easy use of virtual screening. Each year, many successful cases of virtual screening against various targets including protein and nucleotides have been reported in several major review papers (Whitty and Kumaravel 2006; Seifert and Lang 2008; Villoutreix et al. 2009).

H.-J. Park (✉) • S.-J. Park
School of Pharmacy, Sungkyunkwan University, Suwon 440-746, Korea
e-mail: hyunju85@skku.edu

J.D. Dinman (ed.), *Biophysical Approaches to Translational Control of Gene Expression*,   235
Biophysics for the Life Sciences 1, DOI 10.1007/978-1-4614-3991-2_12,
© Springer Science+Business Media New York 2013

Table 12.1 Widely used docking algorithms

Name	URL	Short summary
AutoDock	http://autodock.scripps.edu/	Free software of automated small-molecules docking tools (rigid receptor, flexible ligand)
DOCK	http://dock.compbio.ucsf.edu/	Free software, small molecules-various receptor docking (protein, DNA, and RNA), protein–protein interaction
FlexX	http://www.biosolveit.de/flexx/	Fast computer program for predicting protein–ligand interactions
GOLD	http://www.ccdc.cam.ac.uk/ products/life_sciences/gold/	Calculating the docking modes of small molecules in protein binding sites
Glide	http://www.schrodinger.com/	Fast flexible ligand docking program (small molecule-protein)
ICM	http://www.molsoft.com/	Automatic incorporation of flexibility into the ligand and receptor docking (protein, DNA, and RNA), protein–protein interaction
MORDOR	http://mondale.ucsf.edu/index_ main_frame.html	Docking program using algorithm considering flexibility of both nucleic acid receptor and ligand
Surflex-Dock	http://www.tripos.com/	"Protomol"-guided flexible molecular docking program

The docking process concerns the prediction of ligand conformation and orientation within a targeted binding site (active site) (Kitchen et al. 2004). In order to carry out docking calculations, it is necessary to know the 3-dimensional (3D) structure of a target and the nature of the binding site. 3D structures are identified by X-ray crystallography or NMR experiments and can be downloaded from Protein Data Bank (PDB) (Berman et al. 2000) or predicted by homology modeling using various programs. The next step is to define the binding site by known information or prediction. When input is prepared, chemical compounds present in the database are docked into the defined binding site of the selected target receptor. There are two purposes of docking studies. One is to predict accurate ligand binding orientation referred as "molecular modeling" and the other is to predict activity or binding affinity referred as "scoring." A docking result is evaluated by ligand binding orientation through visual inspection and by binding affinity using the scoring function. Scoring function is designed to predict the biological activity or binding affinity through the evaluation of interactions between ligands and receptor (Halperin et al. 2002). Among various docking programs (Table 12.1), the most widely used programs are AutoDock (Morris et al. 1998; Huey et al. 2007), DOCK (Ewing et al. 2001), FlexX (Kramer et al. 1999; Stahl 2000), Gold (Jones et al. 1995), Glide (Zhou et al. 2001), Internal Coordinate Mechanics (ICM) (Abagyan and Totrov 1994), and Surflex-Dock (Jain 2003; Kellenberger et al. 2004). AutoDock is an automated flexible docking program designed to predict how small molecules (ligands) bind into the receptor structure. AutoDock is performed with an empirical free energy force field based on a Lamarckian genetic algorithm, to bring about speedy prediction of conformation with calculated free energies of association (Morris et al. 1998). This program has application in X-ray crystallography, structure-based drug design, virtual screening, and protein–protein interaction studies. DOCK was introduced by the Kuntz group at UCSF and uses a rigid body docking algorithm and flexible ligand docking algorithm to dock the ligand into a negative image of the binding pocket (Ewing et al. 2001). FlexX is a fully automatic computer program for predicting protein–ligand interaction. FlexX can predict not only the lowest energy geometry of the complex of ligand with protein but also the binding affinities using an empirical scoring function (Böhm 1994). The descriptions of other widely used docking programs including commercial and free softwares are listed in Table 12.1.

12.2 Computational Programs to Predict 2D and 3D RNA Structures

In recent years, while the number of identified RNA sequences has rapidly increased, the number of known 3D structures has not kept pace with it. For this reason, there is a large gap between the number of known RNA sequences and 3D structures. For example, tRNA is one of the most structurally well characterized RNAs and its 1,101,833 characterized sequences are reported in the Rfam (Gardner et al. 2009), a database of sequence families of structural RNAs; however, only 170 structures are reported. To apply structure-based drug design approaches to the identification of RNA binding ligands, computational programs are required for prediction of RNA structures. Several computer programs have been developed for folding of RNA secondary structures, and modeling of RNA 2D and 3D structures. Those computational tools are summarized in Tables 12.2 and 12.3.

12.3 RNA-Targeted Virtual Screening

Many clinical antibiotics including macrolides, aminoglycosides, and others targeting bacterial ribosomal RNA (rRNA) reveal that RNA is the important target for drug development (Knowles et al. 2002; Hermann 2005). The appearance of drug resistance is the most critical problem in treating bacterial (Neu 1992) and viral infections (Perrin and Telenti 1998). RNAs contain highly conserved structural and functional motifs that may serve as drug targets, so the development of resistance to drugs targeting RNA can be slower than that to drugs targeting protein (Gallego and Varani 2001). In contrast to DNA, which mostly has a double-stranded helix structure, RNA is generally single-stranded and folds into complex 3D structures that provide unique pockets for small molecules (Foloppe et al. 2006), thus making RNA an attractive drug target.

Table 12.2 RNA secondary structure viewers/editors programs

Name	URL	Description
PseudoViewer (Han et al. 1999; Byun and Han 2009)	http://wilab.inha.ac.kr/pseudoviewer/	visualize RNA pseudoknot 2D structure automatically
RNAdraw (Matzura and Wennborg 1996)	http://www.rnadraw.com/	RNA 2D structure prediction, analysis, and visualization
RNA Movies (Evers and Giegerich 1999)	http://bibiserv.techfak.uni-bielefeld.de/rnamovies/	System for the visualization of RNA secondary structure spaces
RNAView/RnamlView (Yang et al. 2003)	http://ndbserver.rutgers.edu	Automatically generate 2D displays of RNA/DNA secondary structures with tertiary interactions
VARNA (Darty et al. 2009)	http://varna.lri.fr	Automated drawing, visualization, and annotation of the secondary structure of RNA
RNA Designer (Andronescu et al. 2004)	http://www.rnasoft.ca/cgi-bin/RNAsoft/RNAdesigner	Design de novo RNA structures with certain structural properties
RnaViz 2 (De Rijk et al. 2003)	http://rnaviz.sourceforge.net/	Drawings of RNA secondary structure with portability and structure annotation
Vienna RNA (Hofacker 2003)	http://www.tbi.univie.ac.at/~ivo/RNA/	Program for the prediction and comparison of RNA secondary structures

In spite of these advantages, RNA has not been focus of structure-based drug design, not only due to lack of information of RNA 3D structures, but also due to the sequence-specific unique features of the binding pockets in RNA. The binding pocket of protein usually lies deep in an internal region, separated from solvent. In RNA targets, the binding pockets are large and flat, located along the surface, and relatively exposed to solvent. Therefore, in using docking algorithms to discover RNA-binding drugs, the physicochemical properties of RNA, such as conformational flexibility, high negative charge, and solvation, should be taken into account more accurately than those for proteins. Despite such differences between protein and RNA targets, classical protein-ligand docking programs have sometimes successfully performed in RNA-targeted virtual screenings. For example, Kuntz first reported a successful virtual screening using DOCK 3.5 program to identify small molecules in the Available Chemicals Directory (ACD) that targeted an RNA double helix (Chen et al. 1997). Following the first study, many research groups have reported successful studies of virtual screening targeting RNA through protein-based docking methods (Filikov et al. 2000; Lind et al. 2002; Kang et al. 2004; Park et al. 2008, 2011). Meanwhile, since most of the available docking methods were developed for protein targets, their compliance with RNA targets has been evaluated extensively, and recently Li et al. demonstrated that two widely-used protein docking programs, GOLD 4.0 and Glide 5.0, are appropriate for structure-based drug design and virtual screening for RNA targets (Detering and Varani 2004; Li et al. 2010).

Table 12.3 Computer program for predicting RNA 3D structure

Name	URL	Short summary
Mode RNA (Rother et al. 2011)	http://genesilico.pl/moderna/	Automatic program for comparative modeling of RNA 3D structures. Require sequence alignment and structural template
MMB (formerly RNA Builder) (Flores et al. 2010)	https://simtk.org/home/rnatoolbox	Automatic program for generating model RNA structures from 2D and template structure by simulating in parallel at multiple levels of details
PARADISE (Assemble and S2S) (Jossinet et al. 2010; Jossinet and Westhof 2005)	http://paradise-ibmc.u-strasbg.fr/	Assemble: automatic program for intuitive graphical interface to study and construct complex 3D RNA structures. S2S: graphical system to easily display, manipulate, and interconnect heterogeneous RNA data like multiple sequence alignments, 2D and 3D structures
RNA2D3D (Martinez et al. 2008)	http://www-lmmb.ncifcrf.gov/	Manual manipulation program for RNA 3D modeling with conversion of RNA 2D structures to 3D
ERNA-3D (Zwieb and Müller 1997)	http://www.erna-3d.de/	Molecular Modeling Expert System to develop for the generation of models of RNA and protein molecules. Need to manual manipulation
MC-Fold/MC-Sym (Parisien and Major 2008)	http://www.major.iric.ca/MC-Pipeline/	A web-hosted service for prediction of RNA 3D structure with input 2D secondary
NAST (Jonikas et al. 2009)	https://simtk.org/home/nast	Predicting RNA 3D model from 2D structure
DMD/iFoldRNA (Sharma et al. 2008)	http://danger.med.unc.edu/tools.php	Web portal for interactive RNA folding simulations. Enable to perform molecular dynamics simulations of RNA using coarse-grained structural models (two-beads/residue)
YUP (Tan et al. 2006)	http://rumour.biology.gatech.edu/YammpWeb/	RNA molecular modeling program based on PYTHON. Offer such methods as Monte Carlo, Molecular Mechanics and Energy Minimization

Several modified algorithms were established for RNA-targeted virtual screening. For example, Morley and Afshar established an empirical scoring function that appropriately describes steric, polar, and charged interactions in RNA-ligand complexes (Morley and Afshar 2004). This scoring function was implanted in RiboDock program and validated for docking screening of a large-size chemical database. Virtual screening with this program successfully identified novel ligands for the bacterial ribosomal A-site (Foloppe et al. 2004). Moitessier et al. applied a unique method to the AutoDock docking process, taking into account inherent RNA flexibility and key water molecules, and this modified docking tool was validated by the docking of aminoglycosides to the ribosomal RNA A-site (Moitessier et al. 2006). In Kuntz et al.'s study, 70 experimental RNA-ligand complexes from PDB were re-docked using the DOCK 6 program, and the resulting docked conformations were rescored with AMBER generalized Born/solvent-accessible surface area (GB/SA) and Poisson–Blotzmann/SA (PB/SA) scoring functions in combination with explicit water molecules and sodium counterions. The success rate for reproducing experimental binding modes was significantly improved by using AMBER GB/SA or PB/SA (Lang et al. 2009).

As small molecules induce RNA conformational changes by binding to structures from preexisting dynamic ensembles (Puglisi et al. 1992; Zhang et al. 2007; Frank et al. 2009; Cruz and Westhof 2009; Fulle and Gohlke 2010), large conformational changes in RNA receptors after binding with small molecules during virtual screening must be accounted for. However, current protocols do not consider this point, limiting the range of target structures for the discovery of small molecules. Guilbert and James therefore developed a flexible docking program called MORDOR, which supports flexibility in the ligand and limited flexibility in the RNA for induced-fit binding. MORDOR performed well not only on 57 test sets of RNA-ligand complexes by retrieving experimental poses within 2.5 Å with a 74% success rate, but also in discovering ligands for novel targets such as human telomerase RNA (Guilbert and James 2008; Pinto et al. 2008). When comparing the practicality of DOCK 6 and MORDOR, DOCK 6 screened about 3–10 times faster than MORDOR, while MORDOR performed better on ligands with a large number of rotatable bonds (Lang et al. 2009). Most recently, Al-Hashimi's group reported a new strategy for virtual screening targeting an RNA dynamic ensemble constructed by combining NMR spectroscopy and computational molecular dynamics (MD) (Stelzer et al. 2011). This strategy takes into account large degrees of RNA conformational adaptation during virtual screening. This approach was applied to a search for small molecules for HIV-1 TAR (transactivation response element) RNA (*vide infra*).

12.4 Successful Application of Virtual Screening to RNA Receptors

12.4.1 Case 1: HIV-1 TAR RNA Hairpin

As one of the best characterized RNA-based regulatory machineries, the interaction of TAR with Tat protein has been focused upon as a target to inhibit HIV-1 replication (Yang 2005). Ligands that inhibit the TAR-Tat interaction can be developed as anti-AIDS drugs. Various molecules for TAR RNA that inhibit Tat binding have been identified, and many of those molecules were discovered by virtual screening of chemical libraries. In earlier studies, James used automatic docking methods (DOCK and ICM), employing flexible docking with Monte Carlo simulation and optimized scoring function, identifying phenothiazine compounds from ACD as

TAR RNA ligands (Filikov et al. 2000; Lind et al. 2002). To circumvent the limitation of incorporating RNA flexibility for structure-based virtual screening, and to find novel scaffolds for TAR-Tat inhibitors, ligand-based virtual screening was conducted. The SQUID fuzzy pharmacophore search method successfully identified a novel heterocyclic compound with an order of magnitude improved activity compared to known phenothiazine compounds (Renner et al. 2005).

In 2011, Al-Hashimi's group developed the most outstanding technology for RNA-targeted virtual screening through intensive generation of an ensemble of TAR RNA conformers and a robust re-docking validation test. Their strategy was effectively applied to virtual screening of a relatively small-sized chemical library containing 51,000 compounds, and they identified netilmicin, a selective HIV-1 TAR RNA binder, that inhibited HIV-1 replication in vivo (Stelzer et al. 2011). Details of their study are described below.

12.4.1.1 Validation of Docking Program

To test the accuracy of docking, a total of 96 small molecule-bound RNA structures downloaded from the PDB were assessed for docking performance. All docking performances were carried out using the ICM docking program (Abagyan and Totrov 1994) and results were evaluated by ICM Score and RMSD between native ligand (extracted from RNA structures) and predicted orientation after docking. The binding energies based on the ICM Score were predicted with high accuracy ($R = 0.71$). In more than half of cases (53%), the predicted conformations matched the X-ray or NMR structures within 2.5 Å RMSD.

12.4.1.2 Preparation of HIV-1 TAR RNA Ensemble

To decide on a suitable RNA ensemble, the accuracies of docking with two sets of RNA ensembles of HIV-1 TAR were compared. One ensemble, named TAR^{NMR-MD}, consisted of 20 conformers generated by SAS selection (select-and-sample strategy, Frank et al. 2009). To construct TAR^{NMR-MD}, HIV-1 TAR structure (PDB id: 1ANR, Aboul-ela et al. 1996) was downloaded and molecular dynamics simulations were performed by measuring NMR residual dipolar couplings (RDC) in elongated RNA. The other RNA ensemble, named TAR^{MD}, consisted of 20 randomly selected snapshots from an 80-ns MD simulation of apo-TAR with backbone RMSDs ranging from 3 to 80 Å. The X-ray structure (PDB id: 397D, Ippolito and Steitz 1998) and 20 NMR structures (PDB id: 1ANR, Aboul-ela et al. 1996) of apo-TAR were downloaded from the PDB. A test set ligand library containing 38 known ligands for TAR RNA was obtained from the published literatures. To test the accuracy of the RNA ensemble, virtual screening of the test set ligands targeting the RNA ensembles, TAR^{NMR-MD} and TAR^{MD}, was conducted, respectively. Virtual screening was conducted using ICM after binding pockets were predicted by the ICM PocketFinder module based on the calculated surface area and volume of cavities on the receptor surface.

The results suggested that TAR^{NMR-MD} improved the accuracy of docking compared with TAR^{MD}. From this result TAR^{NMR-MD} was chosen for further virtual screening to identify small molecule TAR-Tat inhibitors.

12.4.1.3 Virtual Screening Against TAR Dynamic Ensemble

A chemical database consisting of 49,166 compounds was obtained from the Center for Chemical Genomics at the University of Michigan and 2,060 compounds from the author's in-house library. Based on the ICM Score, the top 57 commercially available hit compounds were selected and their binding activities were tested by fluorescence-based assays. This identified six small molecules that bound to TAR with high affinity (K_d=55 nM to 122 μM) and inhibited TAR interaction with Tat (K_i=710 nM to 169 μM) in vitro. Among them, netilmicin (K_d=~1.4 μM) bound to HIV-1 TAR with the highest selectivity over HIV-2 TAR. Netilmicin repressed Tat-mediated transactivation of the HIV-1 promoter through its interaction with TAR in live human T-cells and inhibited HIV-1 replication in the HIV-1 indicator cell line TZM-bl.

12.4.2 Case 2: RNA Pseudoknots

Ribosomal −1 frameshifting (−1 FS) is an essential event during translation for the synthesis of two or more proteins encoded by overlapping reading frames on a single mRNA (Dinman and Berry 2007) in many RNA viruses such as retroviruses, corona-viruses, yeast, plant virus, and even bacteria (Jacks and Varmus 1985; Brierley et al. 1991; Chamorro et al. 1992; Tzeng et al. 1992; ten Dam et al. 1994; Kang and Tinoco 1997; Jacobs et al. 2007). Two cis-acting elements are required to regulate −1 FS. One is a slippery sequence where ribosome-associated tRNAs slip, and the other is RNA secondary structure such as a hairpin or pseudoknot that promotes ribosome pausing. Thermodynamic or kinetic control of RNA secondary structure folding may be important in regulating the efficiency of −1 FS. Human immunodeficiency virus type 1 (HIV-1) utilizes −1 FS to regulate the expression ratio of Gag to Gag-Pol, which is critical for the production of infectious virion particles (Paulus et al. 1999). The RNA stem-loop sequence that is involved in −1 FS of HIV-1 is highly conserved in the main subtypes of HIV-1 (Gareiss and Miller 2009). Mutations of this sequence reduce −1 FS and decrease viral infectivity and replication (Baril et al. 2003; Dulude et al. 2006). In severe acute respiratory syndrome coronavirus (SARS-CoV), replicase genes mainly encode two large replicative polyproteins (pp1a and pp1ab) which are expressed by two partially overlapped open reading frames ORF 1a and ORF 1b. As ORF 1b has no independent translation initiation site, polyprotein pp1ab encoded by ORF 1b is only translated as a fused protein form with ORF 1a through −1 FS. As pp1ab includes RNA-dependent RNA polymerase (RdRp) and other replication components which are important proteins for viral replication,

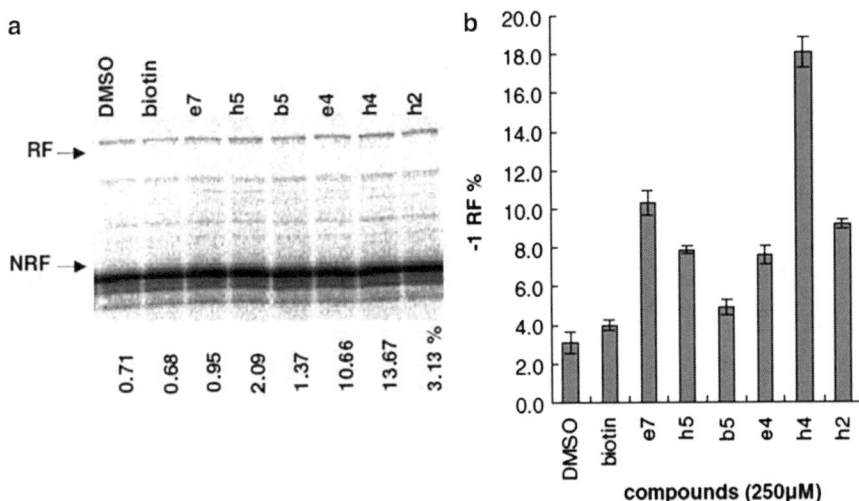

Fig. 12.1 −1 frameshifting efficiencies induced by biotin aptamer RNA pseudoknot in the presence of compounds (250 μM) determined by SDS–PAGE (a) and dual luciferase assay (b). −1 FS % values are shown at the *bottom* of the autoradiogram of SDS–PAGE. Each −1 FS % value from dual luciferase assay is the average of triplicate experiments

−1 FS is essential for the synthesis of enzymatic proteins. The stability of the RNA pseudoknot that induces −1 FS in the SARS-CoV (SARS-pseudoknot) also has a dramatic effect on −1 FS efficiency. Therefore, the RNA secondary structure in the −1 FS site has emerged as an attractive target for drug development (Baranov et al 2005; Plant et al 2005; Su et al. 2005).

Park first conducted virtual screening against the RNA pseudoknot in the −1 FS site to discover ligands that change −1 FS efficiency (Park et al. 2008). In this pilot study, they used the −1 FS system containing biotin aptamer RNA pseudoknot as the RNA secondary structure element. Biotin RNA aptamer was the only ligand-bound RNA pseudoknot structure (PDB id: 1F27) determined by X-crystallography (Nix et al. 2000). Park et al. used the conventional 2D and 3D pharmacophore search program Unity (Martin 1992) for primary database filtering and the FlexX docking program for final docking screening. RNA flexibility was not considered and ligand flexibility was given during FlexX run. Out of about 80,000 compounds in the chemical DB, they obtained 37 hits which increased −1 FS. Compound **h4** showed the highest activity in the in vitro transcription and translation coupled assay (Fig. 12.1). The FlexX-docked pose of **h4** is shown in Fig. 12.2. The docking mode of **h4** is similar to that of biotin; however, **h4** forms a stronger interaction with the receptor RNA (Fig. 12.2a). Compound **h4** forms hydrogen bonds with O4' and the 2-carbonyl oxygen of uracil ring of U7 which is one of the critical residues for interaction with biotin in the X-ray structure, and an additional hydrogen bond with ribose O2' atom of A16 (Fig. 12.2b). These interactions may alter the stability of the RNA pseudoknot and increase the stalled time of the ribosome on the slippery site, thus increasing the rates of −1 FS.

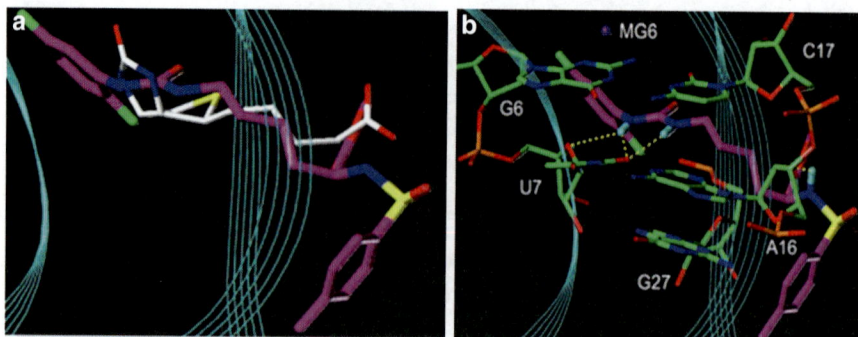

Fig. 12.2 (**a**) Overlay of FlexX-docked pose of **h4** and X-ray pose of biotin in the aptamer RNA pseudoknot. The ligand is rendered in capped stick. Carbon atoms of **h4** are *magenta* and those of biotin are *white. Cyan* lined-ribbon represents the backbone of RNA. (**b**) Docked model of **h4** in complex with biotin-pseudoknot complex. The residues in the active site are rendered in stick. Carbon atoms of pseudoknot are *green, oxygen red, nitrogen blue*, and *phosphorus orange. Yellow dashed lines* are hydrogen bonds

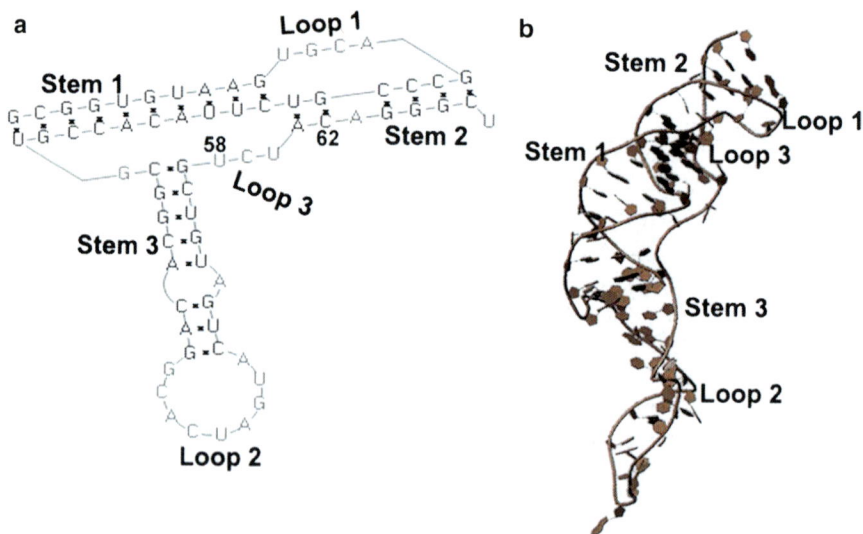

Fig. 12.3 (**a**) Two-dimensional model of SARS-pseudoknot generated by the PSEUDOVIEWER program. (**b**) Three-dimensional structural model of the SARS-pseudoknot used in this study. It was optimized by molecular dynamics simulation using the Amber 8.0 program. *Brown ribbon* renders the phosphate backbone of the RNA pseudoknot

Based on these successful results, Park applied the virtual screening strategy to discover ligands for the SARS-pseudoknot, even though its 3D structure was not completely determined. It was known that the SARS-pseudoknot has a unique 3 stem-3 loop structure. They built a 3D model using the RNA pseudoknot predicting program (PSEUDOVIEWER) and Sybyl molecular modeling software, and then

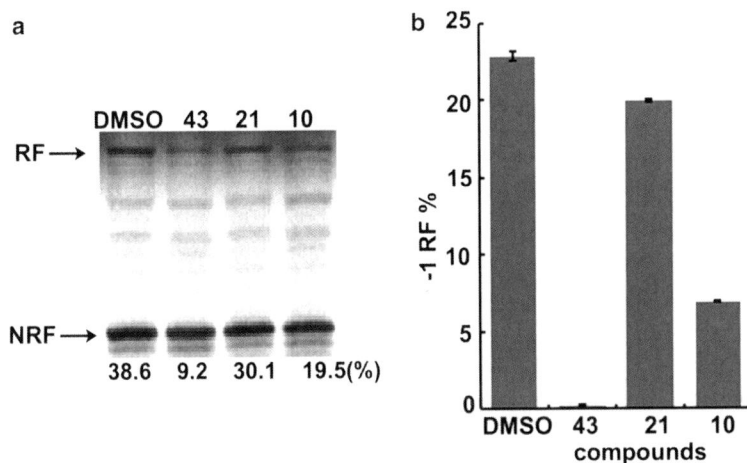

Fig. 12.4 Measurements of −1 FS efficiencies by in vitro TNT assay. (**a**) The −1 FS efficiencies (%) in the presence of **43**, **21**, and **10** obtained from SDS-PAGE analysis. The nonframeshifting product (NRF) is the renilla luciferase protein, and the frameshifting product (RF) is a firefly luciferase-renilla luciferase fusion protein. (**b**) The −1 FS efficiencies obtained from dual luciferase assays

optimized the RNA structure by AMBER molecular dynamics simulation (Case et al. 2005) (Fig. 12.3). Using the DOCK 4.0 program, flexible docking of a commercially-available chemical DB (Leadquest) was conducted. The chemical DB was the same as that previously used for screening against the biotin aptamer RNA pseudoknot. A total of 35 compounds inhibited −1 FS, and three structurally analogous compounds (**43**, **21**, and **10** in Fig. 12.4) reduced −1 FS selectively. Compound **43** was the most active, decreasing −1 FS efficiency by 80%. In HEK 293 cells, **43** inhibited −1 FS in a concentration-dependent manner with an IC_{50} value of approximately 0.45 μM (Park et al. 2011). The docking model of **43** in complex with the SARS-pseudoknot is shown in Fig. 12.5, and reveals that **43** interacts with various residues including key residues to maintain −1 frameshifting efficiency by hydrogen bonds. Compound **43** interacts with the carbonyl oxygen atom (O2) of the U58 uracil base by hydrogen bond, and the nitrogen atom in the thiazole ring of **43** forms a hydrogen bond (2.8 Å) with the 2′-OH group of ribose of C62. The hydrogen bond between the thiazole moiety and the receptor pseudoknot was identified as one of the key intermolecular interactions.

12.4.3 Case 3: Riboswitches (Metabolite-Sensing mRNAs)

Riboswitches are highly organized domains within 5′-UTRs of mRNAs and undergo alternate conformational switches. Riboswitches consist of two domains, an aptamer domain that is a binding site for an effector metabolite, and an expression platform that prompts changes in gene expression. Upon metabolite binding, one of the riboswitch conformers is stabilized, and is capable of controlling expression of a

Fig. 12.5 Docked model of
43: SARS-psuedoknot
complex generated by DOCK
4.0. Several residues in the
binding site are rendered in
capped stick (*brown carbon*),
43 in ball and stick (*green
carbon*), and *red dashed lines*
indicate hydrogen bonds

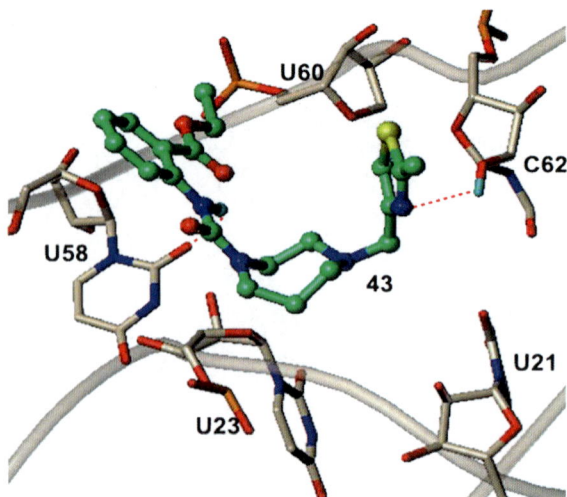

downstream gene either at the transcriptional or translational level (Miranda-Ríos et al. 2001; Mironov et al. 2002; Gilbert and Batey 2005) in bacteria (Winkler et al. 2002; Mandal et al. 2003) as well as in some plants and fungi (Sudarsan et al. 2003; Bocobza and Aharoni 2008). Therefore, the aptamer domain is an essential element of gene regulation. Riboswitches were originally discovered as an antibiotic target (Blount and Breaker 2006; Lee et al. 2009; Kim et al. 2009) and also have the potential to be developed into designer riboswitches for genetics studies (Suess and Weigand 2008). About 20 classes of riboswitches have been reported and a large number (probably up to 100) of new classes may await discovery. Among them, guanine riboswitches are one of the validated targets for development of new antibacterial drugs. X-ray crystal structures of more than 10 riboswitches have been determined, and thus structure-based drug design approaches can be applied to this target (Serganov 2010). Daldrop et al. tried virtual screening to discover novel ligands targeting the purine riboswitch using the program DOCK3.5.54 with minor modification of the scoring function (Daldrop et al. 2011). The *Bacillus subtilis xpt-pbuX* guanine riboswitch carrying a C74U mutation (called GRA, PDB id: 2G9C) (Gilbert et al. 2006) in complex with pyrimidine-2,4,6-triamine was used as a target receptor for docking.

12.4.3.1 Validation of Docking Program

All docking and virtual screening studies were carried out using a slightly modified version of the DOCK3.5.54 program that incorporated RNA-specific parameters to calculate van der Waals and electrostatic energies. This group first tested whether native ligand was correctly reproduced in terms of its binding geometry after self-docking. The RMSD between native ligand and the docking result was measured.

RMSD was less than 0.34 Å, which showed that the docked model was close to native. Secondly, the accuracy of the prediction was tested by docking with known ligands and decoys. The test set consisted of eight known ligands, with binding affinities ranged from 0.01 to 100 µM. For all 15 decoys, no binding was detected at up to 300 µM; except for guanine that was tested up to its solubility limit. All compounds in the test set docked into the active site and the results were analyzed by sorting the docking scores. Seven out of the eight top-scoring compounds were true ligands and all eight compounds with lowest docking energy scores were decoys.

12.4.3.2 Virtual Screening

They used their own in-house database, which included a commercially available 2,592 unique compounds. To evaluate the accuracy of binding prediction, all compounds in the database were docked into the active site and ordered by docking score. According to this list, the true positive rate (fraction of known compounds, ligands, and decoy) was plotted against the false positive (fraction of unassigned database compounds) to get a receiver operation characteristic (ROC) curve. Accuracy was calculated by measuring the area under curve (AUC) of a ROC curve value. The AUC value of ligands was 0.98 and that of decoy was 0.75, suggesting that known ligands were perfectly predicted by virtual screening and decoys were also enriched compared to random (AUC 0.5). From this result, five compounds were selected to examine their binding affinity and modes. Three compounds out of five were analogs of known ligands and two compounds were novel scaffolds. To determine the binding affinities, fluorescence assays and isothermal titration calorimetry (ITC) were used. Four out of five chosen compounds bound to GRA with affinities in the micromolar range.

12.5 Concluding Remarks

The rapidly increasing number of RNA crystal structures in the PDB, and the biological function studies on a variety of RNA structures have provided a basis for structure-based virtual screening targeting RNA. Due to several features of RNA (conformational flexibility, high negative charge, and solvation) that differ from those of proteins, researchers have observed that the conventional protein-ligand docking programs have limitations in accurately predicting RNA-ligand interactions. In some studies, active molecules have been fortuitously obtained by protein-targeted docking programs. However, continuing progress has been made in the development of docking algorithms or scoring functions optimized for RNA receptors, which makes various RNA targets more amenable for structure-based drug design. As an example, consideration of receptor flexibility is important not only for RNA-based receptors, but has also been a critical issue for protein receptors for a long time. Massive efforts have been made to incorporate protein flexibility into docking process, but it exponentially inflates the potential search space and became impractical. Al-Hashimi's approaches to use RNA dynamic ensemble (Stelzer et al. 2011) can be practically applied to other RNA targets. RNA receptors are usually smaller than protein receptors in size, so the process to generate RNA ensembles is not extremely computationally expensive.

Structure- or ligand-based virtual screening has identified a plethora of RNA binding ligands from in-house and commercially available chemical databases, originally designed and prepared for protein targets. Actually, some commercial chemical databases possess compounds which only satisfy "Lipinski's rule of five (Lipinski et al. 1997)". In comparing the activities of small molecule RNA binders identified by virtual screening, we realized that hits with nanomolar activity were very rare, and overall their activities are in the high micromolar range. These results were probably caused by limitations of currently available chemical database, not only by those of protein-friendly docking algorithms. In protein-targeted virtual screening of chemical databases, many hits with nanomolar or submicromolar activity were identified. Thus, another possible way to improve the hit rate of RNA-targeted virtual screening is to prepare an RNA-focused chemical database. Considering the physicochemical properties of RNA binders, a modified "Lipinski's rule of five" needs to be applied to database filtering during virtual screening (Aboulela 2010).

Acknowledgements This work is supported by Basic Science Research Program through the National Research Foundation of Korea (NRF) funded by the Ministry of Education, Science and Technology (Grant 2011-0014385).

References

Abagyan R, Totrov M (1994) Biased probability Monte Carlo conformational searches and electrostatic calculations for peptides and proteins. J Mol Biol 235(3):983–1002

Aboul-ela F (2010) Strategies for the design of RNA-binding small molecules. Future Med Chem 2(1):93–119

Aboul-ela F, Karn J, Varani G (1996) Structure of HIV-1 TAR RNA in the absence of ligands reveals a novel conformation of the trinucleotide bulge. Nucleic Acids Res 24(20):3974–3981

Andronescu M, Fejes AP, Hutter F, Hoos HH, Condon A (2004) A new algorithm for RNA secondary structure design. J Mol Biol 336(3):607–624

Baranov PV, Henderson CM, Anderson CB, Gesteland RF, Atkins JF, Howard MT (2005) Programmed ribosomal frameshifting in decoding the SARS-CoV genome. Virology 332(2):498–510

Baril M, Dulude D, Gendron K, Lemay G, Brakier-Gingras L (2003) Efficiency of a programmed −1 ribosomal frameshift in the different subtypes of the human immunodeficiency virus type 1 group M. RNA 9:1246–12453

Berman HM, Bhat TN, Bourne PE, Feng Z, Gilliland G, Weissig H, Westbrook J (2000) The Protein Data Bank and the challenge of structural genomics. Nat Struct Biol 7(suppl A):957–959

Blount KF, Breaker RR (2006) Riboswitches as antibacterial drug targets. Nat Biotechnol 24(12):1558–1564

Bocobza SE, Aharoni A (2008) Switching the light on plant riboswitches. Trends Plant Sci 13(10):526–533

Böhm HJ (1994) The development of a simple empirical scoring function to estimate the binding constant for a protein-ligand complex of known three-dimensional structure. J Comput Aided Mol Des 8(3):243–256

Brierley I, Rolley NJ, Jenner AJ, Inglis SC (1991) Mutational analysis of the RNA pseudoknot component of a coronavirus ribosomal frameshifting signal. J Mol Biol 220(4):889–902

Byun Y, Han K (2009) PseudoViewer3: generating planar drawings of large-scale RNA structures with pseudoknots. Bioinformatics 25(11):1435–1437

Case DA, Cheatham TE III, Darden T, Gohlke H, Luo R, Merz KM Jr, Onufriev A, Simmerling C, Wang B, Woods RJ (2005) The Amber biomolecular simulation programs. J Comput Chem 26(16):1668–1688

Chamorro M, Parkin N, Varmus HE (1992) An RNA pseudoknot and an optimal heptameric shift site are required for highly efficient ribosomal frameshifting on a retroviral messenger RNA. Proc Natl Acad Sci U S A 89(2):713–717

Chen Q, Shafer RH, Kuntz ID (1997) Structure-based discovery of ligands targeted to the RNA double helix. Biochemistry 36(38):11402–11407

Cruz JA, Westhof E (2009) The dynamic landscapes of RNA architecture. Cell 136(4):604–609

Daldrop P, Reyes FE, Robinson DA, Hammond CM, Lilley DM, Batey RT, Brenk R (2011) Novel ligands for a purine riboswitch discovered by RNA-ligand docking. Chem Biol 18(3):324–335

Darty K, Denise A, Ponty Y (2009) VARNA: interactive drawing and editing of the RNA secondary structure. Bioinformatics 25(1):974–1975

De Rijk P, Wuyts J, De Wachter R (2003) RnaViz 2: an improved representation of RNA secondary structure. Bioinformatics 19(2):299–300

Detering C, Varani G (2004) Validation of automated docking programs for docking and database screening against RNA drug targets. J Med Chem 47(17):4188–4201

Dinman JD, Berry MJ (2007) Regulation of termination and recoding. In: Mathews MB, Soneberg N, Hershey JWB (eds) Translational control in biology and medicine, 1st edn. Cold Spring Harbor, New York

Dulude D, Berchiche YA, Gendron K, Brakier-Gingras L, Heveker N (2006) Decreasing the frameshift efficiency translates into an equivalent reduction of the replication of the human immunodeficiency virus type 1. Virology 345:127–136

Evers D, Giegerich R (1999) RNA movies: visualizing RNA secondary structure spaces. Bioinformatics 15(1):32–37

Ewing TJ, Makino S, Skillman AG, Kuntz ID (2001) DOCK 4.0: search strategies for automated molecular docking of flexible molecule databases. J Comput Aided Mol Des 15(5):411–428

Filikov AV, Mohan V, Vickers TA, Griffey RH, Cook PD, Abagyan RA, James TL (2000) Identification of ligands for RNA targets via structure-based virtual screening: HIV-1 TAR. J Comput Aided Mol Des 14(6):593–610

Flores SC, Wan Y, Russell R, Altman RB (2010) Predicting RNA structure by multiple template homology modeling. Pac Symp Biocomput 2010:216–227

Foloppe N, Chen IJ, Davis B, Hold A, Morley D, Howes R (2004) A structure-based strategy to identify new molecular scaffolds targeting the bacterial ribosomal A-site. Bioorg Med Chem 12:935–947

Foloppe N, Matassova N, Aboul-Ela F (2006) Towards the discovery of drug-like RNA ligands? Drug Discov Today 11(21–22):1019–1027

Frank AT, Stelzer AC, Al-Hashimi HM, Andricioaei I (2009) Constructing RNA dynamical ensembles by combining MD and motionally decoupled NMR RDCs: new insights into RNA dynamics and adaptive ligand recognition. Nucleic Acids Res 37(11):3670–3679

Fulle S, Gohlke H (2010) Molecular recognition of RNA: challenges for modelling interactions and plasticity. J Mol Recognit 23(2):220–231

Gallego J, Varani G (2001) Targeting RNA with small-molecule drugs: therapeutic promise and chemical challenges. Acc Chem Res 34(10):836–843

Gardner PP, Daub J, Tate JG, Nawrocki EP, Kolbe DL, Lindgreen S, Wilkinson AC, Finn RD, Griffiths-Jones S, Eddy SR, Bateman A (2009) Rfam: updates to the RNA families database. Nucleic Acids Res 37(Database issue):D136–D140

Gareiss PC, Miller BL (2009) Ribosomal frameshifting: an emerging drug target for HIV. Curr Opin Investig Drugs 10(2):121–128

Gilbert SD, Batey RT (2005) Riboswitches: natural SELEXion. Cell Mol Life Sci 62(21):2401–2404

Gilbert SD, Mediatore SJ, Batey RT (2006) Modified pyrimidines specifically bind the purine riboswitch. J Am Chem Soc 128(44):14214–14215

Guilbert C, James TL (2008) Docking to RNA via root-mean-square-deviation-driven energy minimization with flexible ligands and flexible targets. J Chem Inf Model 48(6):1257–1268

Halperin I, Ma B, Wolfson H, Nussinov R (2002) Principles of docking: an overview of search algorithms and a guide to scoring functions. Proteins 47(4):409–443

Han K, Kim D, Kim HJ (1999) A vector-based method for drawing RNA secondary structure. Bioinformatics 15(4):286–297

Hermann T (2005) Drugs targeting the ribosome. Curr Opin Struct Biol 15(3):355–366

Hofacker IL (2003) Vienna RNA secondary structure server. Nucleic Acids Res 31(13):3429–3431

Huey R, Morris GM, Olson AJ, Goodsell DS (2007) A semiempirical free energy force field with charge-based desolvation. J Comput Chem 28(6):1145–1152

Ippolito JA, Steitz TA (1998) A 1.3-Å resolution crystal structure of the HIV-1 trans-activation response region RNA stem reveals a metal ion-dependent bulge conformation. Proc Natl Acad Sci U S A 95(17):9819–9824

Jacks T, Varmus HE (1985) Expression of the Rous sarcoma virus pol gene by ribosomal frameshifting. Science 230(4731):1237–1242

Jacobs JL, Belew AT, Rakauskaite R, Dinman JD (2007) Identification of functional, endogenous programmed −1 ribosomal frameshift signals in the genome of Saccharomyces cerevisiae. Nucleic Acids Res 35(1):165–174

Jain AN (2003) Surflex: fully automatic flexible molecular docking using a molecular similarity-based search engine. J Med Chem 46(4):499–511

Jones G, Willett P, Glen RC (1995) Molecular recognition of receptor sites using a genetic algorithm with a description of desolvation. J Mol Biol 245(1):43–53

Jonikas MA, Radmer RJ, Laederach A, Das R, Pearlman S, Herschlag D, Altman RB (2009) Coarse-grained modeling of large RNA molecules with knowledge-based potentials and structural filters. RNA 15(2):189–199

Jossinet F, Westhof E (2005) Sequence to structure (S2S): display, manipulate and interconnect RNA data from sequence to structure. Bioinformatics 21(15):3320–3321

Jossinet F, Ludwig TE, Westhof E (2010) Assemble: an interactive graphical tool to analyze and build RNA architectures at the 2D and 3D levels. Bioinformatics 26(16):2057–2059

Kang H, Tinoco I Jr (1997) A mutant RNA pseudoknot that promotes ribosomal frameshifting in mouse mammary tumor virus. Nucleic Acids Res 25(10):1943–1949

Kang X, Shafer RH, Kuntz ID (2004) Calculation of ligand-nucleic acid binding free energies with the generalized-born model in DOCK. Biopolymers 73(2):192–204

Kellenberger E, Rodrigo J, Muller P, Rognan D (2004) Comparative evaluation of eight docking tools for docking and virtual screening accuracy. Proteins 57(2):225–242

Kim JN, Blount KF, Puskarz I, Lim J, Link KH, Breaker R (2009) Design and antimicrobial action of purine analogues that bind Guanine riboswitches. ACS Chem Biol 4(11):915–927

Kitchen DB, Decornez H, Furr JR, Bajorath J (2004) Docking and scoring in virtual screening for drug discovery: methods and applications. Nat Rev Drug Discov 3(11):935–949

Knowles DJ, Foloppe N, Matassova NB, Murchie AI (2002) The bacterial ribosome, a promising focus for structure-based drug design. Curr Opin Pharmacol 2(5):501–506

Kramer B, Rarey M, Lengauer T (1999) Evaluation of the FLEXX incremental construction algorithm for protein-ligand docking. Proteins 37(2):228–241

Lang PT, Brozell SR, Mukherjee S, Pettersen EF, Meng EC, Thomas V, Rizzo RC, Case DA, James TL, Kuntz ID (2009) DOCK 6: combining techniques to model RNA-small molecule complexes. RNA 15(6):1219–1230

Lee ER, Blount KF, Breaker RR (2009) Roseoflavin is a natural antibacterial compound that binds to FMN riboswitches and regulates gene expression. RNA Biol 6(2):187–194

Li Y, Shen J, Sun X, Li W, Liu G, Tang Y (2010) Accuracy assessment of protein-based docking programs against RNA targets. J Chem Inf Model 50(6):1134–1146

Lind KE, Du Z, Fujinaga K, Peterlin BM, James TL (2002) Structure-based computational database screening, in vitro assay, and NMR assessment of compounds that target TAR RNA. Chem Biol 9(2):185–193

Lipinski CA, Lombardo F, Dominy BW, Feeney PJ (1997) Experimental and computational approaches to estimate solubility and permeability in drug discovery and development settings. Adv Drug Deliv Rev 23:3–25

Mandal M, Boese B, Barrick JE, Winkler WC, Breaker RR (2003) Riboswitches control fundamental biochemical pathways in Bacillus subtilis and other bacteria. Cell 113(5):577–586

Martin YC (1992) 3D database searching in drug design. J Med Chem 35(12):2145–2154

Martinez HM, Maizel JV Jr, Shapiro BA (2008) RNA2D3D: a program for generating, viewing, and comparing 3-dimensional models of RNA. J Biomol Struct Dyn 25(6):669–683

Matzura O, Wennborg A (1996) RNAdraw: an integrated program for RNA secondary structure calculation and analysis under 32-bit Microsoft Windows. Comput Appl Biosci 12(3):247–249

McInnes C (2007) Virtual screening strategies in drug discovery. Curr Opin Chem Biol 11(5):494–502

Miranda-Ríos J, Navarro M, Soberón M (2001) A conserved RNA structure (thi box) is involved in regulation of thiamin biosynthetic gene expression in bacteria. Proc Natl Acad Sci U S A 98(17):9736–9741

Mironov AS, Gusarov I, Rafikov R, Lopez LE, Shatalin K, Kreneva RA, Perumov DA, Nudler E (2002) Sensing small molecules by nascent RNA: a mechanism to control transcription in bacteria. Cell 111(5):747–756

Moitessier N, Westhof E, Hanessian S (2006) Docking of aminoglycosides to hydrated and flexible RNA. J Med Chem 49(3):1023–1033

Morley SD, Afshar M (2004) Validation of an empirical RNA-ligand scoring function for fast flexible docking using Ribodock. J Comput Aided Mol Des 18:189–208

Morris GM, Goodsell DS, Halliday RS, Huey R, Hart WE, Belew RK, Olson AJ (1998) Automated docking using a lamarckian genetic algorithm and an empirical binding free energy function. J Comput Chem 19:1639–1662

Neu HC (1992) The crisis in antibiotic resistance. Science 257(5073):1064–1073

Nix J, Sussman D, Wilson C (2000) The 1.3 Å crystal structure of a biotin-binding pseudoknot and the basis for RNA molecular recognition. J Mol Biol 296(5):1235–1244

Parisien M, Major F (2008) The MC-fold and MC-Sym pipeline infers RNA structure from sequence data. Nature 452(7183):51–55

Park SJ, Jung YH, Kim YG, Park HJ (2008) Identification of novel ligands for the RNA pseudoknot that regulate −1 ribosomal frameshifting. Bioorg Med Chem 16(8):4676–4684

Park SJ, Kim YG, Park HJ (2011) Identification of RNA pseudoknot-binding ligand that inhibits the −1 ribosomal frameshifting of SARS-Coronavirus by structure-based virtual screening. J Am Chem Soc 133(26):10094–10100

Paulus C, Hellebrand S, Tessmer U, Wolf H, Kräusslich HG, Wagner R (1999) Competitive inhibition of human immunodeficiency virus type-1 protease by the Gag-Pol transframe protein. J Biol Chem 30:21539–21543

Perrin L, Telenti A (1998) HIV treatment failure: testing for HIV resistance in clinical practice. Science 280(5371):1871–1873

Pinto IG, Guilbert C, Ulyanov NB, Stearns J, James TL (2008) Discovery of ligands for a novel target, the human telomerase RNA, based on flexible-target virtual screening and NMR. J Med Chem 51:7205–7215

Plant EP, Pérez-Alvarado GC, Jacobs JL, Mukhopadhyay B, Hennig M, Dinman JD (2005) A three-stemmed mRNA pseudoknot in the SARS coronavirus frameshift signal. PLoS Biol 3(6):e172

Puglisi JD, Tan R, Calnan BJ, Frankel AD, Williamson JR (1992) Conformation of the TAR RNA-arginine complex by NMR spectroscopy. Science 257(5066):76–80

Renner S, Ludwig V, Boden O, Scheffer U, Göbel M, Schneider G (2005) New inhibitors of the tat-tar RNA interaction found with a "Fuzzy" Pharmacophore model. Chembiochem 6:1119–1125

Rother M, Rother K, Puton T, Bujnicki JM (2011) ModeRNA: a tool for comparative modeling of RNA 3D structure. Nucleic Acids Res 39(10):4007–4022

Seifert MH, Lang M (2008) Essential factors for successful virtual screening. Mini Rev Med Chem 8(1):63–72

Serganov A (2010) Determination of riboswitch structures: light at the end of the tunnel? RNA Biol 7:98–103

Sharma S, Ding F, Dokholyan NV (2008) iFoldRNA: three-dimensional RNA structure prediction and folding. Bioinformatics 24(17):1951–1952

Stahl M (2000) Modifications of the scoring function in FlexX for virtual screening applications. Persp Drug Discov Des 20:83–98

Stelzer AC, Frank AT, Kratz JD, Swanson MD, Gonzalez-Hernandez MJ, Lee J, Andricioaei I, Markovitz DM, Al-Hashimi HM (2011) Discovery of selective bioactive small molecules by targeting an RNA dynamic ensemble. Nat Chem Biol 7(8):553–559

Su MC, Chang CT, Chu CH, Tsai CH, Chang KY (2005) An atypical RNA pseudoknot stimulator and an upstream attenuation signal for −1 ribosomal frameshifting of SARS coronavirus. Nucleic Acids Res 33(13):4265–4275

Sudarsan N, Barrick JE, Breaker RR (2003) Metabolite-binding RNA domains are present in the genes of eukaryotes. RNA 9(6):644–647

Suess B, Weigand JE (2008) Engineered riboswitches: overview, problems and trends. RNA Biol 5:24–29

Tan RKZ, Petrov AS, Harvey SC (2006) YUP: a molecular simulation program for coarse-grained and multiscaled models. J Chem Theory Comput 2(3):529–540

ten Dam E, Brierley I, Inglis S, Pleij C (1994) Identification and analysis of the pseudoknot-containing gag-pro ribosomal frameshift signal of simian retrovirus-1. Nucleic Acids Res 22(12):2304–2310

Tzeng TH, Tu CL, Bruenn JA (1992) Ribosomal frameshifting requires a pseudoknot in the Saccharomyces cerevisiae double-stranded RNA virus. J Virol 66(2):999–1006

Villoutreix BO, Eudes R, Miteva MA (2009) Structure-based virtual ligand screening: recent success stories. Comb Chem High Throughput Screen 12(10):1000–1016

Whitty A, Kumaravel G (2006) Between a rock and a hard place? Nat Chem Biol 2(3):112–118

Winkler W, Nahvi A, Breaker RR (2002) Thiamine derivatives bind messenger RNAs directly to regulate bacterial gene expression. Nature 419(6910):952–956

Yang M (2005) Discoveries of Tat-TAR interaction inhibitors for HIV-1. Curr Drug Targets Infect Disord 5(4):433–444

Yang H, Jossinet F, Leontis N, Chen L, Westbrook J, Berman H, Westhof E (2003) Tools for the automatic identification and classification of RNA base pairs. Nucleic Acids Res 31(13):3450–3460

Zhang Q, Stelzer AC, Fisher CK, Al-Hashimi HM (2007) Visualizing spatially correlated dynamics that directs RNA conformational transitions. Nature 450(7173):1263–1267

Zhou R, Friesner RA, Ghosh A, Rizzo RC, Jorgensen WL, Levy RM (2001) New linear interaction method for binding affinity calculations using a continuum solvent model. J Phys Chem B 105:10388–10397

Zwieb C, Müller F (1997) Three-dimensional comparative modeling of RNA. Nucleic Acids Symp Ser 36:69–71

Chapter 13
The "Fifth" RNA Nucleotide: A Role for Ribosomal RNA Pseudouridylation in Control of Gene Expression at the Translational Level

Mary McMahon, Cristian Bellodi, and Davide Ruggero

13.1 Introduction

Ribosomal RNA (rRNA) undergoes numerous types of posttranscriptional RNA modifications including base methylation, ribose methylation, and the isomerization of uridine to pseudouridine (Ψ) (Limbach et al. 1994), a process known as pseudouridylation. Modifications are present in ribosomes of all organisms ranging from bacteria to eukarya and in most species appear to be essential for protein synthesis. Pioneering studies employing electron microscopy, X-ray crystallography, NMR spectroscopy, mass spectrometry, and biochemical analyses on one of the earliest discovered and abundant RNA modification, Ψ, have led to the identification of Ψ residues within functionally important regions of the ribosome, the characterization of the core components implicated in rRNA pseudouridylation, and the elucidation of the functional importance of Ψ residues. Despite the development of several technological advancements to investigate rRNA pseudouridylation, a lack of high-resolution crystallographic analysis of the eukaryotic ribosome has hindered our ability to fully appreciate how conformational changes imposed on the ribosome by Ψ modifications may regulate ribosome activity. In the last decade, however, overwhelming evidence supports a role for rRNA pseudouridylation in the posttranscriptional control of gene expression. The establishment of in vivo systems harboring loss of specific Ψ modifications has shed significant light on the biological function of site-specific rRNA pseudouridylation in regulating global rates of protein synthesis (Baudin-Baillieu et al. 2009; Ganot et al. 1997; Liang et al. 2009). More recently, human genetic studies have also highlighted the importance of rRNA pseudouridylation in regulating gene expression at the translational level with

M. McMahon • C. Bellodi • D. Ruggero (✉)
University of California, San Francisco, CA, USA
e-mail: davide.ruggero@ucsf.edu

J.D. Dinman (ed.), *Biophysical Approaches to Translational Control of Gene Expression*, 253
Biophysics for the Life Sciences 1, DOI 10.1007/978-1-4614-3991-2_13,
© Springer Science+Business Media New York 2013

findings that mutations in the gene encoding the evolutionarily conserved pseudouridine synthase, dyskerin, cause the severe life-threatening disease X-linked Dyskeratosis Congenita (Heiss et al. 1998). We are now at a very exciting moment in the field of translational control in which we are capable of addressing several outstanding questions. These include determining the mechanism by which pseudou-ridylation of rRNA is regulated in eukaryotic cells and how rRNA pseudouridyla-tion may modulate gene expression at the translational level in a cell- and tissue-specific manner. For the purpose of this chapter, we will review some of the major advances that have been made possible by numerous biophysical analyses since the discovery of pseudouridine in the early 1950s. We will outline some of the current challenges in the rapidly expanding field of RNA modifications in the translational control of gene expression and briefly discuss our perspective on how several outstanding questions may be answered in order to fully comprehend the role of rRNA pseudouridylation in fine-tuning gene expression.

13.2   Ribosomal RNA Modifications

The RNA component of the ribosome, rRNA, is the most abundant noncoding RNA in cells and undergoes numerous posttranscriptional site-specific nucleotide modifications (Decatur and Fournier 2003). The two most predominant types of modifications involve the addition of a methyl group to the 2′-hydroxyl group of a ribose residue (2′-O-methylation) and the conversion of uridine to Ψ. Modifications occur in the nucleolus and require numerous small nucleolar RNAs (snoRNAs) and specialized multicomponent enzyme complexes (Kiss 2001). Although the precise function of distinct types of rRNA modifications are not fully understood, it is now becoming clear that both methyl and Ψ modifications cluster within important regions of the ribosome (Decatur and Fournier 2002). Findings that the majority of rRNA modifications are highly conserved and increase in number from archaea to eukarya suggest an important functional role for modifications within ribosomes. Understanding the chemical properties of modified nucleotides and the function of components involved in modifications have been critical for advancing our knowl-edge of how distinct rRNA modifications arise; however, much remains to be deter-mined regarding the functional importance of modified rRNA. In this section, we will highlight some important findings regarding rRNA modifications, with a par-ticular focus on rRNA pseudouridylation.

13.2.1   A Brief Overview of rRNA Methylation

The addition of a methyl group (CH_3) to rRNA can occur either at the 2′ hydroxyl position of the ribose (2′-O-methyl, designated Nm, where N is the nucleotide) or at numerous positions on nucleotide bases (designated mN) (Motorin and Helm 2011).

Ribose and base methylations occur in highly conserved regions of rRNA and in many cases cluster near pseudouridine residues (Bachellerie and Cavaille 1997; Decatur and Fournier 2002). Methyltransfer reactions to RNA nucleotides are catalyzed by RNA-methyltransferases, which use *S*-adenosylmethonine (SAM) as a methyl group donor. Groundbreaking analyses by Maden and colleagues since the early 1970s have significantly contributed to our current understanding of methyl sites. By employing radioactive labeling, T1 RNase digestion and separation of rRNA they demonstrated that methylation occurs rapidly upon transcription of rRNA in the nucleolus (Maden et al. 1972; Salim et al. 1970). Subsequently, a detailed list of methyl modifications in *Saccharomyces cerevisiae* provided further evidence that both ribose and base methyl modifications are present in precursor rRNA (pre-rRNA) and mature rRNA species, with some additional base methylations occurring in mature rRNA species (Brand et al. 1977). More recent studies reveal that methylation residues are clustered in functionally important regions of the ribosome, including the peptidyl transferase center (PTC) of the large ribosomal subunit (LSU) and the decoding center of the small subunit (SSU) (Bachellerie and Cavaille 1997; Decatur and Fournier 2002), suggesting that rRNA methylation may play an important role within ribosomes. Importantly, in bacteria, rRNA methylations promote resistance to ribosome-targeted antibiotics, thus indicating that modifications greatly influence bacterial ribosomes (Doi and Arakawa 2007; Long et al. 2006). In yeast, rRNA methylations in functionally important regions of the ribosome are important for maintaining global rates of protein synthesis (Liang et al. 2009) and a very recent study provides evidence that methylation of rRNA may play a role in translational specificity in mammals (Basu et al. 2011). Indeed, the importance of rRNA methylation in mammals is further highlighted by findings that the rRNA methyltransferase, fibrillarin, is essential for development (Newton et al. 2003).

A variety of biophysical approaches used to study the components responsible for modifications of rRNA, including electron microscopy, X-ray crystallography, NMR spectroscopy, mass spectrometry, and in vivo model systems, have shed significant light on our current understanding of the molecular events that guide different types of rRNA methylation. Stand-alone enzymes carry out the majority of base methylations; however, RNA methyltransferases catalyzing ribose methylation reactions rely on specific guide snoRNAs to select RNA substrates (Cavaille et al. 1996; Kiss-Laszlo et al. 1996). In eukaryotes, 2'-*O*-methylation of rRNA requires the box C/D small nucleolar ribonucleoprotein (snoRNP) complex, which consists of four core protein components, fibrillarin (Nop1 in yeast), NOP58 (Nop58 in yeast), NOP56 (Nop56 in yeast), and 15.5 kDa protein (Snu13 in yeast), and a box C/D snoRNA (Lafontaine and Tollervey 1999; Lyman et al. 1999; Reichow et al. 2007; Schimmang et al. 1989; Watkins et al. 2000). C/D snoRNAs specifically guide the snoRNP complex to the target site of 2'-*O*-methylation (Cavaille et al. 1996; Kiss-Laszlo et al. 1996), whereas the enzymatic activity is initiated by the highly conserved RNA methyltransferase fibrillarin, one of the most abundant proteins in the fibrillar region of the nucleolus, the site of pre-rRNA synthesis and processing (Tollervey et al. 1991, 1993). For a more detailed overview of the C/D snoRNP components and their function, please refer to the following reviews (Filipowicz and Pogacic 2002; Ishitani et al. 2008; Reichow et al. 2007). It is clear that significant advances have been made in uncovering the components implicated in methylation of rRNA. However, how distinct methyl modifications affect ribosome function in eukaryotic cells, and more importantly, how this process is regulated remains poorly understood. For example, from a mechanistic viewpoint, it remains to be addressed whether methylation imposes conformational changes in rRNA, influences ribosome composition, and/or affects RNA–RNA or RNA–protein binding. Complementary to studies investigating rRNA methylation, there has been a wealth of research focused on how Ψ, one of the most abundant base modifications among diverse RNA species including transfer RNA (tRNA), rRNA, and small nuclear RNA (snRNA), modulates RNA function (Ofengand 2002). In the remainder of this chapter, we will focus on the biophysical approaches employed to identify the chemical properties and functions of Ψ residues in modulating gene expression at the translation level.

13.2.2 Pseudouridylation of rRNA

13.2.2.1 Structure and Properties of Pseudouridine

Often referred to as the "fifth RNA nucleotide," Ψ was first identified by chromatographic analysis of tRNA from baking yeast by the Allen laboratory in 1957 (Davis and Allen 1957). The chemical structure of Ψ, a 5-ribosyl isomer of uridine, was subsequently solved and characterized by the Allen and Cohn laboratories (Cohn 1959; Yu and Allen 1959). Distinct from the four canonical RNA nucleotides, Ψ is the only nucleotide with a C–C rather than an N–C glycosyl bond (Fig. 13.1). It has been proposed that the presence of a free N1–H provides an additional hydrogen bond donor site (d), which may be exploited for novel RNA–RNA and RNA–protein interactions. NMR studies indicate that Ψ is predominantly found in the *anti* configuration in RNA, providing the ideal spatial arrangement for the formation of water bridges that connect the N1–H group of the Ψ to the backbone phosphate of the preceding residue (Davis et al. 1998; Yarian et al. 1999). The water bridge confers 5' rigidity, which restricts the mobility of the RNA backbone at the site of pseudouridylation, thus stabilizing the RNA and, in particular, its folding domains (Arnez and Steitz 1994; Davis 1995). Pseudouridylation of RNA also appears to be critical for stabilizing RNA–RNA binding by influencing the rigidity of the sugar-phosphate backbone and enhancing base stacking (reviewed in Charette and Gray 2000). In this respect, it is not surprising that Ψ has been referred to as the "molecular glue" that reinforces necessary RNA conformation (Ofengand 2002). Overall, it seems likely that specific conformational changes in the tertiary structure of rRNA, imposed by Ψ modifications, may stabilize rRNA, affect its binding to RNAs and/or proteins, and therefore impact the overall structure of the ribosome.

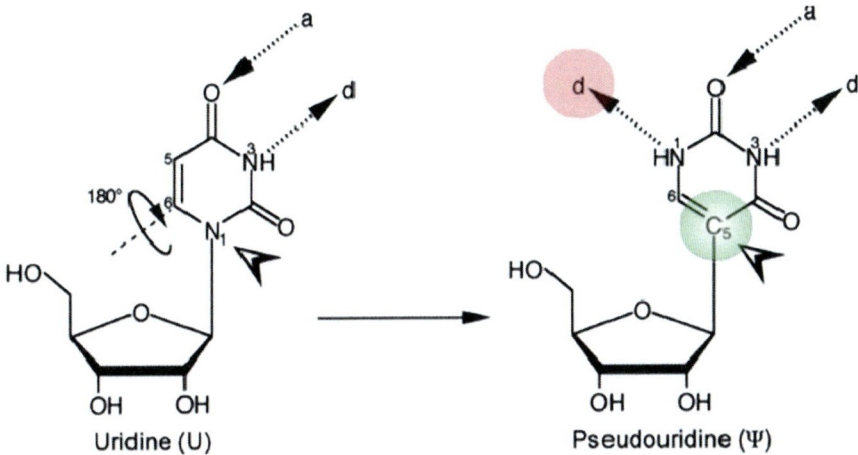

Fig. 13.1 Conversion of uridine to pseudouridine (Ψ). Ψ, one of the most abundant rRNA modifications, is a 5′ ribosyl isomer of uridine (U). Isomerization involves the detachment and rotation (180°) of the uracil base at position N1 (*arrowhead*). The rotation through an N3–C6 diagonal axis (*circular arrow*) results in the formation of a new carbon–carbon (C–C) glycosidic bond at C5, highlighted in *green* (*arrowhead*). The newly synthesized Ψ residue possesses an additional hydrogen bond donor site (d), highlighted in *pink*, increasing the hydrogen bonding capacity of the molecule without affecting the acceptor site (a). Adapted from Charette and Gray (2000)

13.2.2.2 Pseudouridine Synthases

Groundbreaking work and a lifetime of dedication by James Ofengand and colleagues have revolutionized our current understanding of rRNA pseudouridylation (Ofengand 2002). Research aimed at uncovering the functional importance of rRNA pseudouridylation was made possible with the development of methods to detect RNA pseudouridylation, the mapping of Ψ sites on rRNA, and the identification of pseudouridine synthases, the enzymes responsible for pseudouridylation (Arena et al. 1978; Cortese et al. 1974; Johnson and Soll 1970; Samuelsson and Olsson 1990). The first tRNA pseudouridine synthases identified in *Escherichia coli* are classified into five main superfamilies: *rluA*, *rsuA*, *truA*, *truB*, and *truD*, named according to the genes from which the proteins were encoded. Importantly, all subsequent pseudouridine synthases, referred to hereafter as Ψ synthases, were classified based on amino acid sequence homology to these five superfamilies (Gustafsson et al. 1996; Hamma and Ferre-D'Amare 2006; Kaya and Ofengand 2003; Koonin 1996). For example, the archaeal and eukaryotic Ψ synthase Cbf5 belongs to the *truB* superfamily as it contains a TruB domain. (A comprehensive list illustrating the family distribution of Ψ synthases in several organisms can be found in Ofengand (2002). Note: the *truD* superfamily is not included in this list, as they were not identified until 2003 (Kaya and Ofengand 2003).) All Ψ synthase superfamilies display significant sequence diversity, which provide a wealth of information regarding the evolutionary history of Ψ synthases. The *truD* and *truA* superfamilies are highly divergent from all other Ψ synthases, whereas *rsuA* and *rluA* share the highest degree of homology (Koonin 1996). Although the superfamilies display sequence diversity, structural comparisons of proteins from each superfamily reveal that all Ψ synthases share a core consisting of β-sheets and several α-helices, which flank a cleft harboring a catalytically essential aspartate residue (Hoang and Ferre-D'Amare 2001). Most Ψ synthases, with the exception of the archael and eukaryotic *truB* homologues, are capable of selecting target uridine residues without auxiliary guide RNAs, thus acting as stand-alone enzymes (Ofengand 2002). The Ψ synthases of the archael and eukaryotic *truB* family, including Cbf5, have evolved a specific RNA-binding domain, known as a *p*seudo*u*ridine synthase and *a*rcheosine-transglycosy-lases (PUA) domain (Aravind and Koonin 1999), that has paramount importance for the function of these enzymes in the isomerization of substrate RNA.

13.2.2.3   Mechanism of Ψ Synthesis

All Ψ synthases convert uridine to Ψ through a "base-flipping" mechanism that promotes detachment and rotation of the uridine base, a process coordinated by a critical aspartate residue in the active site of the Ψ synthase (Fig. 13.1) (Hoang and Ferre-D'Amare 2001). Mutagenesis studies combined with structural analyses indicate that the aspartate residue is dispensable for the structural integrity of the enzyme active site but is absolutely essential for enzymatic activity (Del Campo et al. 2001; Hoang et al. 2005; Huang et al. 1998; Ramamurthy et al. 1999; Raychaudhuri et al. 1999). This catalytic aspartate uridine faces the uridine at the site of modification and the Ψ synthase accesses the target uridine by "flipping" the uridine residue away from the RNA helical stack. This "base-flipping" mechanism was first demonstrated in *truB*-mediated isomerization of uridine at position 55 within the tRNA T-loop. More precisely, the uridine at the site of pseudouridylation together with the two adjacent residues are "flipped out" from the interior to the exterior of the helical stack, thereby becoming available to the active site of the Ψ synthase (Hoang and Ferre-D'Amare 2001). Importantly, a conserved histidine residue located near the catalytic aspartate mediates the stabilization of the Ψ synthase–RNA complex and facilitates the spatial rearrangement of the substrate uridine (Gu et al. 1998, 1999; Hoang and Ferre-D'Amare 2001). The "base-flipping" mechanism is also performed by DNA methyltransferases, thus suggesting a conserved role for this process in modifying nucleic acids (Klimasauskas et al. 1994).

   Although the detailed kinetic and catalytic mechanism(s) responsible for Ψ conversion remain only partially understood, it is known that the reaction must proceed through a series of events that include cleavage of the *N*-glycosyl link, rotation of the detached uracil base, and formation of a *C*-glycosyl bond to reconnect the uracil base to the ribose (Spedaliere et al. 2004) (Fig. 13.1). Two models describing the molecular events underlying the pseudouridylation reaction have been proposed. The first involves a Michael-type attack by the catalytic aspartate on the carbon at position six (C6) of the uracil base. According to the "Michael mechanism," once the glycosidic bond is broken, the base will detach and be free to rotate around the ester bond of the catalytic aspartate. The rotation repositions C5 of the uracil near C1 of the ribose to form a new glycosidic bond (Gu et al. 1999; Kammen et al. 1988; Ramamurthy et al. 1999). An alternative "acylal mechanism" has also been proposed in which an acylal intermediate is formed by ion pairing or nucleophilic attack at C1 of the ribose by the conserved aspartate residue. The uracilate anion then rotates in the active site to reposition C5 of the base near C1 of the ribose to create the glycosidic bond (Conrad et al. 1999; Huang et al. 1998; Mueller 2002). These two proposed mechanisms are not necessarily exclusive and it is likely that the reaction underlying the catalytic activity of Ψ synthases may integrate characteristics of both models. Since the discovery of Ψ in 1957, significant advances have been made towards understanding the chemical properties and mechanisms of pseudouridylation. In the current technological age, it is highly probable that several outstanding questions regarding the catalytic mechanism of the reaction will be answered. In addition, it is conceivable that studies will be extended to include the analysis of eukaryotic enzymes. In this chapter, we will discuss how Ψ modifications occur, the contribution of rRNA pseudouridylation towards control of gene expression at the level of translation, and the implication of reduced rRNA Ψ levels in human disease.

13.3 The Pseudouridine Synthesis Machinery

The Ψ synthase dyskerin/NAP57 (Cbf5 in yeast and archaea) functions as part of a highly complex evolutionarily conserved H/ACA *small nucleolar ribonucleoprotein* complex termed an H/ACA snoRNP (Kiss et al. 2010). The formation of a functional H/ACA snoRNP complex also requires three additional binding partners, namely GAR1, NOP10, and NHP2, as well as an H/ACA small RNA to guide pseudouridylation of substrate RNA. Importantly, Cbf5/dyskerin associates with two classes of small RNAs, H/ACA snoRNAs that select target uridine sites on substrate rRNA in the nucleolus as well as H/ACA small cajal body RNAs (scaRNAs), which guide pseudouridylation of snRNAs in cajal bodies. In mammals, TERC, the RNA component of telomerase, is also classified as a scaRNA harboring an H/ACA domain. H/ACA snoRNPs have evolved structural components unique from that of the box C/D snoRNPs, which are required for methylation of rRNA (see Sect. 13.2.1), demonstrating that different types of rRNA modifications are evolutionarily distinct and may have diverse functions. Extraordinarily, 8–10 % of total uridine residues in mammalian 18S and 28S rRNA are subject to isomerization (Lestrade and Weber 2006), indicating that this process exhibits high intrinsic specificity. Recent structural findings, obtained utilizing X-ray crystallography, NMR spectroscopy, and mass spectrometry combined with biochemical studies, into the role of each snoRNP component have advanced our current understanding of the stepwise assembly of the snoRNP complex and have provided significant insights into how pseudouridylation of rRNA may occur. These studies will be discussed in this section.

Box H/ACA
snoRNA structure

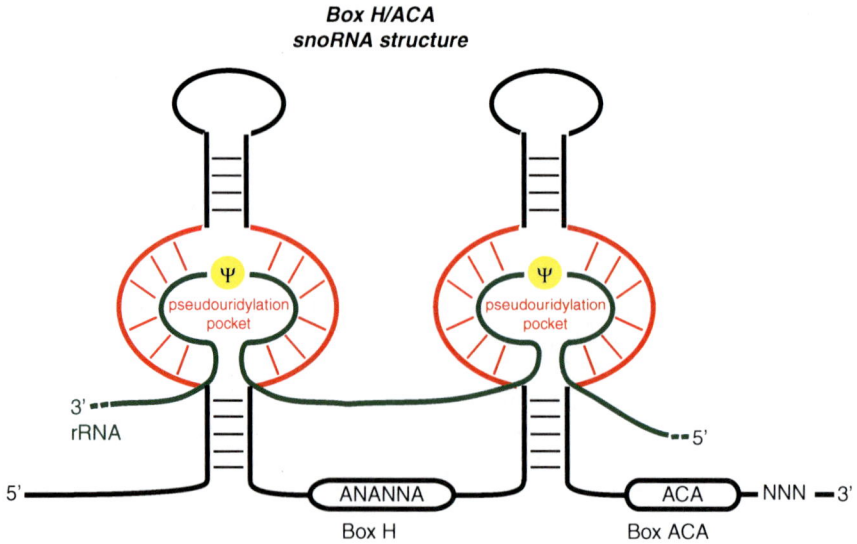

Fig. 13.2 Structure of an H/ACA snoRNA. Schematic representation of a box H/ACA snoRNA containing several evolutionarily conserved structural elements including a box H (ANANNA) and box ACA motif, a hairpin structure, and a pseudouridylation pocket (*red*). The pseudouridylation pocket base pairs to regions in rRNA (*green*) flanking the target uridine. The modified Ψ residue is illustrated in *yellow*

13.3.1 H/ACA SnoRNAs

13.3.1.1 The Structure of H/ACA SnoRNAs

The box H/ACA snoRNAs are responsible for the site-specific conversion of more than 100 uridine residues to Ψ on eukaryotic rRNA (Ganot et al. 1997; Kiss et al. 2004; Lestrade and Weber 2006; Ni et al. 1997). Eukaryotic H/ACA snoRNAs contain several evolutionarily conserved structural elements including the box H and box ACA motifs, a hairpin structure, and pseudouridylation pockets (Ganot et al. 1997; Ni et al. 1997) (Fig. 13.2). The secondary structure of H/ACA snoRNAs typically consists of 60–75 nucleotide-long hairpins that are linked together by a hinge and are followed by a short tail, often referred to as a hairpin-hinge-hairpin-tail arrangement (Bortolin et al. 1999). The conserved box H motif lies in the hinge connecting the two hairpins and consists of an ANANNA sequence, where "N" is any nucleotide. The trinucleotide ACA is located approximately three nucleotides from the 3′ end of the snoRNA sequence. Importantly, the target uridine on rRNA is always positioned 14–16 nucleotides upstream of either the box H motif or the ACA motif, a region referred to as a pseudouridylation pocket (Bortolin et al. 1999). The pseudouridylation pocket is complementary in sequence to the 3–10 nucleotides on either side of the target uridine on rRNA. This binding arrangement is essential for substrate recognition and isomerization. In addition, the presence of two pseudouridine pockets can potentially allow one H/ACA snoRNA to guide modification of two uridine residues, often on different rRNA species, for example as in the case of H/ACA snoRNA U69 (RNU69) (Lestrade and Weber 2006). While it is clear that the isomerization of target uridine residues on rRNA is mediated by sequence-specific H/ACA snoRNAs, it remains poorly understood whether H/ACA snoRNAs are transcriptionally and/or posttranscriptionally regulated. Recently, some insights into how H/ACA snoRNA expression may be controlled have come to light with computational analysis of the genomic organization of H/ACA snoRNAs.

13.3.1.2   Genomic Organization

Genomic searches have revealed that eukaryotic snoRNA-encoding genes are frequently duplicated within distinct genomic regions (Luo and Li 2007; Weber 2006). The vast majority of H/ACA snoRNA gene units reside within intronic regions of protein-encoding host genes and lack independent promoter elements. SnoRNA genes can be arranged as a single intronic unit or as part of a cluster within intronic DNA, both of which are subsequently processed from the pre-mRNA into monocistronic or polycistronic pre-snoRNA, often in a splicing-dependent and RNA polymerase II-dependent manner (Richard and Kiss 2006). A substantial number of H/ACA snoRNA genes cluster within intronic regions of host genes encoding proteins implicated in ribosome biogenesis and ribosome function (Maxwell and Fournier 1995). These findings suggest that a coordinated and regulated expression of H/ACA snoRNAs and components required for ribosome function may occur in cells; however, it remains to be determined if a common transcription factor(s) may coordinately regulate their expression. Furthermore, bioinformatic searches illustrate that a large number of genes hosting snoRNA coding units belong to a family of mRNAs known as 5' TOP mRNAs (Bachellerie et al. 2002), which harbor a polypyrimidine tract at their 5' transcription start sites, that is thought to regulate the expression of these mRNAs (Levy et al. 1991). Interestingly, the promoters of genes harboring mono- or polycistronic snoRNA genes also contain regulatory elements (Dieci et al. 2007; Leader et al. 1997; Nabavi and Nazar 2008), indicating that the expression of snoRNAs genes may be highly controlled. Two human H/ACA snoRNA genes, namely snoRNA36 and snoRNA56, which guide site-specific pseudouridylation of 18S and 28S rRNA, respectively, are located within introns of the gene encoding dyskerin, the mammalian Ψ synthase. It seems likely that the transcription of these two H/ACA snoRNAs may be coupled with transcription of the gene encoding dyskerin. Interestingly, snoRNA genes are also found within nonprotein coding genes and within introns of genes not implicated in ribosome biogenesis (Smith and Steitz 1998; Tycowski and Steitz 2001), indicating that the regulation of snoRNA expression may be more complex than originally thought. Clearly, much remains to be discovered in terms of understanding the expression patterns of specific H/ACA snoRNAs within cells and tissues and the upstream regulatory signals governing their maturation and activity.

13.3.1.3 Orphan H/ACA SnoRNAs

In addition to the H/ACA snoRNAs implicated in the modification of rRNA, recent genomic surveys have also identified a large number of "orphan" H/ACA snoRNA genes, so-called because their putative target residues have not been identified (Huttenhofer et al. 2001; Vitali et al. 2003). While it still remains to be determined if orphan H/ACA snoRNAs are implicated in RNA pseudouridylation, a recent study in yeast demonstrated that pseudouridylation of spliceosomal snRNA by H/ACA scaRNAs can be induced at sites of imperfect sequence identity (Wu et al. 2011). This finding opens the possibility that orphan H/ACA snoRNAs (in addition to H/ACA snoRNAs with assigned target residues) may contribute to the modification of specific uridine residues by imperfect sequence complementarity. It can also be envisioned that noncanonical substrate RNAs, such as mRNAs, may be targets of H/ACA snoRNPs containing orphan H/ACA snoRNAs. Orphan H/ACA snoRNAs may also have additional substrates/roles that are not implicated in RNA modifications, as recent computational analyses suggest that several orphan H/ACA snoRNAs may function as long precursor forms of microRNAs (miRNAs) (Scott et al. 2009). However, whether orphan H/ACA snoRNAs are precursors of functional miRNAs in vivo remains to be explored.

Overall, it is clear that H/ACA snoRNAs are highly specialized RNA elements that modulate pseudouridylation at specific sites on rRNA. Concurrent to research centered on investigating the RNA component of the H/ACA snoRNP, several structural and functional insights into the protein components of this complex have already greatly aided our current understanding of rRNA pseudouridylation in eukaryotic cells and will be discussed in the following sections.

13.3.2 H/ACA SnoRNP Complex Protein Components

13.3.2.1 Cbf5 (Dyskerin in Mammals)

Cbf5 was originally identified in yeast as an essential centromere-binding protein based on its sequence homology to microtubule-binding components (Jiang et al. 1993) and was later demonstrated to be the Ψ synthase responsible for catalyzing the pseudouridylation of rRNA (Lafontaine et al. 1998; Watkins et al. 1998). In line with findings in the bacterial *truB* Ψ synthase, an aspartate at amino acid position 95 of Cbf5 is essential for its pseudouridylation activity (Zebarjadian et al. 1999). Several biochemical analyses in eukaryotes reveal that Cbf5 associates with three core binding partners, Nhp2 (NHP2 in humans), Nop10 (NOP10 in humans), and Gar1 (hGAR1 in humans), which altogether constitute the protein component of the box H/ACA snoRNP (Henras et al. 2004; Pogacic et al. 2000; Watkins et al. 1998; Yang et al. 2000). Importantly, the presence of a conserved RNA-binding domain (PUA domain) within Cbf5 is critical for its ability to bind H/ACA snoRNAs (Fig. 13.3) (Li and Ye 2006; Rashid et al. 2006) and additional studies suggest an

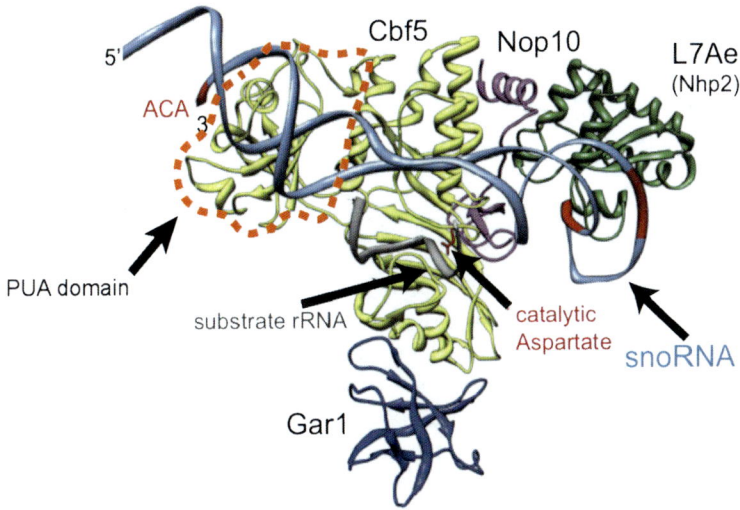

Fig. 13.3 Structure of the H/ACA snoRNP complex. Structural representation of the archael H/ACA snoRNP complex consisting of the core snoRNP component Cbf5 (*yellow*), Nop10 (*purple*), L7Ae (*green*, Nhp2 in yeast), and Gar1 (*dark blue*). The position of the critical aspartate residue (*red*) within the catalytic cleft of Cbf5 lies in close proximity to the substrate rRNA (*grey*). The H/ACA snoRNA (*light blue*) associates with Cbf5, Nop10, L7Ae, and substrate rRNA and positions the target uridine in close proximity to the catalytic cleft of Cbf5. The box ACA domain of the snoRNA (*red*) associates with the PUA domain of Cbf5 (*circled in orange*). Adapted from Hamma and Ferre-D'Amare (2010)

N-terminal extension of this domain is also involved in binding H/ACA snoRNAs (Normand et al. 2006). Importantly, several chaperone proteins including NAF1 (Naf1 in yeast) and SHQ1 (Shq1p in yeast) are also implicated in the stepwise assembly of the H/ACA snoRNP complex and associate with dyskerin in human cells (Darzacq et al. 2006; Grozdanov et al. 2009). In particular, recent evidence in both yeast and human cells indicates that SHQ1 plays a significant role in modulating the assembly of H/ACA snoRNAs with the core snoRNP protein components (Walbott et al. 2011). Specifically, Shq1p acts as a "guide RNA mimic" by binding to the PUA domain of Cbf5, thereby preventing H/ACA small RNA association with Cbf5 prior to complete assembly of the core snoRNP protein components. These findings and others strongly indicate that a hierarchy of component assembly must occur in order to achieve a functionally active H/ACA snoRNP complex. For example, in vitro studies indicate that dyskerin must associate with NOP10 and NHP2 prior to H/ACA snoRNA association (Darzacq et al. 2006). It is conceivable that this stepwise assembly may occur in order to eliminate the formation of a "faulty" or nonoptimally functioning snoRNP. Additional proteins, including the phosphorylated nucleolar protein Nopp140 (NOLC1), also associate with dyskerin and have been implicated in H/ACA snoRNP assembly (Yang et al. 2000); however, the precise role of this protein in modulating H/ACA snoRNP assembly/function remains unclear. Overall, it is evident that dyskerin has a central role as the core H/ACA snoRNP component: it binds and stabilizes H/ACA snoRNAs, it interacts with rRNA substrates, it catalyzes the pseudouridylation of target uridine residues, and it cooperates with the other three protein components, NOP10, NHP2, and hGAR1, to optimize the isomerization activity of the eukaryotic H/ACA snoRNP complex.

13.3.2.2 Gar1

Gar1 was originally identified in yeast as an essential nucleolar protein (Girard et al. 1992) and subsequent experiments demonstrated that Gar1 is required for pseudou-ridylation of rRNA in yeast (Bousquet-Antonelli et al. 1997). Despite extensive co-immunoprecipitation of Gar1 with several H/ACA snoRNAs (Balakin et al. 1996; Dragon et al. 2000; Girard et al. 1992), depletion of Gar1 does not appear to affect snoRNA expression (Girard et al. 1992), suggesting that this protein is not required for H/ACA snoRNA stability. Importantly, hGAR1 associates directly with dysk-erin without interacting with the other snoRNP components in humans (Wang and Meier 2004) (Fig. 13.3). These data specifically implicate hGAR1 in modulating dyskerin function. Current crystal structure models of the archaeal snoRNP indicate that Gar1 is required to enhance the substrate-binding capacity of Cbf5 and sub-strate turnover of the H/ACA snoRNP (Li and Ye 2006; Rashid et al. 2006). A role for Gar1/hGAR1 in promoting substrate turnover rather than H/ACA snoRNA sta-bility or assembly is further supported by the finding that hGAR1 only associates with the fully assembled H/ACA snoRNP (Darzacq et al. 2006).

13.3.2.3 Nhp2

Originally described as an essential high mobility group (HMG)-like protein in yeast (Kolodrubetz and Burgum 1991), Nhp2 was subsequently identified as a Gar1-associated nucleolar protein involved in rRNA pseudouridylation (Henras et al. 1998). Interestingly, Nhp2 contains an RNA-binding domain and shares significant sequence homology with the archael L7Ae ribosomal protein, which associates with both C/D and H/ACA snoRNPs (Rozhdestvensky et al. 2003). It also shares sequence homology with the yeast ribosomal protein Rpl32, which displays specific RNA-binding activities. Nhp2 associates with the hairpin structure of H/ACA snoR-NAs through its central domain (Li and Ye 2006); deletion of Nhp2 affects H/ACA snoRNA levels and significantly hampers rRNA pseudouridylation and cell growth (Henras et al. 1998; Watkins et al. 1998). In vitro studies in human cells indicate that NHP2 does not interact directly with dyskerin but rather associates with NOP10 (Fig. 13.3) (Wang and Meier 2004). Altogether, these findings indicate that NHP2/Nhp2 plays an important role in the stability and association of H/ACA snoRNAs within the snoRNP complex.

13.3.2.4   Nop10

Nop10 was first identified in yeast as an essential Gar1-interacting protein that stabilizes H/ACA snoRNAs and is implicated in pseudouridylation of rRNA (Henras et al. 1998). In vitro biochemical studies in human cells indicate that NOP10 directly binds to dyskerin, and this interaction is required for the association of NHP2 with the H/ACA snoRNP (Wang and Meier 2004). These findings suggest that NOP10 acts as a molecular bridge linking NHP2 to the snoRNP and reinforces the notion that a hierarchy of component assembly must occur in order to generate a functional H/ACA snoRNP complex. Crystal structure analyses of the archael H/ACA snoRNP indicate that Nop10 interacts with a conserved region of Cbf5 in close proximity to its catalytic center (Fig. 13.3) (Rashid et al. 2006). In archaea, it has been demonstrated that Nop10 does not independently interact with RNA (Baker et al. 2005); however in yeast, Nop10 forms two anti-parallel β-sheets, which interact weakly with H/ACA snoRNAs by binding a region close to the 3′ end of the H/ACA snoRNA pseudouridine pocket (Khanna et al. 2006). It therefore seems likely that the association of Nop10 with Cbf5 provides a critical binding surface for the recruitment of Nhp2, which, in turn, stabilizes the association of H/ACA snoRNAs with the snoRNP complex, thus modulating target selection and the activity of the H/ACA snoRNP. It remains to be determined if, in mammalian cells, the components of the H/ACA snoRNP maintain similar functions in vivo.

13.3.3   Structural and Functional Organization of the H/ACA SnoRNP Complex

A number of recent high resolution crystal structure studies in archaea have taken advantage of the evolutionarily conserved structural similarities between archaea and eukaryotes in order to provide functional insights into the eukaryotic H/ACA snoRNP complex (Duan et al. 2009; Hamma et al. 2005; Li and Ye 2006; Rashid et al. 2006). These studies reveal that the catalytic domain of Cbf5 is located in the center of the complex, surrounded by Gar1, L7Ae (Nhp2 homologue), and its own PUA domain (Fig. 13.3). As biochemically demonstrated (Sect. 13.3.2), X-ray crystallography also provides evidence that the binding of Nop10 to Cbf5 results in a conformational change in Nop10, which allows subsequent recruitment of Nhp2 to the snoRNP complex (Rashid et al. 2006). Furthermore, an important thumb loop domain close to the catalytic region of Cbf5 is predicted to help "lock" the substrate RNA into the correct position, stabilizing the structure and favoring the conversion of uridine to Ψ (Hamma et al. 2005; Rashid et al. 2006). The crystal structure of a snoRNP bound to an H/ACA snoRNA has revealed that the ACA motif of the snoRNA binds to the PUA domain of Cbf5, while the hairpin structure associates with Cbf5, Nop10, and L7Ae (Li and Ye 2006) (Fig. 13.3), as previously demonstrated by biochemical analyses (Sect. 13.3.2). Importantly, a complex containing an H/ACA snoRNA, Cbf5, and Nop10 displays basal levels of pseudouridylation activity; however, the four core proteins are required for maximal enzymatic activity (Hamma et al. 2005). Crystallography studies also reveal that a number of interactions take place between the PUA domain of Cbf5, the H/ACA snoRNA hairpin structure, and the two adenines of the box ACA motif (Rashid et al. 2006). These interactions are essential to promote isomerization by maintaining the pseudouridylation pocket of the H/ACA snoRNA in a favorable orientation with substrate RNA (Figs. 13.2 and 13.3). Altogether, these studies in archaea have provided valuable structural insights into the relationships between components of the Ψ synthase apparatus and serve as an excellent model for understanding the tertiary organization of the eukaryotic H/ACA snoRNP complex. However, despite these significant structural insights, much remains to be determined in regards to the molecular mechanisms and the upstream signals that coordinate the assembly of the H/ACA snoRNP complex in vivo.

13.3.4   Regulation of the H/ACA SnoRNP Complex

Despite enormous advancements in understanding the individual function and structural arrangement of the H/ACA snoRNP components (Sects. 13.3.1, 13.3.2, and 13.3.3), there is a substantial lack of information on their transcriptional, translational, and posttranslational regulation. The signals and stimuli promoting the assembly and activity of a functional H/ACA snoRNP complex in eukaryotic cells still remain poorly characterized. Interestingly, dyskerin, like several other rRNA processing and assembly factors, is a direct target of Myc (Alawi and Lee 2007). Myc plays an essential role in the control of cell growth and protein synthesis in cells (van Riggelen et al. 2010). Thus, it is reasonable to speculate that modulating dyskerin levels may be necessary for increased protein synthesis and cell growth. Recent evidence suggests that some of the core H/ACA snoRNP components undergo posttranslational modifications that may be important for the biogenesis and/or function of the H/ACA snoRNP complex. For example, stable-isotope labeling by amino acids in cell culture (SILAC)-based quantitative proteomics indicates that both dyskerin and NHP2 are SUMOylated (Westman et al. 2010) and it would be intriguing to determine whether SUMO modifications regulate their function. It is also conceivable that additional posttranslational modifications of all core components may directly impinge on the assembly and/or subcellular localization of the H/ACA snoRNP thus conferring another layer of complexity in the regulation of the H/ACA snoRNP.

Another important question is centered on whether H/ACA snoRNA expression and subsequent site-specific pseudouridylation of rRNAs are modulated within cells. Can site-specific rRNA pseudouridine residues influence translational control of gene expression in response to stimuli? And if so, what are the molecular mechanisms that underlie this process? A recent finding illustrates that Nopp140, a factor implicated in H/ACA snoRNP assembly, is modulated by p53 (Krastev et al. 2011), a key tumor suppressor protein important for a broad range of cellular events such as cell cycle regulation, apoptosis, and genomic stability, thus suggesting that pseudouridylation may be regulated in response to proliferation, DNA damage, and oncogenic signals. Likewise, it will be interesting to determine if rRNA pseudouridylation is modulated in a tissue-specific manner. Evidence to support this notion comes from findings that some H/ACA snoRNAs display a tissue-specific expression pattern (Castle et al. 2010; Cavaille et al. 2000). Despite an evident lack of information regarding the regulatory mechanisms controlling the activity of H/ACA snoRNPs, studies in both yeast and mouse models have provided important insights into the role of rRNA pseudouridylation in modulating translation and will be discussed in the next section.

13.4   Functional Importance of Pseudouridine Modifications

The biological importance of rRNA Ψ modifications is clear from research employing in vivo models including bacteria, yeast, fruit flies, and mice (Ejby et al. 2007; Giordano et al. 1999; He et al. 2002; Ruggero et al. 2003; Zebarjadian et al. 1999), the majority of which demonstrate an important role for Ψ residues in regulating ribosome biogenesis and protein synthesis to ensure cellular homeostasis. Structural insights obtained using 3D models indicate that Ψ residues are clustered in functionally important regions of the ribosome from eubacteria to eukaryotes and have significantly increased in number throughout evolution (Fig. 13.4) (Decatur and Fournier 2002; Piekna-Przybylska et al. 2008). Why the numbers of Ψ residues

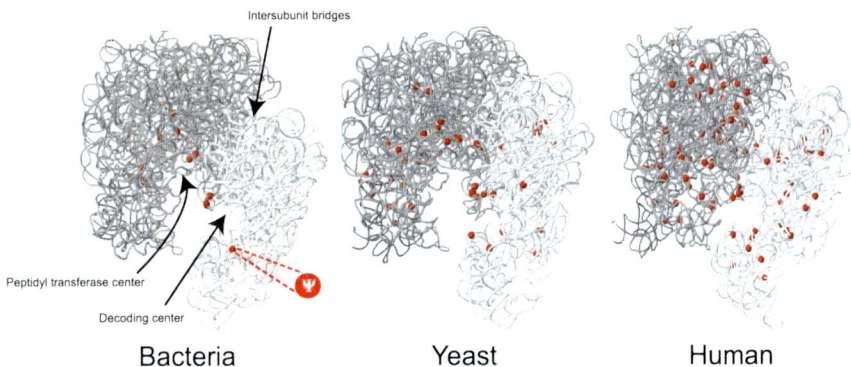

Fig. 13.4 Position of Ψ residues on bacteria, yeast, and human ribosomal subunits. 3D models highlighting the position of Ψ residues (*red*) on the small (*light grey*) and large (*dark grey*) ribosomal subunits from bacteria, yeast, and human obtained from Piekna-Przybylska et al. (2008). The position of the intersubunit bridges, the decoding center, and the peptidyl transferase center are indicated. Note the increase in the number of Ψ residues from bacteria to human and the clustering of Ψ residues in functionally important regions of the ribosome. This model represents only a fraction of Ψ residues present in ribosomal subunits of yeast and human. Please refer to Piekna-Przybylska et al. (2008) for more detailed information

have significantly increased by 4–8-fold from eubacteria to eukaryotes is not fully understood. It is tempting to speculate that multicellular organisms have acquired an increased number of pseudouridines in order to obtain higher quality control mechanisms for modulating gene expression at the translational level in a cell- and tissue-specific manner. Studies in *E. coli* demonstrate that deletion of a number of Ψ synthases, each of which modify distinct uridine residues on 23S rRNA (nine Ψ residues in total identified on 23S rRNA), including *rluA* (Ψ746) and *rluC* (Ψ955, Ψ2504, and Ψ2580) (Conrad et al. 1998; Raychaudhuri et al. 1999), has little effect on growth, whereas absence of *rluD* (Ψ1911, Ψ1915, and Ψ1917) severely inhibits growth (Ejby et al. 2007; Raychaudhuri et al. 1998). Specifically, *rluD* inactivation results in loss of a cluster of three Ψ residues in helix 69 of *E. coli* 23S rRNA, close to the decoding center of the ribosome, and significantly impairs ribosome assembly, biogenesis, and activity (Ejby et al. 2007; Gutgsell et al. 2005). Therefore, it appears that Ψ residues located in distinct regions of rRNA may have differential effects on ribosome function. Furthermore, studies in yeast in which single or multiple H/ACA snoRNAs are depleted have provided additional insights into the role of individual Ψ residues in translation (Liang et al. 2009) (see Sects. 13.4.1 and 13.4.2). An in vivo mouse model characterized by a partial reduction in rRNA pseudouridylation has provided significant insights into the mechanisms by which rRNA pseudouridylation can modulate gene expression in a tissue-specific manner (Yoon et al. 2006). The importance of rRNA pseudouridylation is further highlighted by findings that alterations in DKC1, the gene encoding the Ψ synthase dyskerin, is associated with X-linked Dyskeratosis Congenita (X-DC) and cancer (Heiss et al. 1998; Montanaro et al. 2010) (see Sect. 13.5.1). However, the mechanisms by which reductions in rRNA pseudouridylation may contribute to the specific clinical features present in these diseases are not fully understood. The recent identification of a novel role for Ψ residues in mediating nonsense codon suppression (Karijolich and Yu 2011) indicates that Ψ may posttranscriptionally modulate gene expression at multiple levels. In this section, we will highlight some important findings that demonstrate a role for rRNA pseudouridylation in controlling gene expression at the translational level in eukaryotic cells.

13.4.1   Role for Ψ Modifications in Ribosome Biogenesis

Pseudouridylation occurs rapidly in the nucleolus upon transcription of rRNA (Granneman and Baserga 2004; Maden and Forbes 1972), and studies indicate that global loss of rRNA pseudouridylation in yeast due to genetic deletion of Cbf5 results in pre-rRNA processing defects. Specifically, Cbf5 depletion results in delayed production of 18S and 28S rRNA along with an accumulation of pre-rRNA species (Lafontaine et al. 1998). Further evidence supporting a role for rRNA pseudouridylation in rRNA processing and maturation is evident from studies in yeast lacking Ψ modifications in Helix 69 of 28S rRNA and displaying significant impairments in 28S and 18S rRNA maturation (Liang et al. 2007). A specific single Ψ modification also appears to be important for pre-rRNA processing and rRNA maturation. Namely, depletion of the yeast H/ACA snoRNA snR35, which guides Ψ modification on the hypermodified 1-methyl-3-(3-amino-3-carboxypropyl) Ψ1191 residue ($M^1acp^3\Psi$1191) in 18S rRNA (Brand et al. 1978), leads to defective processing of pre-rRNA and a reduction in mature 18S rRNA species (Liang et al. 2009). Why modification of this residue can have such a substantial effect on 18S rRNA maturation remains unclear. It seems likely that individual Ψ modifications at specific sites on rRNA may affect rRNA maturation by possibly imposing conformational changes on pre-rRNA close to regions where cleavage steps occur. An additional H/ACA snoRNA, snR30 (U17 in humans), is also implicated in processing of rRNA (Morrissey and Tollervey 1993). It has been demonstrated that the ability of this H/ACA snoRNA to bind pre-rRNA is critical for rRNA processing, although it still remains unclear whether this specific H/ACA snoRNA participates in rRNA pseudouridylation.

In light of the fact that rRNA pseudouridylation affects rRNA processing and maturation, it is not surprising that pseudouridylation of rRNA also impinges on the production of mature cytoplasmic ribosome subunits. In particular, a point mutation in Cbf5 at a critical aspartate, D95A, which abolishes rRNA pseudouridylation, severely impairs the production of ribosome subunits (Zebarjadian et al. 1999). In addition, depletion of six Ψ residues within the PTC significantly alters ribosome profiles in yeast: they contain fewer polysomes and display an altered 60S/80S ratio compared to wild-type strains, both of which correlate with a reduction in translation activity (King et al. 2003). Collectively, these studies imply that Ψ residues and/or clusters of specific Ψ modifications in functionally important regions of rRNA play a critical role in ribosome biogenesis and that, when disrupted, significantly reduce overall rates of protein synthesis.

13.4.2 How Does Pseudouridylation of rRNA Control Gene Expression?

The contribution of distinct Ψ residues towards regulating translation has mostly been investigated in yeast by means of genetic deletion of single H/ACA snoRNAs that guide modification at a specific uridine residue (Baudin-Baillieu et al. 2009; Ganot et al. 1997; Liang et al. 2009). Interestingly, these studies demonstrate that in most cases, loss of individual rRNA modifications have no overall effect on global rates of protein synthesis; however, combined loss of a number of modifications within specific regions of the ribosome can significantly diminish overall rates of protein synthesis (Liang et al. 2009). Particularly, loss of Ψ modifications in the decoding center of the SSU results in reduced overall translation levels (Liang et al. 2009). For example, combined loss of the hypermodified $M^1acp^3\Psi$1191 at the P-site and Ψ1187 at the A-site (obtained by depletion of H/ACA snoRNAs snR35 and snR36, respectively) significantly decreases in vivo incorporation rates of (^{35}S) methionine into total protein (25 % decrease), as compared to depletion of individual modifications (7 %

decrease) (Liang et al. 2009). Additional studies in yeast demonstrate that Ψ modifications in the decoding center of the ribosome are critical for reading frame maintenance and stop codon recognition (Baudin-Baillieu et al. 2009). Collectively, these results underscore an important contribution of Ψ modifications, within distinct functional regions of the ribosome, towards modulating protein synthesis.

Importantly, studies in *Drosophila* and mice indicate that a critical threshold level of dyskerin expression in cells is essential for cell and organismal viability. In drosophila, complete loss of *minifly (mfl)* (DKC1 in mammals), the gene encoding the drosophila Ψ synthase, results in larval lethality, whereas drosophila "hypomorphic" mutants harboring a partial loss of function of *mfl* are viable but display a variety of severe developmental defects (Giordano et al. 1999). For example, in the context of the developing imaginal wing disc, it appears that loss of function of *mfl* in specific regions of the wing disc enhances apoptosis, possibly due to cell competition triggered by wild-type cells containing a fully functional Ψ synthase (Tortoriello et al. 2010). Similarly in mice, complete loss of dyskerin, the mammalian rRNA Ψ synthase, results in embryonic lethality (He et al. 2002); however, DKC1 hypomorphic mutant (*Dkc1^m*) mice, which display reduced expression of dyskerin and reductions in both 18S and 28S rRNA pseudouridylation, are viable and recapitulate several pathological features of X-DC including increased tumor susceptibility, dyskeratosis of the skin, and severe anemia (Ruggero et al. 2003) (see Sect. 13.5.1). Conditional deletion of the DKC1 gene in adult mice demonstrates that most tissues cannot tolerate low levels of dyskerin, leading to the progressive elimination of all DKC1-depleted cells in these animals (Ge et al. 2010). Tissues expressing reduced levels of dyskerin exhibit severe impairments in rRNA processing associated with reduced rates of cell proliferation in comparison to wild-type cells (Ge et al. 2010). Collectively, these studies demonstrate that complete loss of dyskerin expression negatively impacts cell growth and that loss of all rRNA pseudouridine residues results in impaired ribosome biogenesis and global reductions in protein synthesis.

Further studies in *Dkc1^m* mice indicate that a partial reduction in dyskerin and rRNA pseudouridylation levels have no overall effect on the rates of global protein synthesis, but rather lead to a selective translational defect in a subset of specific mRNAs (Yoon et al. 2006). An unbiased analysis of mRNAs associated with polysomes (translationally active ribosomes) revealed that primary splenic lymphocytes isolated from *Dkc1^m* mice display a selective defect in translation of a distinct group of mRNAs containing an internal ribosome entry site (IRES) in their 5′ untranslated regions (UTRs) (Yoon et al. 2006). IRES elements are structured RNAs of variable length, originally identified in picornavirus RNAs, that bind to the 40S ribosomal subunit during translation initiation (Hellen and Sarnow 2001; Jang et al. 1988; Pelletier and Sonenberg 1988). In contrast to the general cap-dependent mode of translation, which requires the assembly of proteins known as eukaryotic initiation factors (eIFs) to recruit the 40S ribosomal subunit on the 5′ cap of mRNAs, IRES-dependent translation is an RNA-based mode of translation initiation (Hellen and Sarnow 2001; Holcik and Sonenberg 2005). Specifically, IRES-dependent translation requires more direct contact between the mRNA and ribosome at the expense of some, if not all, of the eIFs. For example, the cricket paralysis virus

(CrPV) IRES assembles with the 40S and 60S ribosomal subunits (80S monosome) on an initiation codon and starts translation without any canonical eIFs (Wilson et al. 2000a, b). Importantly, translation of the CrPV IRES was severely impaired in *Dkc1^m* mouse cells, strongly indicating that defective IRES-dependent translation is due to an intrinsic defect in ribosomes displaying a reduction in rRNA pseudouridylation isolated from *Dkc1^m* mice (Yoon et al. 2006). Significantly, high-resolution cryo-electron microscopy has resolved the structure of the CrPV IRES when it associates with the ribosome, highlighting a direct role in accurate folding of ribosomal subunits upon interaction with mRNAs during IRES-mediated translation initiation (Costantino et al. 2008). Thus, it seems likely that a reduction in Ψ modifications in rRNA may either impair the ability of ribosomal subunits to directly bind the IRES elements or affect the conformational changes in ribosomal subunits upon binding. In line with this hypothesis, a filter-binding assay was employed to demonstrate that ribosomes isolated from *Dkc1^m* mouse cells are greatly impaired in their ability to bind the CrPV IRES element. Importantly, it has also been demonstrated that rRNA Ψ modifications have an evolutionarily conserved role in the recruitment of IRES elements from yeast to human (Jack et al. 2011).

In contrast to the current understanding of viral IRES-mediated translation, the molecular mechanisms underlying cellular IRES-dependent translation are still poorly understood. IRES-dependent translation is a mechanism that modulates gene expression at the translational level during specific cellular events such as mitosis, quiescence, hypoxia, nutrient deprivation, apoptosis, and an accurate switch from cap- to IRES-dependent translation has been shown to be critical in maintaining cellular homeostasis during these circumstances (Barna et al. 2008; Holcik and Sonenberg 2005; Krichevsky et al. 1999; Lang et al. 2002; Sherrill et al. 2004). Therefore, a defect in Ψ modifications and IRES-mediated translation may contribute to specific pathological features present in X-DC (see Sect. 13.5.1). However, it is still not completely known how defects in rRNA pseudouridylation impinge on translation of specific cellular mRNAs that have been identified with an unbiased ribosomal profiling analysis (Yoon et al. 2006). Specifically, downregulation of the mammalian Ψ synthase dyskerin was found to affect the IRES-dependent translation of distinct IRES-containing mRNAs including the cell cycle regulators p27 and p53, as well as anti-apoptotic factors such as XIAP and Bcl-xL (Bellodi et al. 2010a; Yoon et al. 2006). While it is possible that impaired rRNA pseudouridylation due to reduction of dyskerin activity may affect the direct engagement of the IRES element, it is also plausible that the absence of an additional hydrogen bond donor site upon loss of Ψ residues may disrupt critical RNA–protein interactions. For example, in this regard several studies have highlighted the importance of RNA-binding proteins known as *I*RES *trans-a*cting *f*actors (ITAFs) in modulating IRES-mediated translation of several cellular mRNAs (Lewis and Holcik 2008; Spriggs et al. 2005). Therefore, it is possible that loss of Ψ modifications may influence the association of ITAFs with the ribosome, thereby altering the expression of distinct mRNAs. Overall, it is clear that rRNA pseudouridylation is important for the recruitment of mRNAs containing IRES elements to the ribosome; however, the specific Ψ residues responsible for this recognition and the precise mechanism involved still remain to be identified.

In addition to affecting IRES-mediated translation of specific mRNAs, recent findings also demonstrate an evolutionarily conserved role for rRNA pseudouridylation in faithfully maintaining the translational reading frame. Specifically, impaired pseudouridylation of rRNA promoted increased programmed ribosomal frameshifting (PRF) of both viral and cellular mRNAs containing a PRF signal (Jack et al. 2011). In this context, it seems likely that the level of rRNA pseudouridylation may regulate, at the translational level, the expression of messengers containing a PRF signal, in perhaps a cell- and/or tissue-specific manner. Altogether, findings from in vivo systems illustrate that pseudouridylation of rRNA plays an important role in general protein synthesis as well as in controlling the expression of specific mRNAs. Current research efforts are now underway to uncover the regulatory signals that promote pseudouridylation at specific sites on rRNA, the molecular mechanisms by which specific Ψ residues modulate translation, and the mRNA targets that may be affected by site-specific alterations in rRNA pseudouridylation.

13.4.3   A New Role for Ψ Residues in Nonsense Codon Suppression

A recent study has uncovered an unexpected role for Ψ in nonsense codon suppression (Karijolich and Yu 2011). Using in vitro and in vivo systems, the authors demonstrate that nonsense codon suppression can be achieved by targeted pseudouridylation of UAA, UAG, or UGA stop codons. Surprisingly, ΨAA and ΨAG result in the incorporation of either serine or threonine amino acids, whereas pseudouridylation of UGA (ΨGA) leads to the incorporation of tyrosine or phenylalanine. These findings demonstrate highly specialized recognition of pseudouridylated stop codons by aminoacyl-tRNAs. Although the mechanism responsible for Ψ-mediated nonsense suppression remains unclear, it is possible that the unique chemical properties of Ψ may interfere with termination site recognition by inhibiting the function of release factors (RFs) to halt translation (Karijolich and Yu 2011). However, it is not known if pseudouridylation of stop codons occurs in nature and, if so, which signals trigger this process. From a therapeutic perspective, the potential of targeting Ψ residues to mediate nonsense codon suppression have broad implications for the development of novel therapies to treat genetic disorders arising from premature translation termination (Frischmeyer and Dietz 1999). Importantly, the role of pseudouridylation in nonsense codon suppression raises the possibility that uridine residues in cellular mRNAs may also undergo pseudouridylation; however, this hypothesis requires further investigation. As mentioned earlier (Sect. 13.3.1), it is reasonable to postulate that orphan H/ACA snoRNAs, which do not appear to guide pseudouridylation of identified Ψ residues on rRNA, may indeed guide isomerization of uridine residues on mRNA.

13.5   Defective rRNA Modifications in Human Diseases

The development of a faithful mouse model of X-DC has provided significant insights into the role of rRNA pseudouridylation in the control of gene expression at the translational level (Ruggero et al. 2003). More importantly, human genetic studies also highlight an important role for the Ψ synthase dyskerin in translation control. Specifically, the inherited disease X-linked Dyskeratosis Congenita (X-DC) is caused by mutations in the gene encoding the Ψ synthase dyskerin (Heiss et al. 1998) and cells derived from X-DC patients exhibit specific translation defects that explain, in part, some of the severe pathological features of the disease (Bellodi et al. 2010a; Yoon et al. 2006). In addition to X-DC, the contribution of "aberrant" protein synthesis toward the pathogenesis of human diseases has recently undergone a paradigm shift with the identification of an entire class of inherited human syndromes termed "ribosomopathies" (Narla and Ebert 2010). Patients suffering from these diseases, which are associated with mutations in distinct components of the translational apparatus including rRNA synthesis and processing factors, ribosome assembly factors, and ribosomal proteins, share several common clinical features such as bone marrow failure (BMF), skeletal defects, and increased susceptibility to cancer. It remains to be determined if common translational impairments may be responsible for some of the recurrent pathological features present in ribosomopathies (Ganapathi and Shimamura 2008). Research focused on identifying the downstream targets affected by aberrant translational control of gene expression will undoubtedly provide the rationale for new therapies and treatment options for patients suffering from these diseases. In this section, we will highlight some important findings on how reductions in rRNA modifications may contribute to human diseases.

13.5.1   X-Linked Dyskeratosis Congenita

X-DC is a rare inherited human syndrome invariably associated with point mutations in the DKC1 gene locus on chromosome Xq28 (Heiss et al. 1998). From a clinical standpoint, X-DC is characterized by a wide spectrum of pathological features including a triad of cutaneous defects: abnormal skin pigmentation, nail dystrophy, and oral leukoplakia; a variety of non-cutaneous dental, gastrointestinal, genitourinary, neurological, ophthalmic, pulmonary, and skeletal defects; and severe hematological dysfunctions (Dokal 2000; Kirwan and Dokal 2008). BMF associated with peripheral cytopenia, in one or more hematopoietic cell lineages, represents the primary cause of early mortality in X-DC patients along with high susceptibility to solid and hematological malignancies (Alter et al. 2009; Marsh et al. 1992; Montanaro 2010). The clinical features of the disease normally appear in childhood with up to 90 % of patients showing signs of BMF by the age of 30 years. Notably, X-DC is the most severe form of Dyskeratosis Congenita, although clinically milder forms of the disease have also been described. These include autosomal-dominant DC (AD-DC), caused predominantly by mutations in genes implicated in telomere maintenance, including the telomerase RNA component (TERC) (Vulliamy et al. 2001) and the telomerase reverse transcriptase (TERT) (Armanios et al. 2005), and autosomal-recessive DC (AR-DC), resulting from mutations in the H/ACA snoRNP components NHP2 and NOP10 (Vulliamy et al. 2008; Walne et al. 2007). As mentioned earlier (see Sect. 13.3), in addition to associating with H/ACA snoRNAs, dyskerin also associates with H/ACA scaRNAs including TERC (Meier 2006), implicating it in a broad range of cellular processes such as rRNA pseudouridylation, splicing, and telomere maintenance. Importantly, defective telomere maintenance is evident in all forms of DC and several studies indicate that telomere shortening may contribute to some of the clinical features observed in X-DC (Batista et al. 2011; Bessler et al. 2004; Dokal 2000; Mitchell et al. 1999; Vulliamy et al. 2006). Thus, given the multifunctional role of dyskerin within cells, it is plausible to speculate that the severity of X-DC may stem from multiple molecular and cellular defects.

Strikingly, more than 50 X-DC-associated missense mutations in DKC1 have been annotated to date (http://telomerase.asu.edu/diseases.html), the majority of which cluster in the N and C termini of the protein. Tertiary structure analysis indicates that these two regions form the PUA domain of the protein (Rashid et al. 2006), which is critical for its binding to box H/ACA snoRNAs (see Sect. 13.3.3). Thus, it seems likely that mutations within the PUA domain of dyskerin may disrupt H/ACA snoRNA stability and function leading to site-specific reductions in rRNA pseudouridylation. Indeed, introduction of two DKC1 mutations frequently found in X-DC patients into murine embryonic stem cells results in reduced expression of specific H/ACA snoRNAs along with reductions in pseudouridylation of the corresponding sites on rRNA (Mochizuki et al. 2004). These findings support the possibility that X-DC patients may also harbor reductions in a subset of H/ACA snoRNAs and subsequently decreased Ψ modifications at corresponding target residues. While investigations to explore this possibility are currently underway, it would be informative to assess whether specific loss of Ψ residues in distinct regions of rRNA may contribute to particular disease phenotypes. Important functional insights into the mechanisms underlying some of the pathological features present in X-DC have been uncovered utilizing $Dkc1^m$ mice (Ruggero et al. 2003). In these mice, reductions in rRNA pseudouridylation are present prior to disease onset when telomeres are unperturbed, suggesting that rRNA pseudouridylation may contribute to some of the clinical features present in X-DC. Importantly, the first translational targets identified downstream of reductions in rRNA pseudouridylation levels (see Sect. 13.4.2) were also recapitulated in primary X-DC patient cells (Bellodi et al. 2010a; Yoon et al. 2006), indicating that defects in IRES-dependent translation may contribute to the disease.

How does impaired IRES-dependent translation in X-DC potentially contribute to the tissue-specific pathological features observed in this disease? IRES-dependent translation is a fine-tuning mechanism that can regulate gene expression during key cellular conditions including quiescence, survival, and differentiation (Iglesias-Serret et al. 2003; Krichevsky et al. 1999; Miskimins et al. 2001; Pyronnet et al.

2001; Stoneley and Willis 2004). Two tumor suppressors, which regulate cell cycle (p53 and p27) and differentiation (p27), are deregulated in primary $Dkc1^m$ mouse and X-DC patient cells. Several studies have highlighted the importance of p53 in controlling cell quiescence and proliferation, key features of hematopoietic stem cells (HSCs) and lineage-committed hematopoietic cells (Kastan et al. 1991; Liu et al. 2009; Okazuka et al. 2005; Shaulsky et al. 1991; Slatter et al. 2010). Both cell types are essential for the production of mature blood cells and when deregulated may contribute to BMF. Additionally, p27 regulates multiple aspects of hematopoiesis. Specifically, in vivo studies demonstrate that p27 is important for maintaining quiescence, controlling the number of hematopoietic progenitor cells, and restricting the proliferative capacity of naïve CD8+ T cells (Tsukiyama et al. 2001; Wolfraim and Letterio 2005). Collectively, these findings provide strong evidence that deregulations in IRES-dependent translation of both p53 and p27 may contribute to many aspects of BMF, the primary cause of X-DC patient lethality.

In addition, defective IRES-dependent translation in primary X-DC patient cells may provide a possible functional explanation for the increased cancer susceptibility in X-DC. Dyskerin and rRNA pseudouridylation play an important role in a specific translational switch between cap- and IRES-mediated translation that occurs during oncogene-induced senescence (OIS) (Bellodi et al. 2010a), which is one of the first tumor suppressive barriers that restrict cancer progression (Serrano et al. 1997). This translational switch allows the expression of specific mRNAs, such as the tumor suppressor p53, which is necessary and sufficient to promote OIS. Importantly, during this switch, p53 IRES-mediated translation is significantly impaired in the absence of adequate rRNA pseudouridylation, rendering $Dkc1^m$ mouse cells more susceptible to oncogenic transformation (Bellodi et al. 2010a). A role for rRNA modification defects in tumorigenesis is also supported by findings that reductions in dyskerin levels are associated with downregulation of p53 protein levels and activity in human breast cancer cells (Montanaro et al. 2010). Furthermore, a novel point mutation in DKC1 that is associated with reduced levels of rRNA pseudouridylation has recently been identified in a pituitary cancer patient. Interestingly, in this pituitary tumor, decreased dyskerin activity correlates with a reduction in p27 protein expression at the posttranscriptional level (Bellodi et al. 2010b). Collectively, these findings have provided functional insights into how impairments in rRNA pseudouridylation contribute to highly specific translation defects that may contribute, at least in part, to the increased cancer susceptibility observed in X-DC patients as well as patients harboring DKC1 somatic mutations.

Consistent with findings that ablation of dyskerin expression is detrimental to life (Sect. 13.4.2), it is not surprising that two mutations found in the catalytic domain of the gene encoding dyskerin are associated with a severe variant of X-DC known as Hoyeraal-Hreidarsson syndrome (HHS), an exceedingly rare condition characterized by growth retardation, microcephaly, cerebellar hypoplasia, and aplastic anemia (Hoyeraal et al. 1970; Hreidarsson et al. 1988; Yaghmai et al. 2000). The severe clinical phenotypes observed in HHS may result from a decrease in global protein synthesis due to a more acute inactivation of dyskerin catalytic activity. The genetic lesions underlying more than 50 % of DC cases remain unknown, and it will be important to uncover whether mutations in additional genes that also coordinate the assembly and function of the H/ACA snoRNP complex or in genes associated with ribosome function may underlie the pathological features of these patients.

13.5.2   Other Syndromes

13.5.2.1   Bowen-Conradi Syndrome

A mutation in the *NEP1* (Nucleolar Essential Protein 1, also known as *EMG1*) gene locus on chromosome 12p13 has been identified in Bowen-Conradi syndrome (BCS) (Armistead et al. 2009), a severe autosomal recessive developmental disorder (Hunter et al. 1979). In both yeast and human NEP1, a SPOUT-RNA methyltransferase is required for ribosome biogenesis and function (Eschrich et al. 2002; Liu and Thiele 2001). Recent studies have identified NEP1 as a pseudouridine N1-methyltransferase targeting Ψ1191 in the decoding center of yeast 18S rRNA (Meyer et al. 2011). This site corresponds to Ψ1248 in human 18S rRNA, which also contains the hypermodified $M^1acp^3\Psi$. NEP1 specifically methylates Ψ1191 at base position N1 in the nucleus, while the addition of the 3-amino-3-carboxypropyl group to yield $M^1acp^3\Psi$ occurs in the cytoplasm (Brand et al. 1978). Importantly, depletion of H/ACA snoRNA 35 (snr35), which catalyzes the pseudouridylation of Ψ1191 in yeast, severely impairs ribosome function (Liang et al. 2009). Additional studies in yeast reveal that Nep1 may support the association of Rps19 to the SSU (Buchhaupt et al. 2006). As previously mentioned, in this respect it seems plausible that specific rRNA modifications may be essential for enhancing RNA–protein interactions. Interestingly, BCS is characterized by impaired prenatal and postnatal growth, psychomotor retardation and death in early childhood (Hunter et al. 1979), clinical features similar to those observed in Diamond Blackfan Anemia (DBA) patients. Importantly, 25 % of DBA patients harbor a mutation in the gene encoding RPS19 (Ito et al. 2010). Is it possible that a partial loss of ribosomal proteins from ribosomal subunits may affect translational control of specific mRNAs important for distinct disease phenotypes? Recent data suggest that RPL38 mutant mice display severe developmental defects due to impaired translation of hox mRNAs (Kondrashov et al. 2011). Thus, it is reasonable to speculate that impaired control of gene expression at the level of translation may be implicated in ribosomopathies and may in part contribute to some of the severe clinical features associated with these diseases.

13.5.2.2   Mitochondrial Myopathy, Lactic Acidosis, and Sideroblastic Anemia

In addition to mutations in the gene encoding the Ψ synthase dyskerin in X-DC, it is interesting to note that a defect in another Ψ synthase is also implicated in human disease. All patients with the autosomal recessive disorder Myopathy, Lactic

Acidosis, and Sideroblastic Anemia (MLASA) harbor missense mutations in the gene encoding the pseudouridine synthase 1 (PUS1) (Bykhovskaya et al. 2004). In both yeast and mouse, PUS1 has been demonstrated to be a tRNA Ψ synthase. Pathological features of this mitochondrial disorder include myopathy, sideroblastic anemia, lactic acidosis, and mental retardation, among other severe symptoms. Importantly, loss of both cytoplasmic and mitochondrial tRNA pseudouridylation is evident in lymphoblasts from MLASA patients (Patton et al. 2005). Altogether, these findings support a role for reduced pseudouridylation in different species of RNA in human diseases.

13.6   Conclusions and Perspectives

Decades of research have provided a multitude of information regarding the unique chemical properties of Ψ residues. We have uncovered the position of Ψ residues within the ribosome, characterized the components involved in rRNA pseudouridylation, and provided insights into how Ψ modifications control gene expression at the translational level. Numerous multidisciplinary approaches, in particular the combination of biochemical analyses with X-ray crystallography, NMR spectroscopy, mass spectrometry, and in vivo model systems, have altogether determined an important contribution of rRNA pseudouridylation towards ribosome biogenesis and control of protein synthesis. However, several outstanding questions still remain to be addressed regarding the precise catalytic mechanisms of pseudouridylation, the kinetics of pseudouridylation, and the specific mechanisms by which Ψ modifications modulate translational control.

It will be important to determine if defective rRNA pseudouridylation may trigger an rRNA quality control mechanism. For example, nonfunctional rRNA decay (NRD) has been described in yeast as a mechanism to monitor mature rRNA in fully assembled ribosomal subunits (LaRiviere et al. 2006). Whether the NRD pathway may also eliminate ribosomal subunits lacking rRNA Ψ levels below a critical threshold, which would render these ribosomes completely inactive, remains to be investigated. Future directions in the field should also include studies aimed at determining if rRNA pseudouridylation is regulated in cells and if so, whether this can occur both temporally and spatially at the organismal level. Indeed, emerging data highlighting distinct expression patterns of H/ACA snoRNAs suggest that pseudouridylation of rRNA may be modulated in a tissue-specific manner. Additionally, understanding how rRNA pseudouridylation impinges on the structure of the ribosome remains to be further explored. It seems possible that rRNA modifications may affect ribosome composition and function by influencing how ribosomes bind to distinct proteins such as ITAFs, for example, which modulate translational specificity. Furthermore, it remains unknown whether rRNA Ψ modifications may control the engagement of ribosomal subunits to highly structured cis-acting regulatory elements in the 5′ and 3′ UTRs of distinct mRNAs, in addition to IRES elements and frameshift signals (Fig. 13.5). Similarly, it can also

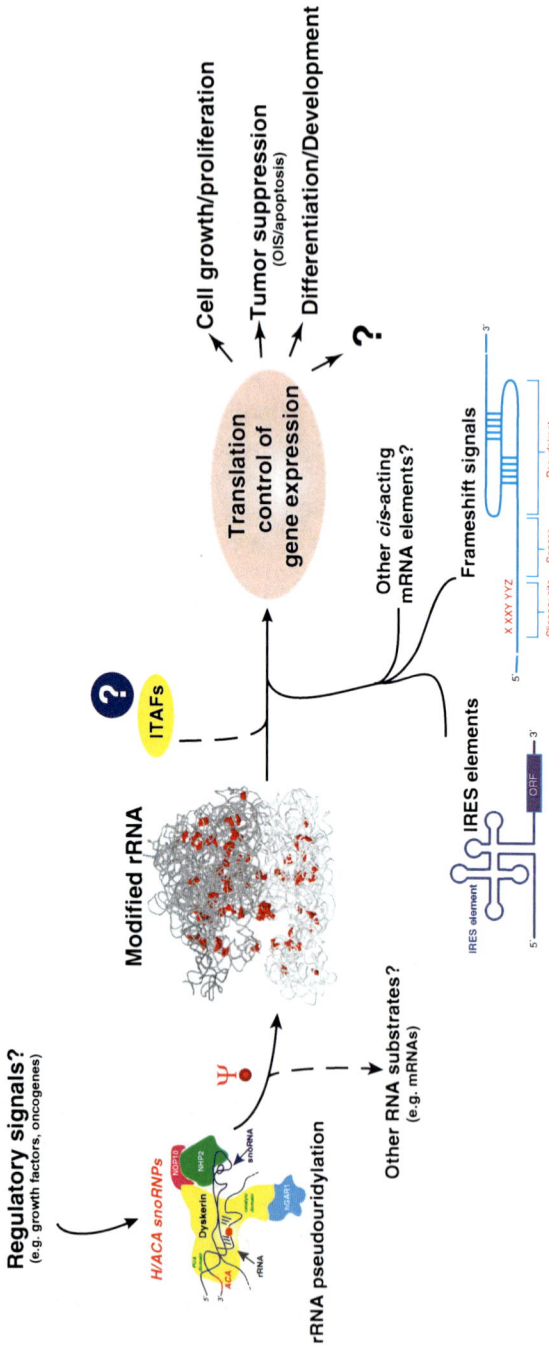

Fig. 13.5 rRNA pseudouridylation fine-tunes control of gene expression at the translational level. Schematic diagram representing the possible mechanisms by which pseudouridylation of rRNA may impinge on the translational control of gene expression. rRNA pseudouridylation is carried out by the H/ACA snoRNP complex; however, the regulatory signals modulating this process remain poorly understood. The incorporation of Ψ residues (*red*) at specific sites on rRNA affects the association of the ribosome with highly structured RNA elements including IRES sequences, programmed ribosomal frameshifting signals, and may also affect the recruitment of additional *cis*-acting mRNA regulatory elements yet to be identified. Ψ residues on rRNA may affect the recruitment of ITAFs and additional factors that enhance translation of specific mRNAs. Recent studies also suggest that other RNA substrates such as mRNA may also be targeted for pseudouridylation by the H/ACA snoRNPs

be envisioned that pseudouridylation of mRNAs may also regulate gene expression at the posttranscriptional level, particularly in light of findings that pseudouridylation can promote nonsense codon suppression (Karijolich and Yu 2011).

An emerging role for rRNA modifications in regulating specific modes of translation has recently been highlighted by findings that methylation of rRNA is also implicated in IRES-dependent translation of specific cellular mRNAs (Basu et al. 2011). Importantly, loss of rRNA methylation at specific sites in zebrafish leads to severe developmental defects. Specifically, site-specific reductions in rRNA methylation due to impaired function of three C/D snoRNAs is associated with tissue-specific morphological abnormalities, including deformities in the midbrain-hindbrain region, delayed pigmentation of the eyes, jaw structure abnormalities, and underdeveloped internal organs in zebrafish (Higa-Nakamine et al. 2012). It is tempting to speculate that specific rRNA residues are critical for modulating translational control of distinct genes required for particular stages of development. Future research efforts employing an unbiased loss of function analysis of individual snoRNAs are required to thoroughly comprehend the molecular mechanisms by which single site-specific rRNA Ψ or methyl modifications may regulate ribosome activity and translational control of gene expression.

In conclusion, it is evident that modification of rRNA plays a role in modulating gene expression at the translational level. Years of insightful analyses have undoubtedly contributed to our current understanding of the biological importance of RNA pseudouridylation and have laid the groundwork for future research to fully explore how pseudouridylation at specific sites on rRNA may impinge on translation initiation, control of gene expression at the translational level, and, when altered, may lead to human disease.

Acknowledgements We are grateful to members of the Ruggero lab for many helpful discussions and to Kimhouy Tong for critically reviewing and editing this manuscript. We apologize to those whose work we were unable to cite. This work is supported by National Institutes of Health grants R01 HL085572 (D.R.) and R01 CA140456 (D.R.). Davide Ruggero is a Leukemia and Lymphoma Society Research Scholar. Cristian Bellodi is a Leukemia and Lymphoma Society (LLS) and Aplastic Anemia and Myelodysplastic Syndromes (MDS) International Foundation (AA&MDS IF) research fellow.

References

Alawi F, Lee M (2007) DKC1 is an evolutionarily conserved c-Myc target. FASEB J 21:A1155

Alter BP, Giri N, Savage SA, Rosenberg PS (2009) Cancer in dyskeratosis congenita. Blood 113:6549–6557

Aravind L, Koonin EV (1999) Novel predicted RNA-binding domains associated with the translation machinery. J Mol Evol 48:291–302

Arena F, Ciliberto G, Ciampi S, Cortese R (1978) Purification of pseudouridylate synthetase I from *Salmonella typhimurium*. Nucleic Acids Res 5:4523–4536

Armanios M, Chen JL, Chang YP, Brodsky RA, Hawkins A, Griffin CA, Eshleman JR, Cohen AR, Chakravarti A, Hamosh A, Greider CW (2005) Haploinsufficiency of telomerase reverse transcriptase leads to anticipation in autosomal dominant dyskeratosis congenita. Proc Natl Acad Sci U S A 102:15960–15964

Armistead J, Khatkar S, Meyer B, Mark BL, Patel N, Coghlan G, Lamont RE, Liu S, Wiechert J, Cattini PA, Koetter P, Wrogemann K, Greenberg CR, Entian KD, Zelinski T, Triggs-Raine B (2009) Mutation of a gene essential for ribosome biogenesis, EMG1, causes Bowen-Conradi syndrome. Am J Hum Genet 84:728–739

Arnez JG, Steitz TA (1994) Crystal structure of unmodified tRNA(Gln) complexed with glutaminyl-tRNA synthetase and ATP suggests a possible role for pseudo-uridines in stabilization of RNA structure. Biochemistry 33:7560–7567

Bachellerie JP, Cavaille J (1997) Guiding ribose methylation of rRNA. Trends Biochem Sci 22:257–261

Bachellerie JP, Cavaille J, Huttenhofer A (2002) The expanding snoRNA world. Biochimie 84:775–790

Baker DL, Youssef OA, Chastkofsky MI, Dy DA, Terns RM, Terns MP (2005) RNA-guided RNA modification: functional organization of the archaeal H/ACA RNP. Genes Dev 19:1238–1248

Balakin AG, Smith L, Fournier MJ (1996) The RNA world of the nucleolus: two major families of small RNAs defined by different box elements with related functions. Cell 86:823–834

Barna M, Pusic A, Zollo O, Costa M, Kondrashov N, Rego E, Rao PH, Ruggero D (2008) Suppression of Myc oncogenic activity by ribosomal protein haploinsufficiency. Nature 456:971–975

Basu A, Das P, Chaudhuri S, Bevilacqua E, Andrews J, Barik S, Hatzoglou M, Komar AA, Mazumder B (2011) Requirement of rRNA methylation for 80S ribosome assembly on a cohort of cellular internal ribosome entry sites. Mol Cell Biol 31(22):4482–4499

Batista LF, Pech MF, Zhong FL, Nguyen HN, Xie KT, Zaug AJ, Crary SM, Choi J, Sebastiano V, Cherry A, Giri N, Wernig M, Alter BP, Cech TR, Savage SA, Reijo Pera RA, Artandi SE (2011) Telomere shortening and loss of self-renewal in dyskeratosis congenita induced pluripotent stem cells. Nature 474:399–402

Baudin-Baillieu A, Fabret C, Liang XH, Piekna-Przybylska D, Fournier MJ, Rousset JP (2009) Nucleotide modifications in three functionally important regions of the *Saccharomyces cerevisiae* ribosome affect translation accuracy. Nucleic Acids Res 37:7665–7677

Bellodi C, Kopmar N, Ruggero D (2010a) Deregulation of oncogene-induced senescence and p53 translational control in X-linked dyskeratosis congenita. EMBO J 29:1865–1876

Bellodi C, Krasnykh O, Haynes N, Theodoropoulou M, Peng G, Montanaro L, Ruggero D (2010b) Loss of function of the tumor suppressor DKC1 perturbs p27 translation control and contributes to pituitary tumorigenesis. Cancer Res 70:6026–6035

Bessler M, Wilson DB, Mason PJ (2004) Dyskeratosis congenita and telomerase. Curr Opin Pediatr 16:23–28

Bortolin ML, Ganot P, Kiss T (1999) Elements essential for accumulation and function of small nucleolar RNAs directing site-specific pseudouridylation of ribosomal RNAs. EMBO J 18:457–469

Bousquet-Antonelli C, Henry Y, G'Elugne JP, Caizergues-Ferrer M, Kiss T (1997) A small nucleolar RNP protein is required for pseudouridylation of eukaryotic ribosomal RNAs. EMBO J 16:4770–4776

Brand RC, Klootwijk J, Van Steenbergen TJ, De Kok AJ, Planta RJ (1977) Secondary methylation of yeast ribosomal precursor RNA. Eur J Biochem 75:311–318

Brand RC, Klootwijk J, Planta RJ, Maden BE (1978) Biosynthesis of a hypermodified nucleotide in *Saccharomyces carlsbergensis* 17S and HeLa-cell 18S ribosomal ribonucleic acid. Biochem J 169:71–77

Buchhaupt M, Meyer B, Kotter P, Entian KD (2006) Genetic evidence for 18S rRNA binding and an Rps19p assembly function of yeast nucleolar protein Nep1p. Mol Genet Genomics 276:273–284

Bykhovskaya Y, Casas K, Mengesha E, Inbal A, Fischel-Ghodsian N (2004) Missense mutation in pseudouridine synthase 1 (PUS1) causes mitochondrial myopathy and sideroblastic anemia (MLASA). Am J Hum Genet 74:1303–1308

Castle JC, Armour CD, Lower M, Haynor D, Biery M, Bouzek H, Chen R, Jackson S, Johnson JM, Rohl CA, Raymond CK (2010) Digital genome-wide ncRNA expression, including SnoRNAs, across 11 human tissues using polyA-neutral amplification. PLoS One 5:e11779

Cavaille J, Nicoloso M, Bachellerie JP (1996) Targeted ribose methylation of RNA in vivo directed by tailored antisense RNA guides. Nature 383:732–735

Cavaille J, Buiting K, Kiefmann M, Lalande M, Brannan CI, Horsthemke B, Bachellerie JP, Brosius J, Huttenhofer A (2000) Identification of brain-specific and imprinted small nucleolar RNA genes exhibiting an unusual genomic organization. Proc Natl Acad Sci U S A 97:14311–14316

Charette M, Gray MW (2000) Pseudouridine in RNA: what, where, how, and why. IUBMB Life 49:341–351

Cohn WE (1959) 5-Ribosyl uracil, a carbon-carbon ribofuranosyl nucleoside in ribonucleic acids. Biochim Biophys Acta 32:569–571

Conrad J, Sun D, Englund N, Ofengand J (1998) The rluC gene of *Escherichia coli* codes for a pseudouridine synthase that is solely responsible for synthesis of pseudouridine at positions 955, 2504, and 2580 in 23S ribosomal RNA. J Biol Chem 273:18562–18566

Conrad J, Niu L, Rudd K, Lane BG, Ofengand J (1999) 16S ribosomal RNA pseudouridine synthase RsuA of *Escherichia coli*: deletion, mutation of the conserved Asp102 residue, and sequence comparison among all other pseudouridine synthases. RNA 5:751–763

Cortese R, Kammen HO, Spengler SJ, Ames BN (1974) Biosynthesis of pseudouridine in transfer ribonucleic acid. J Biol Chem 249:1103–1108

Costantino DA, Pfingsten JS, Rambo RP, Kieft JS (2008) tRNA-mRNA mimicry drives translation initiation from a viral IRES. Nat Struct Mol Biol 15:57–64

Darzacq X, Kittur N, Roy S, Shav-Tal Y, Singer RH, Meier UT (2006) Stepwise RNP assembly at the site of H/ACA RNA transcription in human cells. J Cell Biol 173:207–218

Davis DR (1995) Stabilization of RNA stacking by pseudouridine. Nucleic Acids Res 23:5020–5026

Davis FF, Allen FW (1957) Ribonucleic acids from yeast which contain a fifth nucleotide. J Biol Chem 227:907–915

Davis DR, Veltri CA, Nielsen L (1998) An RNA model system for investigation of pseudouridine stabilization of the codon-anticodon interaction in tRNALys, tRNAHis and tRNATyr. J Biomol Struct Dyn 15:1121–1132

Decatur WA, Fournier MJ (2002) rRNA modifications and ribosome function. Trends Biochem Sci 27:344–351

Decatur WA, Fournier MJ (2003) RNA-guided nucleotide modification of ribosomal and other RNAs. J Biol Chem 278:695–698

Del Campo M, Kaya Y, Ofengand J (2001) Identification and site of action of the remaining four putative pseudouridine synthases in *Escherichia coli*. RNA 7:1603–1615

Dieci G, Fiorino G, Castelnuovo M, Teichmann M, Pagano A (2007) The expanding RNA polymerase III transcriptome. Trends Genet 23:614–622

Doi Y, Arakawa Y (2007) 16S ribosomal RNA methylation: emerging resistance mechanism against aminoglycosides. Clin Infect Dis 45:88–94

Dokal I (2000) Dyskeratosis congenita in all its forms. Br J Haematol 110:768–779

Dragon F, Pogacic V, Filipowicz W (2000) In vitro assembly of human H/ACA small nucleolar RNPs reveals unique features of U17 and telomerase RNAs. Mol Cell Biol 20:3037–3048

Duan J, Li L, Lu J, Wang W, Ye K (2009) Structural mechanism of substrate RNA recruitment in H/ACA RNA-guided pseudouridine synthase. Mol Cell 34:427–439

Ejby M, Sorensen MA, Pedersen S (2007) Pseudouridylation of helix 69 of 23S rRNA is necessary for an effective translation termination. Proc Natl Acad Sci U S A 104:19410–19415

Eschrich D, Buchhaupt M, Kotter P, Entian KD (2002) Nep1p (Emg1p), a novel protein conserved in eukaryotes and archaea, is involved in ribosome biogenesis. Curr Genet 40:326–338

Filipowicz W, Pogacic V (2002) Biogenesis of small nucleolar ribonucleoproteins. Curr Opin Cell Biol 14:319–327

Frischmeyer PA, Dietz HC (1999) Nonsense-mediated mRNA decay in health and disease. Hum Mol Genet 8:1893–1900

Ganapathi KA, Shimamura A (2008) Ribosomal dysfunction and inherited marrow failure. Br J Haematol 141:376–387

Ganot P, Bortolin ML, Kiss T (1997) Site-specific pseudouridine formation in preribosomal RNA is guided by small nucleolar RNAs. Cell 89:799–809

Ge J, Rudnick DA, He J, Crimmins DL, Ladenson JH, Bessler M, Mason PJ (2010) Dyskerin ablation in mouse liver inhibits rRNA processing and cell division. Mol Cell Biol 30:413–422

Giordano E, Peluso I, Senger S, Furia M (1999) Minifly, a *Drosophila* gene required for ribosome biogenesis. J Cell Biol 144:1123–1133

Girard JP, Lehtonen H, Caizergues-Ferrer M, Amalric F, Tollervey D, Lapeyre B (1992) GAR1 is an essential small nucleolar RNP protein required for pre-rRNA processing in yeast. EMBO J 11:673–682

Granneman S, Baserga SJ (2004) Ribosome biogenesis: of knobs and RNA processing. Exp Cell Res 296:43–50

Grozdanov PN, Roy S, Kittur N, Meier UT (2009) SHQ1 is required prior to NAF1 for assembly of H/ACA small nucleolar and telomerase RNPs. RNA 15:1188–1197

Gu X, Yu M, Ivanetich KM, Santi DV (1998) Molecular recognition of tRNA by tRNA pseudouridine 55 synthase. Biochemistry 37:339–343

Gu X, Liu Y, Santi DV (1999) The mechanism of pseudouridine synthase I as deduced from its interaction with 5-fluorouracil-tRNA. Proc Natl Acad Sci U S A 96:14270–14275

Gustafsson C, Reid R, Greene PJ, Santi DV (1996) Identification of new RNA modifying enzymes by iterative genome search using known modifying enzymes as probes. Nucleic Acids Res 24:3756–3762

Gutgsell NS, Deutscher MP, Ofengand J (2005) The pseudouridine synthase RluD is required for normal ribosome assembly and function in *Escherichia coli*. RNA 11:1141–1152

Hamma T, Ferre-D'Amare AR (2006) Pseudouridine synthases. Chem Biol 13:1125–1135

Hamma T, Ferre-D'Amare AR (2010) The box H/ACA ribonucleoprotein complex: interplay of RNA and protein structures in post-transcriptional RNA modification. J Biol Chem 285:805–809

Hamma T, Reichow SL, Varani G, Ferre-D'Amare AR (2005) The Cbf5-Nop10 complex is a molecular bracket that organizes box H/ACA RNPs. Nat Struct Mol Biol 12:1101–1107

He J, Navarrete S, Jasinski M, Vulliamy T, Dokal I, Bessler M, Mason PJ (2002) Targeted disruption of Dkc1, the gene mutated in X-linked dyskeratosis congenita, causes embryonic lethality in mice. Oncogene 21:7740–7744

Heiss NS, Knight SW, Vulliamy TJ, Klauck SM, Wiemann S, Mason PJ, Poustka A, Dokal I (1998) X-linked dyskeratosis congenita is caused by mutations in a highly conserved gene with putative nucleolar functions. Nat Genet 19:32–38

Hellen CU, Sarnow P (2001) Internal ribosome entry sites in eukaryotic mRNA molecules. Genes Dev 15:1593–1612

Henras A, Henry Y, Bousquet-Antonelli C, Noaillac-Depeyre J, Gelugne JP, Caizergues-Ferrer M (1998) Nhp2p and Nop10p are essential for the function of H/ACA snoRNPs. EMBO J 17:7078–7090

Henras AK, Capeyrou R, Henry Y, Caizergues-Ferrer M (2004) Cbf5p, the putative pseudouridine synthase of H/ACA-type snoRNPs, can form a complex with Gar1p and Nop10p in absence of Nhp2p and box H/ACA snoRNAs. RNA 10:1704–1712

Higa-Nakamine S, Suzuki T, Uechi T, Chakraborty A, Nakajima Y, Nakamura M, Hirano N, Kenmochi N (2012) Loss of ribosomal RNA modification causes developmental defects in zebrafish. Nucleic Acids Res 40(1):391–398

Hoang C, Ferre-D'Amare AR (2001) Cocrystal structure of a tRNA Psi55 pseudouridine synthase: nucleotide flipping by an RNA-modifying enzyme. Cell 107:929–939

Hoang C, Hamilton CS, Mueller EG, Ferre-D'Amare AR (2005) Precursor complex structure of pseudouridine synthase TruB suggests coupling of active site perturbations to an RNA-sequestering peripheral protein domain. Protein Sci 14:2201–2206

Holcik M, Sonenberg N (2005) Translational control in stress and apoptosis. Nat Rev Mol Cell Biol 6:318–327

Hoyeraal HM, Lamvik J, Moe PJ (1970) Congenital hypoplastic thrombocytopenia and cerebral malformations in two brothers. Acta Paediatr Scand 59:185–191

Hreidarsson S, Kristjansson K, Johannesson G, Johannsson JH (1988) A syndrome of progressive pancytopenia with microcephaly, cerebellar hypoplasia and growth failure. Acta Paediatr Scand 77:773–775

Huang L, Pookanjanatavip M, Gu X, Santi DV (1998) A conserved aspartate of tRNA pseudouridine synthase is essential for activity and a probable nucleophilic catalyst. Biochemistry 37:344–351

Hunter AG, Woerner SJ, Montalvo-Hicks LD, Fowlow SB, Haslam RH, Metcalf PJ, Lowry RB (1979) The Bowen-Conradi syndrome—a highly lethal autosomal recessive syndrome of microcephaly, micrognathia, low birth weight, and joint deformities. Am J Med Genet 3:269–279

Huttenhofer A, Kiefmann M, Meier-Ewert S, O'Brien J, Lehrach H, Bachellerie JP, Brosius J (2001) RNomics: an experimental approach that identifies 201 candidates for novel, small, non-messenger RNAs in mouse. EMBO J 20:2943–2953

Iglesias-Serret D, Pique M, Gil J, Pons G, Lopez JM (2003) Transcriptional and translational control of Mcl-1 during apoptosis. Arch Biochem Biophys 417:141–152

Ishitani R, Yokoyama S, Nureki O (2008) Structure, dynamics, and function of RNA modification enzymes. Curr Opin Struct Biol 18:330–339

Ito E, Konno Y, Toki T, Terui K (2010) Molecular pathogenesis in Diamond-Blackfan anemia. Int J Hematol 92:413–418

Jack K, Bellodi C, Landry DM, Niederer RO, Meskauskas A, Musalgaonkar S, Kopmar N, Krasnykh O, Dean AM, Thompson SR, Ruggero D, Dinman JD (2011) rRNA pseudouridylation defects affect ribosomal ligand binding and translational fidelity from yeast to human cells. Mol Cell 44:660–666

Jang SK, Krausslich HG, Nicklin MJ, Duke GM, Palmenberg AC, Wimmer E (1988) A segment of the 5′ nontranslated region of encephalomyocarditis virus RNA directs internal entry of ribosomes during in vitro translation. J Virol 62:2636–2643

Jiang W, Middleton K, Yoon HJ, Fouquet C, Carbon J (1993) An essential yeast protein, CBF5p, binds in vitro to centromeres and microtubules. Mol Cell Biol 13:4884–4893

Johnson L, Soll D (1970) In vitro biosynthesis of pseudouridine at the polynucleotide level by an enzyme extract from *Escherichia coli*. Proc Natl Acad Sci U S A 67:943–950

Kammen HO, Marvel CC, Hardy L, Penhoet EE (1988) Purification, structure, and properties of *Escherichia coli* tRNA pseudouridine synthase I. J Biol Chem 263:2255–2263

Karijolich J, Yu YT (2011) Converting nonsense codons into sense codons by targeted pseudouridylation. Nature 474:395–398

Kastan MB, Radin AI, Kuerbitz SJ, Onyekwere O, Wolkow CA, Civin CI, Stone KD, Woo T, Ravindranath Y, Craig RW (1991) Levels of p53 protein increase with maturation in human hematopoietic cells. Cancer Res 51:4279–4286

Kaya Y, Ofengand J (2003) A novel unanticipated type of pseudouridine synthase with homologs in bacteria, archaea, and eukarya. RNA 9:711–721

Khanna M, Wu H, Johansson C, Caizergues-Ferrer M, Feigon J (2006) Structural study of the H/ACA snoRNP components Nop10p and the 3′ hairpin of U65 snoRNA. RNA 12:40–52

King TH, Liu B, McCully RR, Fournier MJ (2003) Ribosome structure and activity are altered in cells lacking snoRNPs that form pseudouridines in the peptidyl transferase center. Mol Cell 11:425–435

Kirwan M, Dokal I (2008) Dyskeratosis congenita: a genetic disorder of many faces. Clin Genet 73:103–112

Kiss T (2001) Small nucleolar RNA-guided post-transcriptional modification of cellular RNAs. EMBO J 20:3617–3622

Kiss AM, Jady BE, Bertrand E, Kiss T (2004) Human box H/ACA pseudouridylation guide RNA machinery. Mol Cell Biol 24:5797–5807

Kiss T, Fayet-Lebaron E, Jady BE (2010) Box H/ACA small ribonucleoproteins. Mol Cell 37:597–606

Kiss-Laszlo Z, Henry Y, Bachellerie JP, Caizergues-Ferrer M, Kiss T (1996) Site-specific ribose methylation of preribosomal RNA: a novel function for small nucleolar RNAs. Cell 85:1077–1088

Klimasauskas S, Kumar S, Roberts RJ, Cheng X (1994) HhaI methyltransferase flips its target base out of the DNA helix. Cell 76:357–369

Kolodrubetz D, Burgum A (1991) Sequence and genetic analysis of NHP2: a moderately abundant high mobility group-like nuclear protein with an essential function in *Saccharomyces cerevisiae*. Yeast 7:79–90

Kondrashov N, Pusic A, Stumpf CR, Shimizu K, Hsieh AC, Xue S, Ishijima J, Shiroishi T, Barna M (2011) Ribosome-mediated specificity in hox mRNA translation and vertebrate tissue patterning. Cell 145:383–397

Koonin EV (1996) Pseudouridine synthases: four families of enzymes containing a putative uridine-binding motif also conserved in dUTPases and dCTP deaminases. Nucleic Acids Res 24:2411–2415

Krastev DB, Slabicki M, Paszkowski-Rogacz M, Hubner NC, Junqueira M, Shevchenko A, Mann M, Neugebauer KM, Buchholz F (2011) A systematic RNAi synthetic interaction screen reveals a link between p53 and snoRNP assembly. Nat Cell Biol 13:809–818

Krichevsky AM, Metzer E, Rosen H (1999) Translational control of specific genes during differentiation of HL-60 cells. J Biol Chem 274:14295–14305

Lafontaine DLJ, Tollervey D (1999) Nop58p is a common component of the box C+D snoRNPs that is required for snoRNA stability. RNA 5:455–467

Lafontaine DL, Bousquet-Antonelli C, Henry Y, Caizergues-Ferrer M, Tollervey D (1998) The box H+ACA snoRNAs carry Cbf5p, the putative rRNA pseudouridine synthase. Genes Dev 12:527–537

Lang KJ, Kappel A, Goodall GJ (2002) Hypoxia-inducible factor-1alpha mRNA contains an internal ribosome entry site that allows efficient translation during normoxia and hypoxia. Mol Biol Cell 13:1792–1801

LaRiviere FJ, Cole SE, Ferullo DJ, Moore MJ (2006) A late-acting quality control process for mature eukaryotic rRNAs. Mol Cell 24:619–626

Leader DJ, Clark GP, Watters J, Beven AF, Shaw PJ, Brown JW (1997) Clusters of multiple different small nucleolar RNA genes in plants are expressed as and processed from polycistronic pre-snoRNAs. EMBO J 16:5742–5751

Lestrade L, Weber MJ (2006) snoRNA-LBME-db, a comprehensive database of human H/ACA and C/D box snoRNAs. Nucleic Acids Res 34:D158–D162

Levy S, Avni D, Hariharan N, Perry RP, Meyuhas O (1991) Oligopyrimidine tract at the 5′ end of mammalian ribosomal protein mRNAs is required for their translational control. Proc Natl Acad Sci U S A 88:3319–3323

Lewis SM, Holcik M (2008) For IRES trans-acting factors, it is all about location. Oncogene 27:1033–1035

Li L, Ye K (2006) Crystal structure of an H/ACA box ribonucleoprotein particle. Nature 443:302–307

Liang XH, Liu Q, Fournier MJ (2007) rRNA modifications in an intersubunit bridge of the ribosome strongly affect both ribosome biogenesis and activity. Mol Cell 28:965–977

Liang XH, Liu Q, Fournier MJ (2009) Loss of rRNA modifications in the decoding center of the ribosome impairs translation and strongly delays pre-rRNA processing. RNA 15: 1716–1728

Limbach PA, Crain PF, McCloskey JA (1994) Summary: the modified nucleosides of RNA. Nucleic Acids Res 22:2183–2196

Liu PC, Thiele DJ (2001) Novel stress-responsive genes EMG1 and NOP14 encode conserved, interacting proteins required for 40S ribosome biogenesis. Mol Biol Cell 12:3644–3657

Liu Y, Elf SE, Miyata Y, Sashida G, Huang G, Di Giandomenico S, Lee JM, Deblasio A, Menendez S, Antipin J, Reva B, Koff A, Nimer SD (2009) p53 regulates hematopoietic stem cell quiescence. Cell Stem Cell 4:37–48

Long KS, Poehlsgaard J, Kehrenberg C, Schwarz S, Vester B (2006) The Cfr rRNA methyltransferase confers resistance to phenicols, lincosamides, oxazolidinones, pleuromutilins, and streptogramin A antibiotics. Antimicrob Agents Chemother 50:2500–2505

Luo Y, Li S (2007) Genome-wide analyses of retrogenes derived from the human box H/ACA snoRNAs. Nucleic Acids Res 35:559–571

Lyman SK, Gerace L, Baserga SJ (1999) Human Nop5/Nop58 is a component common to the box C/D small nucleolar ribonucleoproteins. RNA 5:1597–1604

Maden BE, Forbes J (1972) Standard and non standard products in combined T(1) plus pancreatic RNAase fingerprints of HeLa cell rRNA and its precursors. FEBS Lett 28:289–292

Maden BE, Salim M, Williamson R, Shepherd J (1972) Chemical studies on ribosomal ribonucleic acid and ribosome formation in HeLa cells. Biochem J 129:30P

Marsh JC, Will AJ, Hows JM, Sartori P, Darbyshire PJ, Williamson PJ, Oscier DG, Dexter TM, Testa NG (1992) "Stem cell" origin of the hematopoietic defect in dyskeratosis congenita. Blood 79:3138–3144

Maxwell ES, Fournier MJ (1995) The small nucleolar RNAs. Annu Rev Biochem 64:897–934

Meier UT (2006) How a single protein complex accommodates many different H/ACA RNAs. Trends Biochem Sci 31:311–315

Meyer B, Wurm JP, Kotter P, Leisegang MS, Schilling V, Buchhaupt M, Held M, Bahr U, Karas M, Heckel A, Bohnsack MT, Wohnert J, Entian KD (2011) The Bowen-Conradi syndrome protein Nep1 (Emg1) has a dual role in eukaryotic ribosome biogenesis, as an essential assembly factor and in the methylation of Psi1191 in yeast 18S rRNA. Nucleic Acids Res 39:1526–1537

Miskimins WK, Wang G, Hawkinson M, Miskimins R (2001) Control of cyclin-dependent kinase inhibitor p27 expression by cap-independent translation. Mol Cell Biol 21:4960–4967

Mitchell JR, Wood E, Collins K (1999) A telomerase component is defective in the human disease dyskeratosis congenita. Nature 402:551–555

Mochizuki Y, He J, Kulkarni S, Bessler M, Mason PJ (2004) Mouse dyskerin mutations affect accumulation of telomerase RNA and small nucleolar RNA, telomerase activity, and ribosomal RNA processing. Proc Natl Acad Sci U S A 101:10756–10761

Montanaro L (2010) Dyskerin and cancer: more than telomerase. The defect in mRNA translation helps in explaining how a proliferative defect leads to cancer. J Pathol 222:345–349

Montanaro L, Calienni M, Bertoni S, Rocchi L, Sansone P, Storci G, Santini D, Ceccarelli C, Taffurelli M, Carnicelli D, Brigotti M, Bonafe M, Trere D, Derenzini M (2010) Novel dyskerin-mediated mechanism of p53 inactivation through defective mRNA translation. Cancer Res 70:4767–4777

Morrissey JP, Tollervey D (1993) Yeast snR30 is a small nucleolar RNA required for 18S rRNA synthesis. Mol Cell Biol 13:2469–2477

Motorin Y, Helm M (2011) RNA nucleotide methylation. Wiley Interdiscip Rev RNA 2:611–631

Mueller EG (2002) Chips off the old block. Nat Struct Biol 9:320–322

Nabavi S, Nazar RN (2008) U3 snoRNA promoter reflects the RNA's function in ribosome biogenesis. Curr Genet 54:175–184

Narla A, Ebert BL (2010) Ribosomopathies: human disorders of ribosome dysfunction. Blood 115:3196–3205

Newton K, Petfalski E, Tollervey D, Caceres JF (2003) Fibrillarin is essential for early development and required for accumulation of an intron-encoded small nucleolar RNA in the mouse. Mol Cell Biol 23:8519–8527

Ni J, Tien AL, Fournier MJ (1997) Small nucleolar RNAs direct site-specific synthesis of pseudouridine in ribosomal RNA. Cell 89:565–573

Normand C, Capeyrou R, Quevillon-Cheruel S, Mougin A, Henry Y, Caizergues-Ferrer M (2006) Analysis of the binding of the N-terminal conserved domain of yeast Cbf5p to a box H/ACA snoRNA. RNA 12:1868–1882

Ofengand J (2002) Ribosomal RNA pseudouridines and pseudouridine synthases. FEBS Lett 514:17–25

Okazuka K, Wakabayashi Y, Kashihara M, Inoue J, Sato T, Yokoyama M, Aizawa S, Aizawa Y, Mishima Y, Kominami R (2005) p53 prevents maturation of T cell development to the immature CD4-CD8+ stage in Bcl11b-/- mice. Biochem Biophys Res Commun 328:545–549

Patton JR, Bykhovskaya Y, Mengesha E, Bertolotto C, Fischel-Ghodsian N (2005) Mitochondrial myopathy and sideroblastic anemia (MLASA): missense mutation in the pseudouridine synthase 1 (PUS1) gene is associated with the loss of tRNA pseudouridylation. J Biol Chem 280:19823–19828

Pelletier J, Sonenberg N (1988) Internal initiation of translation of eukaryotic mRNA directed by a sequence derived from poliovirus RNA. Nature 334:320–325

Piekna-Przybylska D, Decatur WA, Fournier MJ (2008) The 3D rRNA modification maps database: with interactive tools for ribosome analysis. Nucleic Acids Res 36:D178–D183

Pogacic V, Dragon F, Filipowicz W (2000) Human H/ACA small nucleolar RNPs and telomerase share evolutionarily conserved proteins NHP2 and NOP10. Mol Cell Biol 20:9028–9040

Pyronnet S, Dostie J, Sonenberg N (2001) Suppression of cap-dependent translation in mitosis. Genes Dev 15:2083–2093

Ramamurthy V, Swann SL, Paulson JL, Spedaliere CJ, Mueller EG (1999) Critical aspartic acid residues in pseudouridine synthases. J Biol Chem 274:22225–22230

Rashid R, Liang B, Baker DL, Youssef OA, He Y, Phipps K, Terns RM, Terns MP, Li H (2006) Crystal structure of a Cbf5-Nop10-Gar1 complex and implications in RNA-guided pseudouridylation and dyskeratosis congenita. Mol Cell 21:249–260

Raychaudhuri S, Conrad J, Hall BG, Ofengand J (1998) A pseudouridine synthase required for the formation of two universally conserved pseudouridines in ribosomal RNA is essential for normal growth of *Escherichia coli*. RNA 4:1407–1417

Raychaudhuri S, Niu L, Conrad J, Lane BG, Ofengand J (1999) Functional effect of deletion and mutation of the *Escherichia coli* ribosomal RNA and tRNA pseudouridine synthase RluA. J Biol Chem 274:18880–18886

Reichow SL, Hamma T, Ferre-D'Amare AR, Varani G (2007) The structure and function of small nucleolar ribonucleoproteins. Nucleic Acids Res 35:1452–1464

Richard P, Kiss T (2006) Integrating snoRNP assembly with mRNA biogenesis. EMBO Rep 7:590–592

Rozhdestvensky TS, Tang TH, Tchirkova IV, Brosius J, Bachellerie JP, Huttenhofer A (2003) Binding of L7Ae protein to the K-turn of archaeal snoRNAs: a shared RNA binding motif for C/D and H/ACA box snoRNAs in Archaea. Nucleic Acids Res 31:869–877

Ruggero D, Grisendi S, Piazza F, Rego E, Mari F, Rao PH, Cordon-Cardo C, Pandolfi PP (2003) Dyskeratosis congenita and cancer in mice deficient in ribosomal RNA modification. Science 299:259–262

Salim M, Williamson R, Maden BE (1970) Methylated oligonucleotides from HeLa cell ribosomal and nucleolar RNA. FEBS Lett 12:109–113

Samuelsson T, Olsson M (1990) Transfer RNA pseudouridine synthases in *Saccharomyces cerevisiae*. J Biol Chem 265:8782–8787

Schimmang T, Tollervey D, Kern H, Frank R, Hurt EC (1989) A yeast nucleolar protein related to mammalian fibrillarin is associated with small nucleolar RNA and is essential for viability. EMBO J 8:4015–4024

Scott MS, Avolio F, Ono M, Lamond AI, Barton GJ (2009) Human miRNA precursors with box H/ACA snoRNA features. PLoS Comput Biol 5:e1000507

Serrano M, Lin AW, McCurrach ME, Beach D, Lowe SW (1997) Oncogenic ras provokes premature cell senescence associated with accumulation of p53 and p16INK4a. Cell 88:593–602

Shaulsky G, Goldfinger N, Peled A, Rotter V (1991) Involvement of wild-type p53 in pre-B-cell differentiation in vitro. Proc Natl Acad Sci U S A 88:8982–8986

Sherrill KW, Byrd MP, Van Eden ME, Lloyd RE (2004) BCL-2 translation is mediated via internal ribosome entry during cell stress. J Biol Chem 279:29066–29074

Slatter TL, Ganesan P, Holzhauer C, Mehta R, Rubio C, Williams G, Wilson M, Royds JA, Baird MA, Braithwaite AW (2010) p53-mediated apoptosis prevents the accumulation of progenitor B cells and B-cell tumors. Cell Death Differ 17:540–550

Smith CM, Steitz JA (1998) Classification of gas5 as a multi-small-nucleolar-RNA (snoRNA) host gene and a member of the 5′-terminal oligopyrimidine gene family reveals common features of snoRNA host genes. Mol Cell Biol 18:6897–6909

Spedaliere CJ, Ginter JM, Johnston MV, Mueller EG (2004) The pseudouridine synthases: revisiting a mechanism that seemed settled. J Am Chem Soc 126:12758–12759

Spriggs KA, Bushell M, Mitchell SA, Willis AE (2005) Internal ribosome entry segment-mediated translation during apoptosis: the role of IRES-trans-acting factors. Cell Death Differ 12:585–591

Stoneley M, Willis AE (2004) Cellular internal ribosome entry segments: structures, trans-acting factors and regulation of gene expression. Oncogene 23:3200–3207

Tollervey D, Lehtonen H, Carmo-Fonseca M, Hurt EC (1991) The small nucleolar RNP protein NOP1 (fibrillarin) is required for pre-rRNA processing in yeast. EMBO J 10:573–583

Tollervey D, Lehtonen H, Jansen R, Kern H, Hurt EC (1993) Temperature-sensitive mutations demonstrate roles for yeast fibrillarin in pre-rRNA processing, pre-rRNA methylation, and ribosome assembly. Cell 72:443–457

Tortoriello G, de Celis JF, Furia M (2010) Linking pseudouridine synthases to growth, development and cell competition. FEBS J 277:3249–3263

Tsukiyama T, Ishida N, Shirane M, Minamishima YA, Hatakeyama S, Kitagawa M, Nakayama K (2001) Down-regulation of p27Kip1 expression is required for development and function of T cells. J Immunol 166:304–312

Tycowski KT, Steitz JA (2001) Non-coding snoRNA host genes in Drosophila: expression strategies for modification guide snoRNAs. Eur J Cell Biol 80:119–125

van Riggelen J, Yetil A, Felsher DW (2010) MYC as a regulator of ribosome biogenesis and protein synthesis. Nat Rev Cancer 10:301–309

Vitali P, Royo H, Seitz H, Bachellerie JP, Huttenhofer A, Cavaille J (2003) Identification of 13 novel human modification guide RNAs. Nucleic Acids Res 31:6543–6551

Vulliamy T, Marrone A, Goldman F, Dearlove A, Bessler M, Mason PJ, Dokal I (2001) The RNA component of telomerase is mutated in autosomal dominant dyskeratosis congenita. Nature 413:432–435

Vulliamy TJ, Marrone A, Knight SW, Walne A, Mason PJ, Dokal I (2006) Mutations in dyskeratosis congenita: their impact on telomere length and the diversity of clinical presentation. Blood 107:2680–2685

Vulliamy T, Beswick R, Kirwan M, Marrone A, Digweed M, Walne A, Dokal I (2008) Mutations in the telomerase component NHP2 cause the premature ageing syndrome dyskeratosis congenita. Proc Natl Acad Sci U S A 105:8073–8078

Walbott H, Machado-Pinilla R, Liger D, Blaud M, Rety S, Grozdanov PN, Godin K, van Tilbeurgh H, Varani G, Meier UT, Leulliot N (2011) The H/ACA RNP assembly factor SHQ1 functions as an RNA mimic. Genes Dev 25:2398–2408

Walne AJ, Vulliamy T, Marrone A, Beswick R, Kirwan M, Masunari Y, Al-Qurashi FH, Aljurf M, Dokal I (2007) Genetic heterogeneity in autosomal recessive dyskeratosis congenita with one subtype due to mutations in the telomerase-associated protein NOP10. Hum Mol Genet 16:1619–1629

Wang C, Meier UT (2004) Architecture and assembly of mammalian H/ACA small nucleolar and telomerase ribonucleoproteins. EMBO J 23:1857–1867

Watkins NJ, Gottschalk A, Neubauer G, Kastner B, Fabrizio P, Mann M, Luhrmann R (1998) Cbf5p, a potential pseudouridine synthase, and Nhp2p, a putative RNA-binding protein, are present together with Gar1p in all H BOX/ACA-motif snoRNPs and constitute a common bipartite structure. RNA 4:1549–1568

Watkins NJ, Segault V, Charpentier B, Nottrott S, Fabrizio P, Bachi A, Wilm M, Rosbash M, Branlant C, Luhrmann R (2000) A common core RNP structure shared between the small nucleolar box C/D RNPs and the spliceosomal U4 snRNP. Cell 103:457–466

Weber MJ (2006) Mammalian small nucleolar RNAs are mobile genetic elements. PLoS Genet 2:e205

Westman BJ, Verheggen C, Hutten S, Lam YW, Bertrand E, Lamond AI (2010) A proteomic screen for nucleolar SUMO targets shows SUMOylation modulates the function of Nop5/Nop58. Mol Cell 39:618–631

Wilson JE, Pestova TV, Hellen CU, Sarnow P (2000a) Initiation of protein synthesis from the A site of the ribosome. Cell 102:511–520

Wilson JE, Powell MJ, Hoover SE, Sarnow P (2000b) Naturally occurring dicistronic cricket paralysis virus RNA is regulated by two internal ribosome entry sites. Mol Cell Biol 20:4990–4999

Wolfraim LA, Letterio JJ (2005) Cutting edge: p27Kip1 deficiency reduces the requirement for CD28-mediated costimulation in naive CD8+ but not CD4+ T lymphocytes. J Immunol 174:2481–2484

Wu G, Xiao M, Yang C, Yu YT (2011) U2 snRNA is inducibly pseudouridylated at novel sites by Pus7p and snR81 RNP. EMBO J 30:79–89

Yaghmai R, Kimyai-Asadi A, Rostamiani K, Heiss NS, Poustka A, Eyaid W, Bodurtha J, Nousari HC, Hamosh A, Metzenberg A (2000) Overlap of dyskeratosis congenita with the Hoyeraal-Hreidarsson syndrome. J Pediatr 136:390–393

Yang Y, Isaac C, Wang C, Dragon F, Pogacic V, Meier UT (2000) Conserved composition of mammalian box H/ACA and box C/D small nucleolar ribonucleoprotein particles and their interaction with the common factor Nopp 140. Mol Biol Cell 11:567–577

Yarian CS, Basti MM, Cain RJ, Ansari G, Guenther RH, Sochacka E, Czerwinska G, Malkiewicz A, Agris PF (1999) Structural and functional roles of the N1- and N3-protons of psi at tRNA's position 39. Nucleic Acids Res 27:3543–3549

Yoon A, Peng G, Brandenburger Y, Zollo O, Xu W, Rego E, Ruggero D (2006) Impaired control of IRES-mediated translation in X-linked dyskeratosis congenita. Science 312:902–906

Yu CT, Allen FW (1959) Studies on an isomer of uridine isolated from ribonucleic acids. Biochim Biophys Acta 32:393–406

Zebarjadian Y, King T, Fournier MJ, Clarke L, Carbon J (1999) Point mutations in yeast CBF5 can abolish in vivo pseudouridylation of rRNA. Mol Cell Biol 19:7461–7472

Chapter 14
Translational Control of Synaptic Plasticity and Memory

Arkady Khoutorsky, Christos Gkogkas, and Nahum Sonenberg

14.1 Introduction

Synaptic plasticity refers to the ability of neurons to change synaptic transmission efficiency in response to specific stimuli. Different stimuli lead to either enhancement or depression of synaptic strength. In mammals, activity-dependent long-lasting enhancement of synaptic strength is called long-term potentiation (LTP), whereas a long-lasting decrease in synaptic efficiency is referred to as long-term depression (LTD). LTP exhibits two temporally distinct phases that are defined by their requirement for new gene expression: early-phase LTP (E-LTP) does not depend on protein synthesis, but relies on modification of preexisting components, whereas late-phase LTP (L-LTP) requires transcription and synthesis of new proteins. LTP is generally considered to be the cellular model for learning and memory, as both LTP and memory share similar molecular and cellular mechanisms. Similarly to LTP, memory formation presents two temporal phases with respect to their requirement of new gene expression. Short-term memory (STM) lasts minutes to hours and does not require the synthesis of new proteins, while long-term memory (LTM) lasts days or even a lifetime and is heavily dependent on the generation of new proteins (Abel et al. 1997; Kandel 2001).

Newly synthesized proteins contribute to structural changes that accompany long-lasting synaptic plasticity and memory. Synaptic stimulation induces the formation of new spines (Desmond and Levy 1983; Toni et al. 1999), and promotes protein synthesis-dependent enlargement of existing spines (Tanaka et al. 2008). These morphological changes are also observed after behavioral training (Ramirez-Amaya et al. 1999, 2001; Leuner et al. 2003).

A. Khoutorsky • C. Gkogkas • N. Sonenberg (✉)
Department of Biochemistry, Goodman Cancer Centre, McGill University,
Montreal, QC, Canada H3A 1A3
e-mail: nahum.sonenberg@mcgill.ca

J.D. Dinman (ed.), *Biophysical Approaches to Translational Control of Gene Expression*,
Biophysics for the Life Sciences 1, DOI 10.1007/978-1-4614-3991-2_14,
© Springer Science+Business Media New York 2013

Neurons are highly polarized cells with the cell body being separated by hundreds of micrometers from their distal dendrites where synaptic contacts are formed. One strategy to control the proteome at synapses is through protein synthesis in the cell body and subsequent transport of newly synthesized components to the distal locations. However, a substantial amount of data supports the idea that, in neurons, many proteins are translated locally at synapses from preexisting mRNAs (Bramham and Wells 2007). Indeed, dendrites of hippocampal neurons contain all the necessary components to translate new proteins: numerous mRNAs are localized in dendrites (Bramham and Wells 2007), ribosomes (Steward and Levy 1982), translation initiation factors such as eIF4G, eIF4E, eIF4A, 4E-BP, as well as signaling molecules of the MAPK-Mnk-eIF4E, mTOR, and CPEB1/gld2 pathway (Wu et al. 1998; Tang et al. 2002; Asaki et al. 2003). Moreover, LTP-inducing stimulation causes ribosomes to move from dendritic shafts to spines with enlarged synapses (Ostroff et al. 2002). Local translation at synaptic compartments enables fast changes in the synaptic proteome in response to stimuli, and offers better spatial control with a differential repertoire of proteins being translated at neighboring synapses of the same neuron (Steward et al. 1998). Remarkably, brain-derived neurotrophic factor (BDNF) elicits L-LTP at Schaffer collateral-CA1 synapses in hippocampus even when the cell bodies of presynaptic and postsynaptic neurons are physically severed from the axons and dendrites respectively (Kang and Schuman 1996). L-LTP under these conditions was still impaired by protein synthesis inhibitors, indicating that local translation in dendrites is required for L-LTP persistence. When the protein synthesis inhibitor emetine was applied locally to the apical dendrites of CA1 pyramidal cells in hippocampus, L-LTP was impaired at apical but not at basal dendrites (Bradshaw et al. 2003), further supporting a key role of local protein synthesis for long-lasting plasticity. Other forms of synaptic plasticity, such as L-LTP induced by subthreshold electrical stimulation in the presence of β-adrenergic receptors agonist and LTD, are all dependent on local protein synthesis (Huber et al. 2000; Gelinas and Nguyen 2005).

The rate of mRNA translation in eukaryotes is controlled by several signaling pathways (Sonenberg and Hinnebusch 2009). Translation initiation is the rate-limiting step under most circumstances and is a major target for regulation. In this chapter, we review the current knowledge on the mechanisms controlling the rate of protein synthesis in neurons in response to synaptic stimulation or behavioral training.

14.2   Translation in Eukaryotes

All nuclear transcribed eukaryotic mRNAs contain, at their 5′ end, the structure m7GpppN (where N is any nucleotide) termed the "cap," which facilitates ribosome recruitment to the mRNA. Ribosome recruitment requires a group of translation initiation factors, termed eIF4 (eukaryotic initiation factor 4). An important member of this group is eIF4F, which is a three subunit complex (Edery et al. 1983; Grifo et al. 1983) composed of (1) eIF4A, an RNA helicase that is thought to unwind the

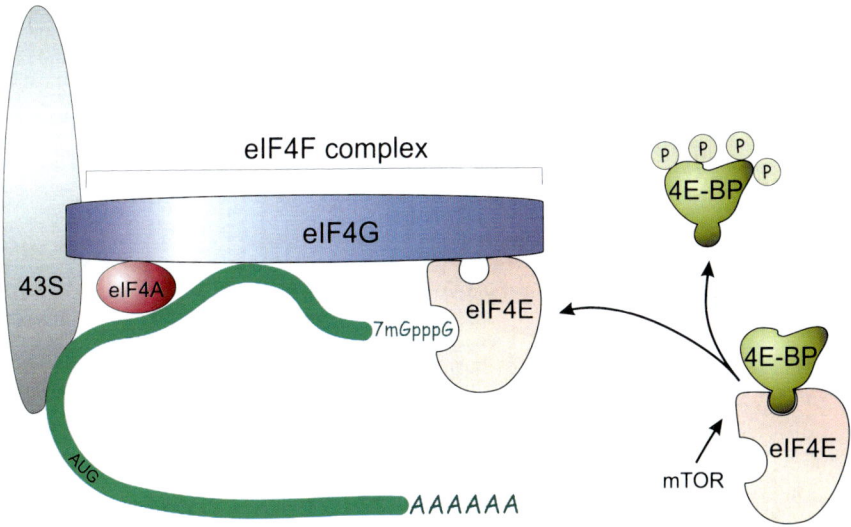

Fig. 14.1 Regulation of translation initiation by eIF4F complex formation The eukaryotic mRNA 5′-cap structure facilitates ribosome recruitment to the mRNA by the eIF4F complex, which consists of three subunits: (1) eIF4E, the cap binding protein; (2) eIF4A, an RNA helicase that unwinds the secondary structure of the mRNA 5′ UTR; and (3) eIF4G, a large scaffolding protein that bridges the mRNA to the 43S preinitiation ribosomal complex. Once bound to the 5′ end of the mRNA, the 43S ribosomal complex traverses the 5′ UTR in a 5′–3′ direction, until it encounters the initiation codon (AUG) where translation begins. 4E-BP inhibits cap-dependent translation initiation by preventing the assembly of the eIF4F complex. 4E-BP and eIF4G compete for binding to the convex dorsal surface of eIF4E. Phosphorylation of 4E-BP by mTOR reduces its affinity for eIF4E and thus leads to eIF4F complex formation and translation initiation

secondary structure of the 5′ UTR of the mRNA, (2) eIF4E, that specifically interacts with the cap structure (Sonenberg et al. 1979), and (3) eIF4G, a large scaffolding protein (of which there are two gene products, eIF4GI and eIF4GII) that binds to both eIF4E and eIF4A and bridges the mRNA to the 43S preinitiation complex. Thus, eIF4G serves as a modular scaffolding protein to assemble the protein machinery that directs the ribosome to the mRNA (Fig. 14.1). Because eIF4E generally exhibits the lowest expression level of all the eukaryotic initiation factors, the cap-recognition step is rate-limiting for translation and a major target for regulation.

A second major translational control mechanism is mediated by the translation initiation factor eIF2 (which is composed of three subunits α, β, and γ) (Sonenberg and Hinnebusch 2009). eIF2 binds Met-tRNA$_i^{Met}$ and GTP to form a ternary complex eIF2·GTP·Met-tRNA$_i^{Met}$. The ternary complex binds the small 40S ribosomal subunit, which is associated with other eIFs to form a 43S ribosomal preinitiation complex. Upon 60S subunit joining, GTP is hydrolyzed by eIF5 leading to the release of eIF2·GDP from the ribosome. An exchange of GDP for GTP on eIF2 is catalyzed by eIF2B and is required to prepare the functional ternary complex for a new round of translational initiation. Phosphorylation of Ser51 in the α subunit of eIF2 blocks the eIF2B-catalyzed GDP to GTP exchange reaction, leading to a decrease in general translation initiation (Dever 2002). eIF2α is phosphorylated by four protein kinases: protein kinase activated by double-stranded RNA (PKR), hemin-regulated inhibitor kinase (HRI), PKR-like endoplasmic reticulum kinase (PERK), and general control non-derepressible-2 kinase (GCN2), each of which is activated by different stress stimuli (Raven and Koromilas 2008).

14.3   Regulation of Translation Initiation

Although eIF4F complex assembly is regulated in several ways, the best-characterized mechanism is by the members of a family of three small molecular weight proteins, the eIF4E-binding proteins (4E-BPs). 4E-BP1, 4E-BP2, and 4E-BP3 specifically inhibit cap-dependent translation initiation by preventing the assembly of the eIF4F complex and, subsequently, ribosome recruitment to the mRNA (Pause et al. 1994; Haghighat et al. 1995). 4E-BPs and eIF4G share a canonical eIF4E-binding site through which they compete for binding to the convex dorsal surface of eIF4E (Fig. 14.1). Non-phosphorylated 4E-BPs bind with high affinity to eIF4E, thus preventing eIF4F complex formation and consequently inhibiting translation (Gingras et al. 1999). Upon phosphorylation, 4E-BPs dissociate from eIF4E, thus allowing eIF4F complex formation and relief of translational inhibition. The major protein kinase that phosphorylates 4E-BPs is the mechanistic/mammalian target of rapamycin (mTOR) (Hay and Sonenberg 2004).

14.3.1   mTOR Pathway

mTOR is a highly evolutionarily conserved serine/threonine kinase that regulates cell homeostasis through key cellular processes, including cell growth and proliferation, translation, autophagy, and cytoskeletal organization (Hay and Sonenberg 2004). The best studied and understood function of mTOR is its role in the control of translation. mTOR is present in two structurally and functionally distinct multiprotein complexes: mTORC1 (mTOR Complex 1) and mTORC2. mTORC1 contains the regulatory associated protein of mTOR (raptor), proline-rich AKT/PKB substrate 40 kD (PRAS40), and several other proteins and is sensitive to the drug rapamycin. Raptor functions as an adaptor protein that binds proteins containing the TOR signaling (TOS) motif and then presents them to the mTOR catalytic domain. Rapamycin is a macrolide that in a complex with the immunophilin FKBP12, inhibits mTOR kinase activity (Kim et al. 2002). Two key substrates of mTORC1 are p70S6 kinase (S6K1/2) and 4E-BPs. The second mTOR complex (mTORC2), which contains the rapamycin-insensitive companion of mTOR (rictor), mSIN1, and several other proteins, is rapamycin insensitive and phosphorylates AKT at Ser473 and protein kinase C alpha (PKCα) (Sarbassov et al. 2004). Though not affected by short rapamycin treatment, prolonged exposure to rapamycin inhibits mTORC2 activity in several cell lines (Sarbassov et al. 2006). mTORC2 mediates spatial control of cell growth by regulating PKCα and the actin cytoskeleton (Sarbassov et al. 2004). mTORC1 is a downstream target of the phosphatidylinositol-3 kinase (PI3K) pathway. PI3K kinase converts phosphatidylinositol-4,5-bisphosphate (PIP2) to phosphatidylinositol-3,4,5-trisphosphate (PIP3), leading to AKT phosphorylation and activation by PI3K-dependent kinase (PDK1) and mTORC2. AKT phosphorylates the tuberous sclerosis complex, consisting of TSC1 and TSC2, at its TSC2 subunit, leading to inhibition of its GTPase activity and, as a result, to activation of the G protein Ras-homolog enriched in brain (Rheb) and mTORC1 (Hay and Sonenberg 2004). Importantly, mTORC1 does not have a significant impact on the global translation rate, but regulates the translation of specific subsets of mRNAs. mTORC1 phosphorylation of the 4E-BPs leads to their dissociation from eIF4E and, as a result to eIF4F complex formation. mTORC1 regulates the translation of mRNAs with extensive secondary structures in their 5′ UTRs. This is achieved via augmenting the helicase activity of eIF4A (part of the eIF4F complex), that is believed to unwind secondary structures in the mRNA 5′ UTR and facilitates the binding and movement of the 40S ribosomal subunit (Parsyan et al. 2011).

Most of the studies on mTOR function in learning and memory predominantly relied on the use of rapamycin. Assessing the role of mTORC1 and mTORC2 by genetic deletion of mTOR, raptor, or rictor has been impeded due to prenatal lethality of the general knock-out (KO) mice (Gangloff et al. 2004; Sarbassov et al. 2004). However, current studies with conditional KO mice are likely to shed light on the role of mTORC1 and mTORC2 in learning and memory.

14.3.2 mTOR in Synaptic Plasticity and Memory Formation

mTOR activity is modulated by numerous neuronal receptors such as N-methyl-D-aspartate (NMDA), alpha-amino-3-hydroxy-5-methyl-4-isoxazolepropionic acid (AMPA), BDNF, metabotropic glutamate (mGlu), and dopaminergic (Hou and Klann 2004; Banko et al. 2005; Schicknick et al. 2008; Zhou et al. 2010). Compelling evidence supports an important role for mTOR in synaptic plasticity and memory formation in different species. The first evidence was derived from studies in *Aplysia* and crayfish (Yanow et al. 1998; Beaumont et al. 2001), and was later supported by numerous reports in vertebrates. Several learning-inducing behavioral tasks, as well as L-LTP and mGluR-LTD-inducing stimulation, trigger the phosphorylation of the mTORC1 downstream targets: 4E-BPs and S6K1/2 (Cammalleri et al. 2003; Hou and Klann 2004; Kelleher et al. 2004; Banko et al. 2005; Qin et al. 2005; Tsokas et al. 2005; Belelovsky et al. 2009). Importantly, activation of the mTOR pathway in response to synaptic activity was observed locally in dendrites, consistent with the idea of local translation in dendritic compartments (Tang and Schuman 2002; Cammalleri et al. 2003; Tsokas et al. 2005; Gobert et al. 2008).

mTOR inhibition by rapamycin blocks L-LTP (Tang et al. 2002), LTD (Hou and Klann 2004), and memory consolidation in mammals in a number of behavioral tasks (Shimizu et al. 2000; Dash et al. 2006; Bekinschtein et al. 2007; Schicknick et al. 2008; Stoica et al. 2011). A recent pharmocogenetic study showed that doses of rapamycin, which are ineffective in blocking L-LTP and memory in wild-type mice, inhibited L-LTP and LTM in mTOR heterozygous mice, thus providing the first genetic evidence for the role of mTOR in synaptic plasticity and memory formation (Stoica et al. 2011). Moreover, genetic mutants of the upstream and downstream components to mTOR have been used to study the role of the mTOR pathway in LTM.

For example, TSC1 and TSC2 heterozygous mice show hyperactivation of the mTOR pathway and alterations in synaptic plasticity and memory. TSC2$^{+/-}$ mice exhibit a lowered threshold for induction of L-LTP and impairment in hippocampus-dependent LTM (Ehninger et al. 2008). Remarkably, brief rapamycin treatment rescued the LTP and memory deficits, supporting the idea that the activity of mTOR in the optimal range is essential for memory formation. TSC1$^{+/-}$ mice similarly demonstrate LTM deficits and impaired social behavior (Goorden et al. 2007).

Conditional deletion of the rapamycin-binding protein FKBP12 in the forebrain of mice (cKO) enhanced mTOR activity and increased S6K1 phosphorylation (Hoeffer et al. 2008). L-LTP was enhanced relative to wild-type mice and was resistant to rapamycin but was sensitive to anisomycin, a potent inhibitor of protein synthesis. FKBP12 cKO mice exhibited enhanced contextual fear memory and obsessive-compulsive behavior in several tasks, a phenotype frequently associated with autism spectrum disorders (ASDs). Collectively, these results show that modulation of mTOR pathway activity has a strong impact on L-LTP and memory and can lead to ASD-associated behavior.

Intensive research has been conducted to elucidate the roles of the mTOR downstream targets, 4E-BPs and S6K1/2, in synaptic plasticity and memory formation. Since no conditional transgenic lines for these proteins are available, the conclusions derived from studies on general 4E-BPs and S6K1/2 knock-out mice should be interpreted with caution, since long-term adaptation and feedback mechanisms could be involved. Since 4E-BP2 is the major 4E-BP paralog in the brain, experiments focused on 4E-BP2 KO mice. These mice displayed spatial learning deficits and LTM impairment in several behavioral tasks (Banko et al. 2005, 2007), but they also exhibit a lowered threshold for the induction of L-LTP, similar to TSC2$^{+/-}$ and GCN2$^{-/-}$ mice (Banko et al. 2005). Strong stimulation led to the obstruction of L-LTP, which was suggested to result from translation hyperactivation in the absence of 4E-BP2 (Banko and Klann 2008). In support of this theory, it was found that although memory was impaired in most behavioral tasks, it was enhanced in an insular cortex-dependent form of LTM—conditional taste aversion (CTA). In this test, animals learn to associate an appetizing taste, such as saccharin, with gastric distress induced by intraperitoneal injection of LiCl.

In accordance with the well-established role of 4E-BPs, a recent study demonstrated the importance of eIF4F complex formation in memory consolidation (Hoeffer et al. 2011). Prevention of eIF4E's interaction with eIF4G by the small molecule 4EGI-1 impaired LTM, while STM remained intact. Interestingly, no effect on reconsolidation was observed, suggesting that the requirement for the eIF4E-eIF4G complex is different for these two memory processes.

The second major downstream targets of mTOR, S6K1 and S6K2, are also implicated in memory formation. Intact L-LTP was measured in slices from S6K1 and S6K2 KO mice. However, slices from S6K1-deficient mice displayed impaired protein synthesis-independent potentiation (E-LTP) (Antion et al. 2008b). Behavioral studies revealed that S6K1-deficient mice express an early-onset contextual fear memory deficit within 1 h of training, a deficit in conditioned taste aversion (CTA), impaired Morris water maze (MWM) acquisition, and hypoactive exploratory behavior. S6K2-deficient mice exhibit decreased contextual fear memory 7 days after training (but not one) and a reduction in latent inhibition of CTA. Surprisingly, mGluR-LTD was enhanced in S6K2 KO mice, whereas mGluR-LTD was intact in S6K1 knockouts (Antion et al. 2008a). Taken together, these results suggest that mTOR signaling to S6K1/2 is not critical for expression of long-lasting synaptic plasticity (L-LTP and LTD), findings that are difficult to explain in light of the memory phenotype in S6K1 KO mice (Antion et al. 2008b). Redundancy with other kinases, compensatory, or negative feedback mechanisms (for example activation of insulin receptor substrate 1 [IRS-1] in the absence of S6Ks) could explain the lack of an L-LTP phenotype. Although it is clear that S6K1/2 play a role in memory formation, more studies using conditional knockout or knock-in mice are required for more accurate examination of their functions. In contrast to the findings in mice, long-term facilitation (LTF) in *Aplysia* was blocked by expression of dominant negative S6K (Weatherill et al. 2010).

Both LTP and mGluR-LTD activate ERK and its downstream targets MAPK signal-integrating kinase/MAPK-interacting kinase 1 and 2 (Mnk1/2), that in turn phosphorylate eIF4E at Ser209 (Kelleher et al. 2004; Banko et al. 2006). This is consistent with the important role of ERK signaling in protein synthesis-dependent synaptic plasticity and memory formation.

14.3.3 Significance of eIF2α Phosphorylation in Memory Formation

eIF2α phosphorylation at Ser51 controls both the rate of general translation and gene-specific translation. eIF2α phosphorylation precludes ternary complex formation and consequently impairs general translation. Paradoxically, translation of mRNAs containing upstream open reading frames (uORFs) is stimulated (Harding et al. 2000; Vattem and Wek 2004). Activating transcription factor 4 (ATF4) contains two uORFs and its translation is stimulated upon eIF2α phosphorylation. ATF4, also known as cAMP-responsive element binding protein-2 (CREB-2), represses cAMP response element (CRE)-dependent transcription (Karpinski et al. 1992) that is widely considered to be critical for establishing long-lasting synaptic potentiation and memory (Bartsch et al. 1995; Pittenger et al. 2002). Stimulation of hippocampal slices with either BDNF, forskolin, tetanic stimulation, or behavioral training in rats reduced eIF2α phosphorylation, raising the possibility that the resulting increase in general translation or the decrease in ATF4 translation might be important for LTP and memory (Takei et al. 2001; Costa-Mattioli et al. 2007). Indeed, genetic reduction of eIF2α phosphorylation in eIF2α$^{+/S51A}$ knock-in mice or GCN2 (one of the major eIF2α kinases) ablation had a strong impact on LTP and memory (Costa-Mattioli et al. 2005, 2007). The threshold for L-LTP was lowered and learning and LTM was enhanced in MWM, associative fear conditioning, and CTA tasks, while STM remained intact. Accordingly, preventing activity-dependent eIF2α dephosphorylation with Sal003, an inhibitor of eIF2α phosphatase, impairs L-LTP and LTM. Interestingly, in slices from ATF4$^{-/-}$ mice, Sal003 was ineffective in reducing L-LTP, indicating that impairment of L-LTP by Sal003 is mediated by ATF4.

Another study examined whether eIF2α's regulation of general translation or gene-specific translation controls the conversion of STM to LTM (Jiang et al. 2010). When PKR (another eIF2α kinase)-mediated eIF2α phosphorylation was conditionally increased in CA1 pyramidal cells in the hippocampus, L-LTP and memory consolidation were impaired. The rate of de novo general translation was not affected, whereas the translation of ATF4 was enhanced and CREB-dependent transcription was suppressed. The authors concluded that the memory consolidation via dephosphorylation of eIF2α depends more on gene-specific translation (decreased ATF4 synthesis) and transcription (increased CREB-dependent transcription) than on general translation. This conclusion was supported by the fact that low doses of anisomycin, that reduced de novo general translation by ~70%, did not impair hippocampal-dependent LTM.

It is still unclear how the activity of the eIF2α kinases are regulated by neuronal activity. It was hypothesized that IMPACT, a GCN2 inhibitor that is present in the brain, is rapidly upregulated by synaptic activity, leading to GCN2 inhibition (Costa-Mattioli et al. 2009).

14.4   Regulation of Translation by Polyadenylation and CPEB

Poly(A) binding protein (PABP) interacts with both the mRNA poly(A) tail and eIF4G, leading to mRNA circularization and translational activation (Derry et al. 2006). The length of the poly(A) tail is controlled by the action of various poly(A) polymerases and deadenylases. Regulation of mRNA poly(A) tail length by neuronal activity emerges as an important mechanism in translational control of synaptic plasticity and memory. Cytoplasmic polyadenylation element (CPE)-binding protein (CPEB) binds CPE sequences at proximal 3′ UTR of mRNA and modulates its poly(A) length. Several CPEB-associated proteins have been identified in *Xenopus* oocytes: Gld2, a poly(A) polymerase, and PARN, a deadenylase and symplekin that serves as a scaffold upon which regulatory factors are assembled (Kim and Richter 2006). In unstimulated state the deadenylase activity of PARN prevents poly(A) tail elongation. Activation of NMDA and metabotropic Glutamate receptors (mGluRs) leads to CPEB phosphorylation by both Aurora A kinase and calcium/

calmodulin-dependent protein kinase II (CaMKII), triggering poly(A) tail elongation of CPE-containing mRNAs (probably by dissociation of PARN from the complex and Gld2 activation) and their enhanced translation (Huang et al. 2002; Atkins et al. 2004; Shin et al. 2004; Kim and Richter 2006). In neurons, CPEB is present in post-synaptic densities (PSD) in hippocampus (Wu et al. 1998). Importantly, several molecules with well established functions in learning and memory, such as CaMKIIα and tissue plasminogen activator (TPA), contain CPE sequences in their 3′ UTRs and their mRNA poly(A) tails are elongated in response to stimulation (Wu et al. 1998; Shin et al. 2004). Visual experience induces elongation of CaMKIIα mRNA poly(A) tail and its subsequent translation in the visual cortex (Wu et al. 1998). The significance of CPEB for protein synthesis-dependent synaptic plasticity and memory was investigated in CPEB-1 KO mice. Theta burst-induced L-LTP was impaired in slices from KO mice, while four trains of high frequency stimulation elicited normal L-LTP, suggesting stimulus specificity in engaging the CPEB pathway (Alarcon et al. 2004). Behavioral studies revealed normal hippocampus-dependent learning and memory; however, the extinction of these memories was impaired (Berger-Sweeney et al. 2006). Though extinction is protein synthesis-dependent, the underlying mechanisms are somewhat different from those of memory consolidation (Herry et al. 2010). Another study found that CPEB controls the translation of the transcription factor c-jun, which in turn regulates the expression of a growth hormone that is also implicated in synaptic plasticity (Zearfoss et al. 2008).

In *Aplysia*, CPEB is translated locally at the activated synapses and is required for persistence of long-term synaptic facilitation. Interestingly, *Aplysia* CPEB (ApCPEB) exhibits prion-like properties, forming a self-sustaining multimer (Si et al. 2003a, b). Formation of CPEB multimers is enhanced by the neurotransmitter serotonin, and is believed to contribute to persistence of synaptic facilitation over long periods of time (Miniaci et al. 2008; Si et al. 2010).

14.5 Role of miRNAs in Synaptic Plasticity and Memory

MicroRNAs (miRNAs) are small (~21 nucleotide) noncoding RNAs that bind to the 3′ UTR of target mRNAs and suppress their expression by mechanisms that are not fully elucidated. Many miRNAs are present in the brain (Kosik 2006) and the expression of some miRNAs is transcriptionally regulated by neuronal activity (Khudayberdiev et al. 2009; Nudelman et al. 2010; Wibrand et al. 2010). Components involved in the regulation of miRNA functions including Dicer, Argonaute, Fragile-X mental retardation protein (FMRP), P-bodies along with numerous miRNAs are present in dendrites, suggesting their possible role in the regulation of protein synthesis-dependent plasticity and memory (Lugli et al. 2005; Kye et al. 2007). For example, in *Aplysia*, miR-124 has been shown to constrain LTF and was rapidly downregulated by serotonin, a neurotransmitter critical for LTF (Rajasethupathy et al. 2009). In *Drosophila*, proteasome-dependent degradation of Armitage, a helicase required for miRNA-mediated gene repression, relieves CaMKII mRNA from translational repression, and allows for its translation at synapses, which is critical for olfactory learning (Ashraf et al. 2006). Armitage's mammalian ortholog, MOV10, is degraded in hippocampal neurons by the proteosome upon synaptic activity, leading to mRNA unsilencing and the dendritic translation of both CaMKIIα and the depalmitoylating enzyme lysophospholipase1 (Lypla1) (Banerjee et al. 2009). Recent reports illustrated the significance of miRNAs in synaptic plasticity and memory in mammals. Forebrain-specific deletion of Dicer decreased brain-specific miRNAs and enhanced memory in several behavioral tasks (Konopka et al. 2010). Expression of plasticity-related genes, such as BDNF and Matrix metallopeptidase 9, was increased in KO mice. Transgenic mice overexpressing miRNA-132 exhibited memory impairment in a novel object recognition task (Hansen et al. 2010). Further recent studies shed light on the functions of two other brain-specific miRNAs, miRNA-134 has been shown to be involved in synaptic plasticity and memory consolidation (Gao et al. 2010), while miRNA-128b regulates memory extinction (Lin et al. 2011).

14.6   Translational Regulation by Elongation Factors

While translational initiation is considered to be the rate-limiting step in translation and the major target for regulation, translational regulation of the elongation process has been implicated during distinct forms of synaptic plasticity and memory formation. Eukaryotic elongation factor 2 (eEF2) regulates the elongation step of translation and eEF2 phosphorylation by the specific eEF2 kinase (eEF2K) at Thr56 slows the rate of elongation and consequently decreases the rate of general translation. Remarkably, the translation of several mRNAs implicated in synaptic plasticity is enhanced. Phosphorylation of eEF2 during mGluR-LTD inhibits general translation, while specifically enhancing the translation of Arc/Arg3.1, CaMKIIα, and MAP1B mRNAs (Davidkova and Carroll 2007; Park et al. 2008). It was hypothesized that the stalling of elongation following eEF2 phosphorylation releases step-limiting initiation factors (such as eIF4E), leading to the enhanced translation of poorly translated mRNAs (Walden et al. 1981). Consistent with this hypothesis, blocking elongation with low doses of cycloheximide effectively reduces general translation, while enhancing the translation of Arc/Arg3.1 mRNAs (Park et al. 2008). eEF2 phosphorylation is regulated by neuronal activity. While it is induced by mGluR-LTD, LTP in dentate gyrus and taste conditioning, it is reduced by fear conditioning training (Belelovsky et al. 2005; Park et al. 2008; Im et al. 2009; Panja et al. 2009). eEF2α KO mice demonstrate impaired mGluR-LTD and enhanced L-LTP (Park et al. 2008) and overexpression of eEF2K in the hippocampus impairs L-LTP and long-term fear memory (Im et al. 2009). These results indicate that the phosphorylation status of eEF2 might have different impacts on synaptic depression (mGluR-LTD) and synaptic potentiation (L-LTP).

Additionally, activity-dependent synaptic spine maturation and dendritic BDNF expression has been shown to depend on mGluR-mediated phosphorylation of eEF2 by eEFK (Verpelli et al. 2010).

14.7 FMRP

Fragile X syndrome (FXS) is the most frequent form of single-gene inherited mental retardation and many cases manifest as ASDs (Levenga et al. 2010). In humans, the FMR1 gene codes for the FMRP, a transcript-specific RNA-binding protein whose action is generally believed to repress translation (Hinds et al. 1993). Various mutations causing loss of FMRP expression result in FXS. FMRP is highly expressed in the mammalian brain (Hinds et al. 1993). Moreover, FMRP is localized to neuronal dendrites and dendritic spines. The RNA binding activity of FMRP is thought to be mediated through three domains: two hnRNP-K-homology (KH) domains and an RGG box (arginine-glycine-glycine) domain. FMRP selectively binds approximately 4% of mammalian mRNAs and is believed to help regulate their function (Hinds et al. 1993). In several studies, it has been clearly shown that FMRP associates with actively translating polyribosomes in neuronal and non-neuronal cells and in brain synaptoneurosomes (Feng et al. 1997b). This association is mediated by the KH domains of FMRP. Interestingly, a missense mutation in the second KH domain has been reported in a rare FXS patient (Feng et al. 1997a). Phosphorylated FMRP associates with stalled ribosomes, whereas nonphosphorylated FMRP cosediments with actively translating ribosomes (Ceman et al. 2003). FMRP phosphorylation is regulated by S6K1 (Narayanan et al. 2008) which phosphorylates FMRP on a conserved serine residue that is required for mRNA binding.

A recent high-throughput study demonstrated that FMRP interacts with the coding region of pre- and post-synaptic transcripts, as well as ASD-associated transcripts (Darnell et al. 2011). The mechanism proposed for FMRP's translational repression of bound transcripts is via the stalling of ribosomal translocation on these target mRNAs (Darnell et al. 2011). Another model suggests that FMRP represses translation through its interaction with BC1 (a neuronal RNA) (Zalfa et al. 2003). BC1 binding to FMRP can lead to increased binding to its target mRNAs or to the inhibition of eIF4A, thus favoring the repression of translation initiation on mRNAs with structured 5′ UTRs (Wang et al. 2002; Lin et al. 2008). However, the FMRP-BC1 interaction is contentious since several studies have failed to recapitulate this finding (Bagni 2008; Iacoangeli et al. 2008). Recently, it was also shown that FMRP interacts with Cytoplasmic FMRP Interacting Protein 1 (CYFIP1), which is an eIF4E-binding protein that competes for eIF4E binding with eIF4G (Napoli et al. 2008). Unlike the 4E-BPs, CYFIP1 does not bind to eIF4E through the canonical eIF4E-binding site.

FMR1 KO animals have provided invaluable insight into the molecular pathways associated with FMRP in the brain. A salient feature of FXS patients is the increased occurrence of dendritic spines with a long, thin morphology (immature spines)

(Irwin et al. 2000; Grossman et al. 2006). This phenotype has also been observed in FMR1 KO mice (Comery et al. 1997). Deletion of FMR1 in mice leads to an increased rate of basal protein synthesis (Qin et al. 2005), enhanced seizure susceptibility (Grossman et al. 2010), impaired social interaction (Spencer et al. 2005; Mineur et al. 2006), altered learning and memory (Mineur et al. 2002; Nielsen et al. 2002; Zhao et al. 2005), and enhanced mGluR-LTD, which is no longer dependent on protein synthesis (Koekkoek et al. 2005; Nosyreva and Huber 2006). In agreement with S6K-dependent phosphorylation of FMRP, S6K2, and S6K1/2 double-knockout mice display enhanced mGluR-LTD, while in S6K1 KO mice mGluR-LTD is not altered (Antion et al. 2008a). This LTD phenotype in FMR1 KO mice supports the mGluR theory of FXS, which posits that FMRP acts as a negative translational regulator downstream of the mGluRs. mGluR-LTD depends on dendritic protein synthesis, which is required for AMPA receptors internalization. In the absence of FMRP, there is increased protein synthesis, which leads to excessive AMPA receptors internalization. Thus the inhibition of mGluRs could be a useful therapy for FXS. Indeed, genetic reduction of mGluR5 in FMR1 KO mice (Dolen et al. 2007) or treatment with the mGluR5 antagonist MPEP (Yan et al. 2005), reverse FMR1 KO-associated phenotypes.

14.8   Aberrant Translation Leads to Neurological Diseases

Dysregulation of translation leads to pathological conditions such as cancer, obesity, and memory impairment. Bidirectional alterations in synaptic protein synthesis leading to increased or decreased levels of plasticity-related proteins have been proposed to impair neural network performance in a fraction of patients with ASD (Sonenberg and Hinnebusch 2007; Costa-Mattioli et al. 2009), and lead to impairment of cognition, impaired language, communication, social interaction, and relationships (Fombonne 2009). Interestingly, in some cases, this can lead to the development of "savant abilities" which are restricted instances of normal or superior skills in a variety of artistic or cognitive domains (Heaton and Wallace 2004).

In several single-gene disorders with high rates of autism, levels of synaptic proteins and connectivity are increased. In FXS intellectual disability, disruptive and autistic-like behavior, epileptic seizures and language impairment are observed (Levenga et al. 2010). Moreover, mutations in genes encoding for proteins upstream of mTOR have been linked to autism in humans (Wiznitzer 2004; Butler et al. 2005). For instance, in Tuberous sclerosis complex (*TSC1/2* genes, inhibitors of the mTOR pathway) 85% of cases display cognitive impairment tightly linked to autistic features, epilepsy and abnormal or absent speech (Curatolo et al. 2010). Similarly, in PTEN hamartoma syndrome (caused by mutations of the *PTEN* gene, an inhibitor of the PI3K/mTOR pathway), neurobehavioral features resembling autism, macrocephaly, and language impairment are diagnosed in a subset of patients (Butler et al. 2005).

Deletion of PTEN or TSC2 in the mouse brain results in macrocephaly (Kwon et al. 2006; Ehninger et al. 2008; Way et al. 2009), which is reminiscent of the high prevalence of macrocephaly amongst children with ASD (Courchesne 2004). Knockout mouse models of PTEN display deficits in social interaction, social learning, and behavioral abnormalities in response to sensory stimuli, anxiety, and learning (Kwon et al. 2006). These mice also have abnormal dendritic and axonal growth and synapse numbers. Similarly, TSC2$^{+/-}$ mice display memory deficits (Ehninger et al. 2008) and an enlargement of neuronal somata and dendritic spines (Tavazoie et al. 2005). Activation of the mTOR pathway leads to growth and branching of dendrites, whereas rapamycin treatment or the overexpression of 4E-BP1 blocks the effect in cultured neurons (Jaworski et al. 2005). Rapamycin prevents and reverses neuronal hypertrophy, resulting in the amelioration of a subset of PTEN-associated abnormal behaviors, thus providing strong evidence that the mTORC1 pathway downstream of PTEN is critical for this complex phenotype (Zhou et al. 2009). Interestingly, brief treatment with rapamycin in adult TSC2$^{+/-}$ mice rescues not only the synaptic plasticity deficit, but also the behavioral deficits in this mouse model. Finally, FKBP12 conditional knockout (cKO) mice display autistic/obsessive-compulsive-like perseveration (Hoeffer et al. 2008).

A link between eIF4E and autism was recently reported. In a boy with classic autism a de novo chromosome translocation between 4q and 5q was mapped to the eIF4E gene (Neves-Pereira et al. 2009). Strikingly, the screening of 120 autism families for mutations revealed two unrelated families in which both autistic siblings and one of the parents harbored the same single nucleotide insertion at position 25 in the basal element of the eIF4E promoter (Neves-Pereira et al. 2009). This mutation results in both increased levels of eIF4E mRNA and protein amounts.

The implication of eIF4E in ASDs is further strengthened by recent findings that CYFIP1, an eIF4E-binding protein, is also associated with ASDs. CYFIP1 is a binding partner of FMRP (Schenck et al. 2001, 2003) and directly binds to eIF4E to repress neuronal translation in an activity-dependent manner (Napoli et al. 2008). Thus, CYFIP1 represses translation by directly binding to eIF4E and preventing the formation of the eIF4F complex, in a manner similar to the 4EBPs. Interestingly, CYFIP1 has been proposed as a risk gene for autism in ASDs, Prader Willie syndrome (PWS), and Angelman syndrome patients through microduplication or microdeletion (Doornbos et al. 2009) of the 15q11.2 chromosomal region or deletions in the 15q.11-q13 (Sahoo et al. 2006). Furthermore, CYFIP1 mRNA levels are reduced in both PWS and FXS patients (Nowicki et al. 2007). This shows that eIF4F complex formation, which regulates translation initiation, can be dysregulated in ASDs.

A common feature of neurodegenerative diseases (such as Parkinson's or Alzheimer's) is the accumulation of aberrant or misfolded proteins that eventually aggregate to form inclusion bodies. This pathological state can be attributed to dysregulated translational control. In Parkinson's disease (PD), resting tremor, slowness of movement, rigidity, and postural instability are linked to the cell-death of dopaminergic neurons in the substantia nigra. In *Drosophila* models of PD, it has been shown that rapamycin can have beneficial effects (preventing loss of neurons, flight muscle degeneration, climbing deficits, and mitochondrial alterations) and that this effect depends on 4E-BP activation but not autophagy induction (Tain et al. 2009). In Alzheimer's disease (AD), profound memory loss and cognitive dysfunction have been linked to increased mTOR activity and signaling in cellular and animal models (Caccamo et al. 2010, 2011). In these AD models, Aβ (beta-amyloid) accumulations induce mTOR hyperactivation and rapamycin treatment decreases Aβ depositions and rescues cognitive deficits. Interestingly, phosphorylation of eIF2α has been shown to increase the translation of BACE1 (beta-site APP cleaving enzyme-1) (O'Connor et al. 2008). BACE1 is elevated in AD and is the rate limiting enzyme for Aβ production. Accordingly, phospho-PKR and phospho-eIF2α immunoreactivity is higher in AD-affected brains compared to age-matched controls (Chang et al. 2002).

14.9 Conclusions

Neuronal activity induces mild increase of general translation, but the translation of certain mRNAs which are critical for synaptic plasticity and memory is significantly stimulated. Although inhibition of mTOR signaling by rapamycin reduces total protein synthesis by only ~12%, in contrast to the 50–90% reduction by the general protein synthesis inhibitor anisomycin (Beretta et al. 1996; Parsons et al. 2006), strikingly, rapamycin impairs memory as strongly as anisomycin. This is explained by the fact that mTOR stimulates the translation of a subset of mRNAs which are critical for memory formation. However, the identity of these mRNAs remains unknown. Similarly, regulation of specific mRNAs by eIF2α dephosphorylation, in contrast to the enhancement of global translation, appears to be critical for synaptic plasticity and memory formation (Costa-Mattioli et al. 2007; Jiang et al. 2010). Translational control at the 3′ UTR by regulation of poly(A) tail length is emerging as an important mechanism for synaptic plasticity and ongoing in vivo research will shed new light on the role of this important mode of translational regulation in memory formation. Most importantly, the aberrant translational control in ASD mouse models and the high percentage of single-gene mutations in the upstream components of the mTOR pathway (TSC1/TSC2 and PTEN) in ASD patients prompted intensive research to identify the mRNAs whose abnormal translation leads to an autistic phenotype.

Acknowledgments We thank Valerie Henderson and Ruifeng Cao for critical reading of the chapter.

References

Abel T, Nguyen PV, Barad M, Deuel TA, Kandel ER, Bourtchouladze R (1997) Genetic demonstration of a role for PKA in the late phase of LTP and in hippocampus-based long-term memory. Cell 88(5):615–626
Alarcon JM, Hodgman R, Theis M, Huang YS, Kandel ER, Richter JD (2004) Selective modulation of some forms of schaffer collateral-CA1 synaptic plasticity in mice with a disruption of the CPEB-1 gene. Learn Mem 11(3):318–327

Antion MD, Hou L, Wong H, Hoeffer CA, Klann E (2008a) mGluR-dependent long-term depression is associated with increased phosphorylation of S6 and synthesis of elongation factor 1A but remains expressed in S6K-deficient mice. Mol Cell Biol 28(9):2996–3007

Antion MD, Merhav M, Hoeffer CA, Reis G, Kozma SC, Thomas G, Schuman EM, Rosenblum K, Klann E (2008b) Removal of S6K1 and S6K2 leads to divergent alterations in learning, memory, and synaptic plasticity. Learn Mem 15(1):29–38

Asaki C, Usuda N, Nakazawa A, Kametani K, Suzuki T (2003) Localization of translational components at the ultramicroscopic level at postsynaptic sites of the rat brain. Brain Res 972(1–2):168–176

Ashraf SI, McLoon AL, Sclarsic SM, Kunes S (2006) Synaptic protein synthesis associated with memory is regulated by the RISC pathway in Drosophila. Cell 124(1):191–205

Atkins CM, Nozaki N, Shigeri Y, Soderling TR (2004) Cytoplasmic polyadenylation element binding protein-dependent protein synthesis is regulated by calcium/calmodulin-dependent protein kinase II. J Neurosci 24(22):5193–5201

Bagni C (2008) On BC1 RNA and the fragile X mental retardation protein. Proc Natl Acad Sci U S A 105(17):E19

Banerjee S, Neveu P, Kosik KS (2009) A coordinated local translational control point at the synapse involving relief from silencing and MOV10 degradation. Neuron 64(6):871–884

Banko JL, Klann E (2008) Cap-dependent translation initiation and memory. Prog Brain Res 169:59–80

Banko JL, Poulin F, Hou L, DeMaria CT, Sonenberg N, Klann E (2005) The translation repressor 4E-BP2 is critical for eIF4F complex formation, synaptic plasticity, and memory in the hippocampus. J Neurosci 25(42):9581–9590

Banko JL, Hou L, Poulin F, Sonenberg N, Klann E (2006) Regulation of eukaryotic initiation factor 4E by converging signaling pathways during metabotropic glutamate receptor-dependent long-term depression. J Neurosci 26(8):2167–2173

Banko JL, Merhav M, Stern E, Sonenberg N, Rosenblum K, Klann E (2007) Behavioral alterations in mice lacking the translation repressor 4E-BP2. Neurobiol Learn Mem 87(2):248–256

Bartsch D, Ghirardi M, Skehel PA, Karl KA, Herder SP, Chen M, Bailey CH, Kandel ER (1995) Aplysia CREB2 represses long-term facilitation: relief of repression converts transient facilitation into long-term functional and structural change. Cell 83(6):979–992

Beaumont V, Zhong N, Fletcher R, Froemke RC, Zucker RS (2001) Phosphorylation and local presynaptic protein synthesis in calcium- and calcineurin-dependent induction of crayfish long-term facilitation. Neuron 32(3):489–501

Bekinschtein P, Katche C, Slipczuk LN, Igaz LM, Cammarota M, Izquierdo I, Medina JH (2007) mTOR signaling in the hippocampus is necessary for memory formation. Neurobiol Learn Mem 87(2):303–307

Belelovsky K, Elkobi A, Kaphzan H, Nairn AC, Rosenblum K (2005) A molecular switch for translational control in taste memory consolidation. Eur J Neurosci 22(10):2560–2568

Belelovsky K, Kaphzan H, Elkobi A, Rosenblum K (2009) Biphasic activation of the mTOR pathway in the gustatory cortex is correlated with and necessary for taste learning. J Neurosci 29(23):7424–7431

Beretta L, Svitkin YV, Sonenberg N (1996) Rapamycin stimulates viral protein synthesis and augments the shutoff of host protein synthesis upon picornavirus infection. J Virol 70(12):8993–8996

Berger-Sweeney J, Zearfoss NR, Richter JD (2006) Reduced extinction of hippocampal-dependent memories in CPEB knockout mice. Learn Mem 13(1):4–7

Bradshaw KD, Emptage NJ, Bliss TV (2003) A role for dendritic protein synthesis in hippocampal late LTP. Eur J Neurosci 18(11):3150–3152

Bramham CR, Wells DG (2007) Dendritic mRNA: transport, translation and function. Nat Rev Neurosci 8(10):776–789

Butler MG, Dasouki MJ, Zhou XP, Talebizadeh Z, Brown M, Takahashi TN, Miles JH, Wang CH, Stratton R, Pilarski R et al (2005) Subset of individuals with autism spectrum disorders and extreme macrocephaly associated with germline PTEN tumour suppressor gene mutations. J Med Genet 42(4):318–321

Caccamo A, Majumder S, Richardson A, Strong R, Oddo S (2010) Molecular interplay between mammalian target of rapamycin (mTOR), amyloid-beta, and Tau: effects on cognitive impairments. J Biol Chem 285(17):13107–13120

Caccamo A, Maldonado MA, Majumder S, Medina DX, Holbein W, Magri A, Oddo S (2011) Naturally secreted amyloid-beta increases mammalian target of rapamycin (mTOR) activity via a PRAS40-mediated mechanism. J Biol Chem 286(11):8924–8932

Cammalleri M, Lutjens R, Berton F, King AR, Simpson C, Francesconi W, Sanna PP (2003) Time-restricted role for dendritic activation of the mTOR-p70S6K pathway in the induction of late-phase long-term potentiation in the CA1. Proc Natl Acad Sci U S A 100(24):14368–14373

Ceman S, O'Donnell WT, Reed M, Patton S, Pohl J, Warren ST (2003) Phosphorylation influences the translation state of FMRP-associated polyribosomes. Hum Mol Genet 12(24):3295–3305

Chang RC, Wong AK, Ng HK, Hugon J (2002) Phosphorylation of eukaryotic initiation factor-2alpha (eIF2alpha) is associated with neuronal degeneration in Alzheimer's disease. Neuroreport 13(18):2429–2432

Comery TA, Harris JB, Willems PJ, Oostra BA, Irwin SA, Weiler IJ, Greenough WT (1997) Abnormal dendritic spines in fragile X knockout mice: maturation and pruning deficits. Proc Natl Acad Sci U S A 94(10):5401–5404

Costa-Mattioli M, Gobert D, Harding H, Herdy B, Azzi M, Bruno M, Bidinosti M, Ben Mamou C, Marcinkiewicz E, Yoshida M et al (2005) Translational control of hippocampal synaptic plasticity and memory by the eIF2alpha kinase GCN2. Nature 436(7054):1166–1173

Costa-Mattioli M, Gobert D, Stern E, Gamache K, Colina R, Cuello C, Sossin W, Kaufman R, Pelletier J, Rosenblum K et al (2007) eIF2alpha phosphorylation bidirectionally regulates the switch from short- to long-term synaptic plasticity and memory. Cell 129(1):195–206

Costa-Mattioli M, Sossin WS, Klann E, Sonenberg N (2009) Translational control of long-lasting synaptic plasticity and memory. Neuron 61(1):10–26

Courchesne E (2004) Brain development in autism: early overgrowth followed by premature arrest of growth. Ment Retard Dev Disabil Res Rev 10(2):106–111

Curatolo P, Napolioni V, Moavero R (2010) Autism spectrum disorders in tuberous sclerosis: pathogenetic pathways and implications for treatment. J Child Neurol 25(7):873–880

Darnell JC, Van Driesche SJ, Zhang C, Hung KY, Mele A, Fraser CE, Stone EF, Chen C, Fak JJ, Chi SW et al (2011) FMRP stalls ribosomal translocation on mRNAs linked to synaptic function and autism. Cell 146(2):247–261

Dash PK, Orsi SA, Moore AN (2006) Spatial memory formation and memory-enhancing effect of glucose involves activation of the tuberous sclerosis complex-mammalian target of rapamycin pathway. J Neurosci 26(31):8048–8056

Davidkova G, Carroll RC (2007) Characterization of the role of microtubule-associated protein 1B in metabotropic glutamate receptor-mediated endocytosis of AMPA receptors in hippocampus. J Neurosci 27(48):13273–13278

Derry MC, Yanagiya A, Martineau Y, Sonenberg N (2006) Regulation of poly(A)-binding protein through PABP-interacting proteins. Cold Spring Harb Symp Quant Biol 71:537–543

Desmond NL, Levy WB (1983) Synaptic correlates of associative potentiation/depression: an ultrastructural study in the hippocampus. Brain Res 265(1):21–30

Dever TE (2002) Gene-specific regulation by general translation factors. Cell 108(4):545–556

Dolen G, Osterweil E, Rao BS, Smith GB, Auerbach BD, Chattarji S, Bear MF (2007) Correction of fragile X syndrome in mice. Neuron 56(6):955–962

Doornbos M, Sikkema-Raddatz B, Ruijvenkamp CA, Dijkhuizen T, Bijlsma EK, Gijsbers AC, Hilhorst-Hofstee Y, Hordijk R, Verbruggen KT, Kerstjens-Frederikse WS et al (2009) Nine patients with a microdeletion 15q11.2 between breakpoints 1 and 2 of the Prader-Willi critical region, possibly associated with behavioural disturbances. Eur J Med Genet 52(2–3):108–115

Edery I, Humbelin M, Darveau A, Lee KA, Milburn S, Hershey JW, Trachsel H, Sonenberg N (1983) Involvement of eukaryotic initiation factor 4A in the cap recognition process. J Biol Chem 258(18):11398–11403

Ehninger D, Han S, Shilyansky C, Zhou Y, Li W, Kwiatkowski DJ, Ramesh V, Silva AJ (2008) Reversal of learning deficits in a Tsc2+/− mouse model of tuberous sclerosis. Nat Med 14(8):843–848

Feng Y, Absher D, Eberhart DE, Brown V, Malter HE, Warren ST (1997a) FMRP associates with polyribosomes as an mRNP, and the I304N mutation of severe fragile X syndrome abolishes this association. Mol Cell 1(1):109–118

Feng Y, Gutekunst CA, Eberhart DE, Yi H, Warren ST, Hersch SM (1997b) Fragile X mental retardation protein: nucleocytoplasmic shuttling and association with somatodendritic ribosomes. J Neurosci 17(5):1539–1547

Fombonne E (2009) Epidemiology of pervasive developmental disorders. Pediatr Res 65(6): 591–598

Gangloff YG, Mueller M, Dann SG, Svoboda P, Sticker M, Spetz JF, Um SH, Brown EJ, Cereghini S, Thomas G et al (2004) Disruption of the mouse mTOR gene leads to early postimplantation lethality and prohibits embryonic stem cell development. Mol Cell Biol 24(21):9508–9516

Gao J, Wang WY, Mao YW, Graff J, Guan JS, Pan L, Mak G, Kim D, Su SC, Tsai LH (2010) A novel pathway regulates memory and plasticity via SIRT1 and miR-134. Nature 466(7310):1105–1109

Gelinas JN, Nguyen PV (2005) Beta-adrenergic receptor activation facilitates induction of a protein synthesis-dependent late phase of long-term potentiation. J Neurosci 25(13):3294–3303

Gingras AC, Raught B, Sonenberg N (1999) eIF4 initiation factors: effectors of mRNA recruitment to ribosomes and regulators of translation. Annu Rev Biochem 68:913–963

Gobert D, Topolnik L, Azzi M, Huang L, Badeaux F, Desgroseillers L, Sossin WS, Lacaille JC (2008) Forskolin induction of late-LTP and up-regulation of 5′ TOP mRNAs translation via mTOR, ERK, and PI3K in hippocampal pyramidal cells. J Neurochem 106(3):1160–1174

Goorden SM, van Woerden GM, van der Weerd L, Cheadle JP, Elgersma Y (2007) Cognitive deficits in Tsc1+/− mice in the absence of cerebral lesions and seizures. Ann Neurol 62(6):648–655

Grifo JA, Tahara SM, Morgan MA, Shatkin AJ, Merrick WC (1983) New initiation factor activity required for globin mRNA translation. J Biol Chem 258(9):5804–5810

Grossman AW, Elisseou NM, McKinney BC, Greenough WT (2006) Hippocampal pyramidal cells in adult Fmr1 knockout mice exhibit an immature-appearing profile of dendritic spines. Brain Res 1084(1):158–164

Grossman AW, Aldridge GM, Lee KJ, Zeman MK, Jun CS, Azam HS, Arii T, Imoto K, Greenough WT, Rhyu IJ (2010) Developmental characteristics of dendritic spines in the dentate gyrus of Fmr1 knockout mice. Brain Res 1355:221–227

Haghighat A, Mader S, Pause A, Sonenberg N (1995) Repression of cap-dependent translation by 4E-binding protein 1: competition with p220 for binding to eukaryotic initiation factor-4E. EMBO J 14(22):5701–5709

Hansen KF, Sakamoto K, Wayman GA, Impey S, Obrietan K (2010) Transgenic miR132 alters neuronal spine density and impairs novel object recognition memory. PLoS One 5(11):e15497

Harding HP, Novoa I, Zhang Y, Zeng H, Wek R, Schapira M, Ron D (2000) Regulated translation initiation controls stress-induced gene expression in mammalian cells. Mol Cell 6(5):1099–1108

Hay N, Sonenberg N (2004) Upstream and downstream of mTOR. Genes Dev 18(16):1926–1945

Heaton P, Wallace GL (2004) Annotation: the savant syndrome. J Child Psychol Psychiatry 45(5):899–911

Herry C, Ferraguti F, Singewald N, Letzkus JJ, Ehrlich I, Luthi A (2010) Neuronal circuits of fear extinction. Eur J Neurosci 31(4):599–612

Hinds HL, Ashley CT, Sutcliffe JS, Nelson DL, Warren ST, Housman DE, Schalling M (1993) Tissue specific expression of FMR-1 provides evidence for a functional role in fragile X syndrome. Nat Genet 3(1):36–43

Hoeffer CA, Tang W, Wong H, Santillan A, Patterson RJ, Martinez LA, Tejada-Simon MV, Paylor R, Hamilton SL, Klann E (2008) Removal of FKBP12 enhances mTOR-Raptor interactions, LTP, memory, and perseverative/repetitive behavior. Neuron 60(5):832–845

Hoeffer CA, Cowansage KK, Arnold EC, Banko JL, Moerke NJ, Rodriguez R, Schmidt EK, Klosi E, Chorev M, Lloyd RE et al (2011) Inhibition of the interactions between eukaryotic initiation factors 4E and 4G impairs long-term associative memory consolidation but not reconsolidation. Proc Natl Acad Sci U S A 108(8):3383–3388

Hou L, Klann E (2004) Activation of the phosphoinositide 3-kinase-Akt-mammalian target of rapamycin signaling pathway is required for metabotropic glutamate receptor-dependent long-term depression. J Neurosci 24(28):6352–6361

Huang YS, Jung MY, Sarkissian M, Richter JD (2002) N-methyl-D-aspartate receptor signaling results in Aurora kinase-catalyzed CPEB phosphorylation and alpha CaMKII mRNA polyadenylation at synapses. EMBO J 21(9):2139–2148

Huber KM, Kayser MS, Bear MF (2000) Role for rapid dendritic protein synthesis in hippocampal mGluR-dependent long-term depression. Science 288(5469):1254–1257

Iacoangeli A, Rozhdestvensky TS, Dolzhanskaya N, Tournier B, Schutt J, Brosius J, Denman RB, Khandjian EW, Kindler S, Tiedge H (2008) On BC1 RNA and the fragile X mental retardation protein. Proc Natl Acad Sci U S A 105(2):734–739

Im HI, Nakajima A, Gong B, Xiong X, Mamiya T, Gershon ES, Zhuo M, Tang YP (2009) Post-training dephosphorylation of eEF-2 promotes protein synthesis for memory consolidation. PLoS One 4(10):e7424

Irwin SA, Galvez R, Greenough WT (2000) Dendritic spine structural anomalies in fragile-X mental retardation syndrome. Cereb Cortex 10(10):1038–1044

Jaworski J, Spangler S, Seeburg DP, Hoogenraad CC, Sheng M (2005) Control of dendritic arborization by the phosphoinositide-3'-kinase-Akt-mammalian target of rapamycin pathway. J Neurosci 25(49):11300–11312

Jiang Z, Belforte JE, Lu Y, Yabe Y, Pickel J, Smith CB, Je HS, Lu B, Nakazawa K (2010) eIF2alpha Phosphorylation-dependent translation in CA1 pyramidal cells impairs hippocampal memory consolidation without affecting general translation. J Neurosci 30(7):2582–2594

Kandel ER (2001) The molecular biology of memory storage: a dialogue between genes and synapses. Science 294(5544):1030–1038

Kang H, Schuman EM (1996) A requirement for local protein synthesis in neurotrophin-induced hippocampal synaptic plasticity. Science 273(5280):1402–1406

Karpinski BA, Morle GD, Huggenvik J, Uhler MD, Leiden JM (1992) Molecular cloning of human CREB-2: an ATF/CREB transcription factor that can negatively regulate transcription from the cAMP response element. Proc Natl Acad Sci U S A 89(11):4820–4824

Kelleher RJ III, Govindarajan A, Jung HY, Kang H, Tonegawa S (2004) Translational control by MAPK signaling in long-term synaptic plasticity and memory. Cell 116(3):467–479

Khudayberdiev S, Fiore R, Schratt G (2009) MicroRNA as modulators of neuronal responses. Commun Integr Biol 2(5):411–413

Kim JH, Richter JD (2006) Opposing polymerase-deadenylase activities regulate cytoplasmic polyadenylation. Mol Cell 24(2):173–183

Kim DH, Sarbassov DD, Ali SM, King JE, Latek RR, Erdjument-Bromage H, Tempst P, Sabatini DM (2002) mTOR interacts with raptor to form a nutrient-sensitive complex that signals to the cell growth machinery. Cell 110(2):163–175

Koekkoek SK, Yamaguchi K, Milojkovic BA, Dortland BR, Ruigrok TJ, Maex R, De Graaf W, Smit AE, VanderWerf F, Bakker CE et al (2005) Deletion of FMR1 in Purkinje cells enhances parallel fiber LTD, enlarges spines, and attenuates cerebellar eyelid conditioning in fragile X syndrome. Neuron 47(3):339–352

Konopka W, Kiryk A, Novak M, Herwerth M, Parkitna JR, Wawrzyniak M, Kowarsch A, Michaluk P, Dzwonek J, Arnsperger T et al (2010) MicroRNA loss enhances learning and memory in mice. J Neurosci 30(44):14835–14842

Kosik KS (2006) The neuronal microRNA system. Nat Rev Neurosci 7(12):911–920

Kwon CH, Luikart BW, Powell CM, Zhou J, Matheny SA, Zhang W, Li Y, Baker SJ, Parada LF (2006) Pten regulates neuronal arborization and social interaction in mice. Neuron 50(3):377–388

Kye MJ, Liu T, Levy SF, Xu NL, Groves BB, Bonneau R, Lao K, Kosik KS (2007) Somatodendritic microRNAs identified by laser capture and multiplex RT-PCR. RNA 13(8):1224–1234

Leuner B, Falduto J, Shors TJ (2003) Associative memory formation increases the observation of dendritic spines in the hippocampus. J Neurosci 23(2):659–665

Levenga J, de Vrij FM, Oostra BA, Willemsen R (2010) Potential therapeutic interventions for fragile X syndrome. Trends Mol Med 16(11):516–527

Lin D, Pestova TV, Hellen CU, Tiedge H (2008) Translational control by a small RNA: dendritic BC1 RNA targets the eukaryotic initiation factor 4A helicase mechanism. Mol Cell Biol 28(9):3008–3019

Lin Q, Wei W, Coelho CM, Li X, Baker-Andresen D, Dudley K, Ratnu VS, Boskovic Z, Kobor MS, Sun YE et al (2011) The brain-specific microRNA miR-128b regulates the formation of fear-extinction memory. Nat Neurosci 14(9):1115–1117

Lugli G, Larson J, Martone ME, Jones Y, Smalheiser NR (2005) Dicer and eIF2c are enriched at postsynaptic densities in adult mouse brain and are modified by neuronal activity in a calpain-dependent manner. J Neurochem 94(4):896–905

Mineur YS, Sluyter F, de Wit S, Oostra BA, Crusio WE (2002) Behavioral and neuroanatomical characterization of the Fmr1 knockout mouse. Hippocampus 12(1):39–46

Mineur YS, Huynh LX, Crusio WE (2006) Social behavior deficits in the Fmr1 mutant mouse. Behav Brain Res 168(1):172–175

Miniaci MC, Kim JH, Puthanveettil SV, Si K, Zhu H, Kandel ER, Bailey CH (2008) Sustained CPEB-dependent local protein synthesis is required to stabilize synaptic growth for persistence of long-term facilitation in Aplysia. Neuron 59(6):1024–1036

Napoli I, Mercaldo V, Boyl PP, Eleuteri B, Zalfa F, De Rubeis S, Di Marino D, Mohr E, Massimi M, Falconi M et al (2008) The fragile X syndrome protein represses activity-dependent translation through CYFIP1, a new 4E-BP. Cell 134(6):1042–1054

Narayanan U, Nalavadi V, Nakamoto M, Thomas G, Ceman S, Bassell GJ, Warren ST (2008) S6K1 phosphorylates and regulates fragile X mental retardation protein (FMRP) with the neuronal protein synthesis-dependent mammalian target of rapamycin (mTOR) signaling cascade. J Biol Chem 283(27):18478–18482

Neves-Pereira M, Muller B, Massie D, Williams JH, O'Brien PC, Hughes A, Shen SB, Clair DS, Miedzybrodzka Z (2009) Deregulation of EIF4E: a novel mechanism for autism. J Med Genet 46(11):759–765

Nielsen DM, Derber WJ, McClellan DA, Crnic LS (2002) Alterations in the auditory startle response in Fmr1 targeted mutant mouse models of fragile X syndrome. Brain Res 927(1):8–17

Nosyreva ED, Huber KM (2006) Metabotropic receptor-dependent long-term depression persists in the absence of protein synthesis in the mouse model of fragile X syndrome. J Neurophysiol 95(5):3291–3295

Nowicki ST, Tassone F, Ono MY, Ferranti J, Croquette MF, Goodlin-Jones B, Hagerman RJ (2007) The Prader-Willi phenotype of fragile X syndrome. J Dev Behav Pediatr 28(2):133–138

Nudelman AS, DiRocco DP, Lambert TJ, Garelick MG, Le J, Nathanson NM, Storm DR (2010) Neuronal activity rapidly induces transcription of the CREB-regulated microRNA-132, in vivo. Hippocampus 20(4):492–498

O'Connor T, Sadleir KR, Maus E, Velliquette RA, Zhao J, Cole SL, Eimer WA, Hitt B, Bembinster LA, Lammich S et al (2008) Phosphorylation of the translation initiation factor eIF2alpha increases BACE1 levels and promotes amyloidogenesis. Neuron 60(6):988–1009

Ostroff LE, Fiala JC, Allwardt B, Harris KM (2002) Polyribosomes redistribute from dendritic shafts into spines with enlarged synapses during LTP in developing rat hippocampal slices. Neuron 35(3):535–545

Panja D, Dagyte G, Bidinosti M, Wibrand K, Kristiansen AM, Sonenberg N, Bramham CR (2009) Novel translational control in Arc-dependent long term potentiation consolidation in vivo. J Biol Chem 284(46):31498–31511

Park S, Park JM, Kim S, Kim JA, Shepherd JD, Smith-Hicks CL, Chowdhury S, Kaufmann W, Kuhl D, Ryazanov AG et al (2008) Elongation factor 2 and fragile X mental retardation protein control the dynamic translation of Arc/Arg3.1 essential for mGluR-LTD. Neuron 59(1):70–83

Parsons RG, Gafford GM, Helmstetter FJ (2006) Translational control via the mammalian target of rapamycin pathway is critical for the formation and stability of long-term fear memory in amygdala neurons. J Neurosci 26(50):12977–12983

Parsyan A, Svitkin Y, Shahbazian D, Gkogkas C, Lasko P, Merrick WC, Sonenberg N (2011) mRNA helicases: the tacticians of translational control. Nat Rev Mol Cell Biol 12(4):235–245

Pause A, Belsham GJ, Gingras AC, Donze O, Lin TA, Lawrence JC Jr, Sonenberg N (1994) Insulin-dependent stimulation of protein synthesis by phosphorylation of a regulator of 5′-cap function. Nature 371(6500):762–767

Pittenger C, Huang YY, Paletzki RF, Bourtchouladze R, Scanlin H, Vronskaya S, Kandel ER (2002) Reversible inhibition of CREB/ATF transcription factors in region CA1 of the dorsal hippocampus disrupts hippocampus-dependent spatial memory. Neuron 34(3):447–462

Qin M, Kang J, Burlin TV, Jiang C, Smith CB (2005) Postadolescent changes in regional cerebral protein synthesis: an in vivo study in the FMR1 null mouse. J Neurosci 25(20):5087–5095

Rajasethupathy P, Fiumara F, Sheridan R, Betel D, Puthanveettil SV, Russo JJ, Sander C, Tuschl T, Kandel E (2009) Characterization of small RNAs in Aplysia reveals a role for miR-124 in constraining synaptic plasticity through CREB. Neuron 63(6):803–817

Ramirez-Amaya V, Escobar ML, Chao V, Bermudez-Rattoni F (1999) Synaptogenesis of mossy fibers induced by spatial water maze overtraining. Hippocampus 9(6):631–636

Ramirez-Amaya V, Balderas I, Sandoval J, Escobar ML, Bermudez-Rattoni F (2001) Spatial long-term memory is related to mossy fiber synaptogenesis. J Neurosci 21(18):7340–7348

Raven JF, Koromilas AE (2008) PERK and PKR: old kinases learn new tricks. Cell Cycle 7(9):1146–1150

Sahoo T, Peters SU, Madduri NS, Glaze DG, German JR, Bird LM, Barbieri-Welge R, Bichell TJ, Beaudet AL, Bacino CA (2006) Microarray based comparative genomic hybridization testing in deletion bearing patients with Angelman syndrome: genotype-phenotype correlations. J Med Genet 43(6):512–516

Sarbassov DD, Ali SM, Kim DH, Guertin DA, Latek RR, Erdjument-Bromage H, Tempst P, Sabatini DM (2004) Rictor, a novel binding partner of mTOR, defines a rapamycin-insensitive and raptor-independent pathway that regulates the cytoskeleton. Curr Biol 14(14):1296–1302

Sarbassov DD, Ali SM, Sengupta S, Sheen JH, Hsu PP, Bagley AF, Markhard AL, Sabatini DM (2006) Prolonged rapamycin treatment inhibits mTORC2 assembly and Akt/PKB. Mol Cell 22(2):159–168

Schenck A, Bardoni B, Moro A, Bagni C, Mandel JL (2001) A highly conserved protein family interacting with the fragile X mental retardation protein (FMRP) and displaying selective interactions with FMRP-related proteins FXR1P and FXR2P. Proc Natl Acad Sci U S A 98(15):8844–8849

Schenck A, Bardoni B, Langmann C, Harden N, Mandel JL, Giangrande A (2003) CYFIP/Sra-1 controls neuronal connectivity in Drosophila and links the Rac1 GTPase pathway to the fragile X protein. Neuron 38(6):887–898

Schicknick H, Schott BH, Budinger E, Smalla KH, Riedel A, Seidenbecher CI, Scheich H, Gundelfinger ED, Tischmeyer W (2008) Dopaminergic modulation of auditory cortex-dependent memory consolidation through mTOR. Cereb Cortex 18(11):2646–2658

Shimizu E, Tang YP, Rampon C, Tsien JZ (2000) NMDA receptor-dependent synaptic reinforcement as a crucial process for memory consolidation. Science 290(5494):1170–1174

Shin CY, Kundel M, Wells DG (2004) Rapid, activity-induced increase in tissue plasminogen activator is mediated by metabotropic glutamate receptor-dependent mRNA translation. J Neurosci 24(42):9425–9433

Si K, Giustetto M, Etkin A, Hsu R, Janisiewicz AM, Miniaci MC, Kim JH, Zhu H, Kandel ER (2003a) A neuronal isoform of CPEB regulates local protein synthesis and stabilizes synapse-specific long-term facilitation in aplysia. Cell 115(7):893–904

Si K, Lindquist S, Kandel ER (2003b) A neuronal isoform of the aplysia CPEB has prion-like properties. Cell 115(7):879–891

Si K, Choi YB, White-Grindley E, Majumdar A, Kandel ER (2010) Aplysia CPEB can form prion-like multimers in sensory neurons that contribute to long-term facilitation. Cell 140(3):421–435

Sonenberg N, Hinnebusch AG (2007) New modes of translational control in development, behavior, and disease. Mol Cell 28(5):721–729

Sonenberg N, Hinnebusch AG (2009) Regulation of translation initiation in eukaryotes: mechanisms and biological targets. Cell 136(4):731–745

Sonenberg N, Rupprecht KM, Hecht SM, Shatkin AJ (1979) Eukaryotic mRNA cap binding protein: purification by affinity chromatography on sepharose-coupled m7GDP. Proc Natl Acad Sci U S A 76(9):4345–4349

Spencer CM, Alekseyenko O, Serysheva E, Yuva-Paylor LA, Paylor R (2005) Altered anxiety-related and social behaviors in the Fmr1 knockout mouse model of fragile X syndrome. Genes Brain Behav 4(7):420–430

Steward O, Levy WB (1982) Preferential localization of polyribosomes under the base of dendritic spines in granule cells of the dentate gyrus. J Neurosci 2(3):284–291

Steward O, Wallace CS, Lyford GL, Worley PF (1998) Synaptic activation causes the mRNA for the IEG Arc to localize selectively near activated postsynaptic sites on dendrites. Neuron 21(4):741–751

Stoica L, Zhu PJ, Huang W, Zhou H, Kozma SC, Costa-Mattioli M (2011) Selective pharmacogenetic inhibition of mammalian target of Rapamycin complex I (mTORC1) blocks long-term synaptic plasticity and memory storage. Proc Natl Acad Sci U S A 108(9):3791–3796

Tain LS, Mortiboys H, Tao RN, Ziviani E, Bandmann O, Whitworth AJ (2009) Rapamycin activation of 4E-BP prevents parkinsonian dopaminergic neuron loss. Nat Neurosci 12(9):1129–1135

Takei N, Kawamura M, Hara K, Yonezawa K, Nawa H (2001) Brain-derived neurotrophic factor enhances neuronal translation by activating multiple initiation processes: comparison with the effects of insulin. J Biol Chem 276(46):42818–42825

Tanaka J, Horiike Y, Matsuzaki M, Miyazaki T, Ellis-Davies GC, Kasai H (2008) Protein synthesis and neurotrophin-dependent structural plasticity of single dendritic spines. Science 319(5870):1683–1687

Tang SJ, Schuman EM (2002) Protein synthesis in the dendrite. Philos Trans R Soc Lond B Biol Sci 357(1420):521–529

Tang SJ, Reis G, Kang H, Gingras AC, Sonenberg N, Schuman EM (2002) A rapamycin-sensitive signaling pathway contributes to long-term synaptic plasticity in the hippocampus. Proc Natl Acad Sci U S A 99(1):467–472

Tavazoie SF, Alvarez VA, Ridenour DA, Kwiatkowski DJ, Sabatini BL (2005) Regulation of neuronal morphology and function by the tumor suppressors Tsc1 and Tsc2. Nat Neurosci 8(12):1727–1734

Toni N, Buchs PA, Nikonenko I, Bron CR, Muller D (1999) LTP promotes formation of multiple spine synapses between a single axon terminal and a dendrite. Nature 402(6760):421–425

Tsokas P, Grace EA, Chan P, Ma T, Sealfon SC, Iyengar R, Landau EM, Blitzer RD (2005) Local protein synthesis mediates a rapid increase in dendritic elongation factor 1A after induction of late long-term potentiation. J Neurosci 25(24):5833–5843

Vattem KM, Wek RC (2004) Reinitiation involving upstream ORFs regulates ATF4 mRNA translation in mammalian cells. Proc Natl Acad Sci U S A 101(31):11269–11274

Verpelli C, Piccoli G, Zanchi A, Gardoni F, Huang K, Brambilla D, Di Luca M, Battaglioli E, Sala C (2010) Synaptic activity controls dendritic spine morphology by modulating eEF2-dependent BDNF synthesis. J Neurosci 30(17):5830–5842

Walden WE, Godefroy-Colburn T, Thach RE (1981) The role of mRNA competition in regulating translation. I. Demonstration of competition in vivo. J Biol Chem 256(22):11739–11746

Wang H, Iacoangeli A, Popp S, Muslimov IA, Imataka H, Sonenberg N, Lomakin IB, Tiedge H (2002) Dendritic BC1 RNA: functional role in regulation of translation initiation. J Neurosci 22(23):10232–10241

Way SW, McKenna J III, Mietzsch U, Reith RM, Wu HC, Gambello MJ (2009) Loss of Tsc2 in radial glia models the brain pathology of tuberous sclerosis complex in the mouse. Hum Mol Genet 18(7):1252–1265

Weatherill DB, Dyer J, Sossin WS (2010) Ribosomal protein S6 kinase is a critical downstream effector of the target of rapamycin complex 1 for long-term facilitation in Aplysia. J Biol Chem 285(16):12255–12267

Wibrand K, Panja D, Tiron A, Ofte ML, Skaftnesmo KO, Lee CS, Pena JT, Tuschl T, Bramham CR (2010) Differential regulation of mature and precursor microRNA expression by NMDA and metabotropic glutamate receptor activation during LTP in the adult dentate gyrus in vivo. Eur J Neurosci 31(4):636–645

Wiznitzer M (2004) Autism and tuberous sclerosis. J Child Neurol 19(9):675–679

Wu L, Wells D, Tay J, Mendis D, Abbott MA, Barnitt A, Quinlan E, Heynen A, Fallon JR, Richter JD (1998) CPEB-mediated cytoplasmic polyadenylation and the regulation of experience-dependent translation of alpha-CaMKII mRNA at synapses. Neuron 21(5):1129–1139

Yan QJ, Rammal M, Tranfaglia M, Bauchwitz RP (2005) Suppression of two major Fragile X Syndrome mouse model phenotypes by the mGluR5 antagonist MPEP. Neuropharmacology 49(7):1053–1066

Yanow SK, Manseau F, Hislop J, Castellucci VF, Sossin WS (1998) Biochemical pathways by which serotonin regulates translation in the nervous system of Aplysia. J Neurochem 70(2):572–583

Zalfa F, Giorgi M, Primerano B, Moro A, Di Penta A, Reis S, Oostra B, Bagni C (2003) The fragile X syndrome protein FMRP associates with BC1 RNA and regulates the translation of specific mRNAs at synapses. Cell 112(3):317–327

Zearfoss NR, Alarcon JM, Trifilieff P, Kandel E, Richter JD (2008) A molecular circuit composed of CPEB-1 and c-Jun controls growth hormone-mediated synaptic plasticity in the mouse hippocampus. J Neurosci 28(34):8502–8509

Zhao MG, Toyoda H, Ko SW, Ding HK, Wu LJ, Zhuo M (2005) Deficits in trace fear memory and long-term potentiation in a mouse model for fragile X syndrome. J Neurosci 25(32): 7385–7392

Zhou J, Blundell J, Ogawa S, Kwon CH, Zhang W, Sinton C, Powell CM, Parada LF (2009) Pharmacological inhibition of mTORC1 suppresses anatomical, cellular, and behavioral abnormalities in neural-specific Pten knock-out mice. J Neurosci 29(6):1773–1783

Zhou X, Lin DS, Zheng F, Sutton MA, Wang H (2010) Intracellular calcium and calmodulin link brain-derived neurotrophic factor to p70S6 kinase phosphorylation and dendritic protein synthesis. J Neurosci Res 88(7):1420–1432

Index

J.D. Dinman (ed.), *Biophysical Approaches to Translational Control of Gene Expression*,   311
Biophysics for the Life Sciences 1, DOI 10.1007/978-1-4614-3991-2,
© Springer Science+Business Media New York 2013